John Wood

The Illustrated Natural History

Elibron Classics
www.elibron.com

Elibron Classics series.

ISBN 0-543-81766-0 (paperback)
ISBN 0-543-81765-2 (hardcover)

This Elibron Classics Replica Edition is an unabridged facsimile of the edition published in 1861 by Routledge, Warne & Routledge, London.

THE

ILLUSTRATED

NATURAL HISTORY.

BY THE

REV. J. G. WOOD, M.A. F.L.S. ETC.

AUTHOR OF "SKETCHES AND ANECDOTES OF ANIMAL LIFE," ETC.

WITH FOUR HUNDRED AND EIGHTY ORIGINAL DESIGNS.

BY WILLIAM HARVEY.

New Edition,

CORRECTED AND CONSIDERABLY ENLARGED.

LONDON:

ROUTLEDGE, WARNE, & ROUTLEDGE,

FARRINGDON STREET; AND 56, WALKER STREET, NEW YORK.

1861.

LONDON:
R. CLAY, SON, AND TAYLOR, PRINTERS,
BREAD STREET HILL.

PREFACE.

ALTHOUGH the number of works on Natural History might deter any new writer from venturing on so extensively handled a subject, there is at present no work of a really popular character in which accuracy of information and systematic arrangement are united with brevity and simplicity of treatment.

All the best-known popular works on Natural History are liable to many objections, among which may be named a want of correct classification, the absence of explanations of the meanings and derivations of scientific words, the strange inaccuracy of many of the accompanying illustrations, and of the accounts of many animals. Nor do the conventional anecdotes chronicled in their pages evince that personal experience of the animal race which alone can repress romance and prevent inaccuracy. These deficiencies, it is hoped, are at all events partly supplied in the present work.

The present volume, although exceeding the limits originally contemplated, is but a brief digest of a large mass of materials, derived either from personal experience, from the most recent zoological writers, or from the kindness of many friends, who are familiar with almost every portion of the world, and to whom my best thanks are due. My original intention was to carry the work as far as the Zoophytes, but it grew so rapidly, especially in the first two classes, the Mammals and Birds, that it was found necessary to conclude at the Insects, and even then to give but an exceedingly short and meagre account of them. This was much regretted by me, as my experience had lain so much in the practical entomological part of Natural History, that during the earlier stages of the work I looked forward with some pleasure to giving a very much fuller account of the British Insects than will be found in the last few pages of this volume.

In arrangement, the order of the Catalogue of the British Museum has been followed, with the view of rendering it a useful companion to that most valuable collection, especially for young visitors. In accordance with that catalogue, the volume commences with a short sketch of mankind and of the theories respecting the different races of humanity; and at the same time a few of the distinctions are mentioned which so widely separate man from any other inhabitant of the earth.

As for the Illustrations, they will best speak for themselves. It will, however, be well to observe that they have all been designed expressly for the present work, and that the combined abilities of Messrs. Harvey and Dalziel, as artist and engravers, are a guarantee for their accuracy and perfect execution. For the anatomical and microscopical vignettes, I am myself answerable, as well as for several of the later drawings, such as the Thorny Woodcock-shell, the Leaf Insect, the Rove Beetle, together with parts of a few others, all of which were drawn from actual specimens.

It has been an object with me in the accounts of each animal, to give as far as possible *new* anecdotes. In many cases, the anecdotes related have never been published before, and in many more, they have been extracted from works which, either from their scarcity, their cost, or their nature, would be very unlikely to be placed in the hands of general readers.

I dismiss these pages with almost a feeling of regret, that a task which has to me been a labour of love, has come to an end. Indeed, the only drawback experienced during its progress was its necessary brevity, which constrained me to omit many creatures, not only beautiful and wonderful in form, but interesting in habits. I was also compelled to describe many others so briefly, as to render the account little more than a formal announcement of their name, country, and food.

If, however, the perusal of the following pages should induce any one to look upon the great plan of Creation more as a whole than merely as an aggregation of separate parts, or to notice how wonderfully each creature is adapted for its peculiar station, by Him who has appointed to each its proper position, and assigned to each its own duties, which could not be performed so well by any other creature, or even by the same animal in another place, my end will be attained.

Perhaps, also, this volume may cause some who have hitherto been troubled with a causeless abhorrence of certain creatures against which they have nourished early prejudices, to examine them with a more indulgent—I should perhaps say, a more reverent eye. I say reverent, because it has long given me deep pain when I have heard others stigmatizing as ugly, horrid, or frightful, those beings whom their Maker saw at the beginning of the world, and declared very good. A naturalist will see as much beauty in a snake, spider, or toad, as in any of those animals which we are accustomed to consider models of beauty; and so will those who have before feared or despised them, if they can only persuade themselves to examine them with an unprejudiced eye.

In those three creatures mentioned a few lines above, there is great beauty even on a superficial examination. The movements of the snake are most graceful, and the changing colours of its varied scales leave the imitations of art far behind. The spiders too are beautiful, even in colour; some are bright crimson, some pale pink, some entirely yellow, some banded with broad streaks of alternately velvety black and silvery white; while the eye of the toad is a living gem of beauty. But when we come to look closer,—to watch their habits—to note their instincts—or, by the use of the microscope, to lay open to our view some of the details of their organization,—then indeed are we lost in wonder and amaze at the vastness of creation, which, even in one little, apparently insignificant animal, presents to our eyes innumerable marvels—marvels which increase in number and beauty as our power for perceiving them increases.

MERTON COLLEGE, OXFORD,
December 16, 1852.

PREFACE

TO THE SECOND EDITION.

THE unexpectedly rapid sale of this work, five thousand copies having been sold in little more than eighteen months, has induced Messrs. Routledge to publish a second edition. In the present edition very great pains have been taken to render the book more worthy of the public notice. At least one-third more matter has been introduced, and between thirty and forty new illustrations will be found in its pages. The entire work has been carefully revised, and the trifling errors which existed in the first edition have been corrected. The price has also been considerably lowered from the sum at which the volume was originally published.

I may be excused for giving a short account of the reasons that led to the production of this book.

It was with considerable reluctance that I first undertook a work that necessarily involved so much labour, and it was not without much misgiving on my part, that it first made its public appearance. No one could be more sensible of its deficiencies than myself; added to which, I was fully aware of the impossibility of condensing the requisite amount of information into so small a compass. It was therefore no less gratifying than unexpected, to find that my fears were groundless, and that my labours were not in vain.

From my earliest childhood, the study of Natural History was my favourite pursuit, and in the prosecution of this study, I was invariably disappointed by the character of the books that are generally considered as standard works on the subject. In none of them is there any systematic order, and much less is there any attempt to explain the arrangement of the various links in the chain of animated nature. For this reason the scientific names always appeared to be little but a collection of cacophonous and unmeaning syllables, hardly more intelligible to the unclassical reader than Abracadabra or Aldiborontiphoskiphorniosticos.

Moreover, very little practical acquaintance with the subject was required, to perceive that, in almost every case, the work was nothing but a compilation from various authors, the compiler having thrown together truth and error with perfect impartiality. For example, in one of the best of these works, the panther, the ocelot, and the ounce, are asserted to be varieties of the leopard; while the chimpansee and the orang-outan are considered as the same animal, although their formation, their colour, their habits and their country are all different. The one has a small, rather smooth skull, while the skull of the other is large, and heavily ridged; the fur of one is black, that of the other reddish brown; the one lives at the foot of trees, the other in their branches; the one inhabits Western Africa, while the other is confined to Borneo and Sumatra.

My object, therefore, in undertaking this work, was to produce a book which would be comparatively free from these defects, and to afford to the young inquirer that assistance which I myself had so often sought in vain. That I should perfectly succeed in so arduous a task could not possibly be expected; but that I have at all events partially succeeded, is evident from the large sale of the work in so short a time.

There was also another object that I endeavoured to attain. In the preface to the first edition, I mentioned that the book was intended as a companion to the British Museum. For that purpose it has been used, and has been found to answer as well as so superficial a work could be expected to do.

It is my sincere wish that this new edition may be found to approach nearer to my original plan, and to carry out my design better than the first.

MERTON COLLEGE, OXFORD,
December, 1854.

Systematic Index.

Division I. VERTEBRATA.

Class I. MAMMALIA.

Order I. *PRIMATES*.

Family I. Hominidæ.

Genus I. HOMO. Sapiens, *Man*.

Fam. II. Simiadæ.

TROGLODYTES. Niger, *Chimpansee*.
SIMIA. Satyrus, *Orang-Outan*.
HYLOBATES. Agilis, *Agile Gibbon*.
PRESBYTES. Larvatus, *Kahau*.
——— Entellus, *Entellus*.
CYNOCEPHALUS. Mormon, *Mandrill*.

Fam. III. Cebidæ.

ATELES. Paniscus, *Coaita Spider Monkey*.
MYCETES. Ursinus, *Ursine Howler*.
CALLITHRIX. Torquatus, *Collared Tee Tee*.
JACCHUS. Vulgaris, *Marmoset*.

Fam. IV. Lemuridæ.

LEMUR. Macaco, *Ruffled Lemur*.
LORIS. Gracilis, *Slender Loris*.

Fam. V. Vespertilionidæ.

Sub-fam. *a*. *Phyllostomina*.
VAMPIRUS. Spectrum, *Vampire*.
Sub-fam. *c*. *Vespertilionina*.
PLECOTUS. Auritus, *Long-eared Bat*.

Order II. *FERÆ*.

Fam. I. Felidæ.

Sub. fam. *a*. *Felina*.
LEO. Barbarus, *Lion*.
TIGRIS. Regalis, *Tiger*.
LEOPARDUS. Varius, *Leopard*.
——— Uncia, *Ounce*.
——— Onca, *Jaguar*.
——— Concolor, *Puma*.
——— Pardalis, *Ocelot*.
FELIS. Domestica, *Cat*.
CARACAL. Melanotis, *Caracal*.
LYNCUS. Canadensis, *Canada Lynx*.
GUEPARDA. Jubata, *Chetah*.
Sub-fam. *b*. *Hyænina*.
HYÆNA. Striata, *Striped Hyæna*.
Sub-fam. *c*. *Viverrina*.
VIVERRA. Civetta, *Civet Cat*.
GENETTA. Vulgaris, *Genet*.
HERPESTES. Ichneumon, *Egyptian Ichneumon*.
Sub-fam. *d*. *Canina*.
CANIS. Familiaris, *Dog*.
——— Lupus, *Wolf*.
——— Aureus, *Jackal*.
VULPES. Vulgaris, *Fox*.
Sub-fam. *e*. *Mustelina*.
MARTES. Abietum, *Pine Marten*.
——— Zibellina, *Sable*.
PUTORIUS. Fœtidus, *Polecat*.
MUSTELA. Erminea, *Stoat*.
——— Vulgaris, *Weasel*.
MELLIVORA. Ratel, *Honey Ratel*.
GULO. Luscus, *Glutton*.
MELES. Vulgaris, *Badger*.
LUTRA. Vulgaris, *Otter*.

Fam. II. Ursidæ.

Sub-fam. *a*. *Ursina*.
URSUS. Arctos, *Bear*.
——— Horribilis, *Grizzly Bear*.
THALARCTOS. Maritimus, *Polar Bear*.
Sub-fam. *c*. *Procyonina*.
PROCYON. Lotor, *Racoon*.
Sub-fam. *d*. *Cercoleptina*.
NASUA. Fusca, *Coati-mondi*.
CERCOLEPTES. Caudivolvulus, *Kinkajou*.

Fam. III. Talpidæ.

Sub-fam. *a*. *Talpina*.
TALPA. Europæa, *Mole*.
Sub-fam. *d*. *Erinacina*.
SOREX. Araneus, *Shrew*.

CROSSOPUS. Fodiens, *Water Shrew.*
ERINACEUS. Europæus, *Hedgehog.*

Fam. IV. Macropidæ.
Sub-fam. *b. Macropina.*
MACROPUS. Major, *Kangaroo.*
Sub-fam. *e. Didelphina.*
DIDELPHYS. Virginiana, *Opossum.*

Fam. V. Phocidæ.
Sub-fam. *b. Phocina.*
PHOCA. Vitulina, *Seal.*
MORUNGA. Proboscidea, *Elephant Seal.*
Sub-fam. *c. Trichecina.*
TRICHECUS. Rosmarus, *Walrus.*

Order III. *CETE.*
Fam. I. Balænidæ.
BALÆNA. Mysticetus, *Whale.*
PHYSETER. Macrocephalus, *Cachalot.*

Fam. II. Delphinidæ.
DELPHINUS. Delphis, *Dolphin.*
PHOCÆNA. Communis, *Porpoise.*
MONODON. Monoceros, *Narwhal.*

Order IV. *GLIRES.*
Fam. I. Muridæ.
Sub-fam. *a. Murina.*
MUS. Decumanus, *Rat.*
—— Musculus, *Mouse.*
MICROMYS. Minutus, *Harvest Mouse.*
Sub-fam. *b. Arvicolina.*
CRICETUS. Frumentarius, *Hamster.*
ARVICOLA. Amphibius, *Water Rat.*
Sub-fam. *d. Castorina.*
CASTOR. Fiber, *Beaver.*

Fam. II. Hystricidæ.
Sub-fam. *a. Hystricina.*
HYSTRIX. Cristata. *Porcupine.*
Sub-fam. *c. Dasyproctina.*
DASYPROCTA. Aguti, *Agouti.*
Sub-fam. *d. Hydrochærina.*
HYDROCHÆRUS. Capybara, *Capybara.*

Fam. III. Leporidæ.
LEPUS. Timidus, *Hare.*
—— Variabilis, *Alpine Hare.*
—— Cuniculus, *Rabbit.*

Fam. IV. Jerboidæ.
Sub-fam. *a. Chinchillina.*
CHINCHILLA. Laniger, *Chinchilla.*

Sub-fam. *c. Dipina.*
DIPUS. Ægyptius, *Jerboa.*
Sub-fam. *d. Myoxina.*
MYOXUS. Avellanarius, *Dormouse.*
Sub-fam. *e. Sciurina.*
SCIURUS. Europæus, *Squirrel.*
PTEROMYS. Alpinus, *Flying Squirrel.*
ARCTOMYS. Marmotta, *Marmot.*

Order V. *UNGULATA.*
Fam. I. Bovidæ.
Sub-fam. *a. Bovina.*
BOS. Taurus, *Bull.*
—— —— *Zebu.*
BUBALUS. Buffelus, *Buffalo.*
—— Caffer, *Cape Buffalo.*
BISON. Americanus, *Bison.*
POEPHAGUS. Grunniens, *Yak.*
OVIBOS. Moschatus, *Musk Ox.*
CATOBLEPAS. Gnu, *Gnoo.*
PORTAX. Picta, *Nylghau.*
STREPSICEROS. Kudu, *Koodoo.*
BOSELAPHUS. Oreas, *Eland.*
ORYX. Leucoryx, *Oryx.*
GAZELLA. Euchore, *Springbok.*
—— Ariel, *Gazelle.*
RUPICAPRA. Tragus, *Chamois.*
CAPRA. Ibex, *Ibex.*
—— Hircus, *Goat.*
OVIS. Aries, *Ram.*
Sub-fam. *b. Camelopardina.*
CAMELOPARDALIS. Giraffa, *Giraffe.*
Sub-fam. *c. Camelina.*
CAMELUS. Arabicus, *Camel.*
—— Bactrianus, *Bactrian Camel.*
LLAMA. Pacos, *Llama.*
Sub-fam. *d. Moschina.*
MOSCHUS. Moschiferus, *Musk-deer.*
Sub-fam. *e. Cervina.*
CERVUS. Capreolus, *Roebuck.*
—— Elaphus, *Stag.*
—— Canadensis, *Wapiti.*
AXIS. Maculata, *Axis.*
DAMA. Vulgaris, *Fallow-deer.*
RANGIFER. Tarandus, *Reindeer.*
ALCES, Palmatus, *Elk.*

Fam. II. Equidæ.
EQUUS. Caballus, *Horse.*
ASINUS. Vulgaris, *Ass.*
—— Dzigguetai, *Dziggue-tai.*
—— Zebra, *Zebra.*
—— Quagga, *Quagga.*

Fam. III. Elephantidæ.
Sub-fam. *a. Elephantina.*
ELEPHAS. Indicus, *Indian Elephant.*

ELEPHAS. Africanus, *African Elephant.*
Sub-fam. *b. Tapirina.*
TAPIRUS. Terrestris, *Tapir.*
Sub-fam. *c. Suina.*
SUS. Scrofa, *Boar.*
—— Babyroussa, *Babyroussa.*
DICOTYLES. Tajaçu, *Peccary.*
Sub-fam. *d. Rhinocerina.*
RHINOCEROS. Unicornis, *Rhinoceros.*
—————— Bicornis, *Rhinaster.*
Sub-fam. *e. Hippopotamina.*
HIPPOPOTAMUS. Amphibius, *Hippopotamus.*

Fam. IV. Bradypidæ.
BRADYPUS. Tridactylus, *Sloth.*

Fam. V. Dasypidæ.
Sub-fam. *a. Manina.*
MANIS. Tetradactyla, *Phatagin.*
—— Pentadactyla, *Short-tailed Manis.*
Sub-fam. *b. Dasypina.*
DASYPUS. Sexcinctus, *Armadillo.*
Sub-fam. *c. Myrmecophagina.*
MYRMECOPHAGA. Jubata, *Ant-eater.*
TAMANDUA. Tetradactyla, *Middle Ant-eater.*
CYCLOTHURUS. Didactyla, *Little Ant-eater.*
Sub-fam. *d. Ornithorhynchina.*
ORNITHORHYNCHUS. Paradoxus, *Ornithorhynchus.*

Class II. AVES.
Order I. *ACCIPITRES.*
Sub-order I. ACCIPITRES-DIURNI.
Fam. I. Gypaëtidæ.
GYPAETUS. Barbatus, *Lämmergeyer.*

Fam. II. Sarcorhamphidæ.
SARCORHAMPHOS. Gryphon, *Condor.*
—————— Papa, *King Vulture.*

Fam. III. Vulturidæ.
Sub-fam. *a. Vulturina.*
GYPS. Fulvus, *Griffin Vulture.*

Fam. IV. Falconidæ.
Sub-fam. *a. Aquilinæ.*
AQUILA. Chrysaëtos, *Golden Eagle.*
PANDION. Haliaëtus, *Osprey.*
HALIAËTUS. Leucocephalus, *White-headed Eagle.*
Sub-fam. *c. Buteoninæ.*
BUTEO. Vulgaris, *Buzzard.*

Sub-fam. *d. Milvinæ.*
PERNIS. Apivorus, *Honey Buzzard.*
MILVUS. Regalis, *Kite.*
ELANOÏDES. Furcatus, *Swallow-tailed Falcon.*
Sub-fam *e. Falconinæ.*
FALCO. Gyrfalco, *Gyrfalcon.*
——— Peregrinus, *Peregrine Falcon.*
HYPOTRIORCHIS. Subbuteo, *Hobby.*
—————— Æsalon, *Merlin.*
TINNUNCULUS. Alaudarius, *Kestrel.*
Sub-fam. *f. Accipitrinæ.*
ASTUR. Palumbarius, *Goshawk.*
ACCIPITER. Nisus, *Sparrow Hawk.*
Sub-fam. *g. Circinæ.*
SERPENTARIUS. Reptilivorus, *Secretary Bird.*
CIRCUS. Cyaneus, *Hen-Harrier.*

Sub-order II. ACCIPITRES-NOCTURNI.
Fam. I. Strigidæ.
Sub-fam. *a. Surninæ.*
SURNIA. Ulula, *Hawk Owl.*
NYCTEA. Nivea, *Snowy Owl.*
ATHENE. Cunicularia, *Burrowing Owl.*
Sub-fam. *b. Buboninæ.*
EPHIALTES. Scops, *Scops-eared Owl.*
BUBO. Maximus, *Great-eared Owl.*
Sub-fam. *d. Striginæ.*
STRIX. Flammea, *Barn Owl.*

Order II. *PASSERES.*
Tribe I. FISSIROSTRES.
Sub-tribe I. FISSIROSTRES-NOCTURNÆ.
Fam. I. Caprimulgidæ.
Sub-fam. *a. Caprimulginæ.*
CAPRIMULGUS. Europæus, *Goatsucker.*
Sub-tribe II. FISSIROSTRES-DIURNÆ.
Fam. II. Hirundinidæ.
Sub-fam *a. Cypselinæ.*
CYPSELUS. Melba, *Alpine-Swift.*
———— Apus, *Swift.*
Sub-fam. *b. Hirundininæ.*
HIRUNDO. Rustica, *Chimney Martin.*
COTILE. Riparia, *Sand Martin.*
CHELIDON. Urbica, *Martin*
Fam. III. Coraciidæ.
Sub-fam. *a. Coraciinæ.*
CORACIAS. Garrula, *Roller*

Fam. IV. Trogonidæ.
TROGON. Resplendens, *Resplendent Trogon.*

Fam. V. Alcedinidæ.
Sub-fam. *a. Alcedininæ.*
ALCEDO. Hispida, *King-fisher.*

Fam. VI. Meropidæ.
Sub-fam. *a. Meropinæ.*
MEROPS. Apiaster, *Bee-eater.*

Tribe II. TENUIROSTRES.

Fam. I. Upupidæ.
Sub-fam. *a. Upupinæ.*
UPUPA. Epops, *Hoopoe.*

Fam. II. Trochilidæ.
TROCHILUS. Colubris, *Ruby-throated Humming-bird.*
ORNISMYA. Gouldii, *Gould's Humming-bird.*
——— Sappho, *Bar-tailed Humming-bird.*
——— Cora, *Cora Humming-bird.*
——— Chrysolopha, *Double-crested Humming-bird.*

Fam. III. Certhidæ.
Sub-fam. *a. Certhinæ.*
CERTHIA. Familiaris, *Creeper.*
Sub-fam. *b. Sittinæ.*
SITTA. Europæa, *Nuthatch.*
Sub-fam. *c. Menurinæ.*
TROGLODYTES. Parvulus, *Wren.*

Tribe III. DENTIROSTRES.

Fam. I. Luscinidæ.
Sub-fam. *a. Luscininæ.*
CALAMODYTA. Locustella, *Grasshopper Warbler.*
LUSCINIA. Philomela, *Nightingale.*
SYLVIA. Undata, *Dartford Warbler.*
——— Cinerea, *Whitethroat.*
——— Atricapilla, *Black-cap Warbler.*
——— Hortensis, *Petti-chaps.*
——— Rufa, *Chiff-chaff.*
——— Trochilus, *Willow Wren.*
REGULUS. Cristatus, *Golden-crested Wren.*
Sub-fam. *b. Erythacinæ.*
SAXICOLA. Œnanthe, *Wheat-ear.*
RUTICILLA. Phœnicura, *Red-start.*
ERYTHACUS. Rubecula, *Red-breast.*
Sub-fam. *c. Accentorinæ.*
ACCENTOR. Modularius, *Hedge Accentor.*

Sub-fam. *d. Parinæ.*
PARUS. Major, *Great Tit-mouse.*
——— Cœruleus, *Blue Tit-mouse.*
——— Caudatus, *Long-tailed Titmouse.*
Sub-fam. *e. Motacillinæ.*
MOTACILLA. Yarrellii, *Pied Wagtail.*
——— Flava, *Yellow Wagtail.*
ANTHUS. Pratensis, *Meadow Pipit.*

Fam. II. Turdidæ.
Sub fam. *a. Formicarinæ.*
HYDROBATA. Cinclus, *Dipper.*
Sub-fam. *b. Turdinæ.*
TURDUS. Viscivorus, *Missel-toe Thrush.*
——— Pilaris, *Fieldfare.*
——— Musicus. *Song-Thrush.*
——— Iliacus, *Redwing.*
——— Torquatus, *Ring-Ouzel.*
——— Merula, *Blackbird.*
ORPHEUS. Polyglottus, *Mocking Bird.*
Sub-fam. *c. Oriolinæ.*
ORIOLUS. Galbula.

Fam. III. Muscicapidæ.
Sub-fam. *a. Muscicapinæ.*
MUSCICAPA. Grisola, *Spotted Fly-catcher.*

Fam. IV. Ampelidæ.
Sub-fam. *a. Ampelinæ.*
AMPELIS. Garrulus, *Bohemian Wax-wing.*

Fam. V. Lanidæ.
Sub-fam. *a. Laninæ.*
LANIUS. Excubitor, *Great Grey Shrike.*
——— Collurio, *Red-backed Shrike.*

Tribe IV. CONIROSTRES.

Fam. I. Corvidæ.
Sub-fam. *a. Garrulinæ.*
GARRULUS. Glandarius, *Jay.*
Sub-fam. *b. Corvinæ.*
NUCIFRAGA. Caryocatactes, *Nutcracker.*
PICA. Caudata, *Magpie.*
CORVUS. Corax, *Raven.*
——— Frugilegus, *Rook.*
——— Monedula, *Jack-daw.*
——— Corone, *Crow.*
——— Cornix, *Hooded Crow.*
Sub-fam- *c. Pyrrhocoracinæ.*
CORACIA. Garacula, *Chough.*

Fam. II. Paradiseidæ.
PARADISEA. Apoda, *Emerald Bird of Paradise.*

Fam. III. Sturnidæ.
Sub-fam. *a. Ptilonorhynchinæ.*
PTILONORHYNCHUS. Sericeus, *Satin Bower-Bird.*
Sub-fam. *d. Icterinæ.*
ICTERUS. Baltimorus, *Baltimore Oriole.*
Sub-fam. *g. Sturninæ.*
STURNUS. Vulgaris, *Starling.*

Fam. IV. Fringillidæ.
Sub-fam. *a. Coccothraustinæ.*
COCCOTHRAUSTES. Vulgaris, *Grosbeak.*
Sub-fam. *b. Fringillinæ.*
FRINGILLA. Cœlebs, *Chaffinch.*
——— Carduelis, *Goldfinch.*
——— Cannabina, *Linnet.*
——— Spinus, *Siskin.*
——— Chloris, *Greenfinch.*
CARDUELIS. Canaria, *Canary.*
PASSER. Domesticus, *House Sparrow.*
Sub-fam. *c. Emberizinæ.*
EMBERIZA. Miliaria, *Bunting.*
——— Citrinella, *Yellow Bunting.*
——— Hortulana, *Ortolan.*
Sub-fam. *d. Alaudinæ.*
ALAUDA. Arvensis, *Skylark.*
——— Arborea, *Woodlark.*
Sub-fam. *e. Pyrrhulinæ.*
PYRRHULA. Rubicilla, *Bullfinch.*
Sub-fam. *f. Loxinæ.*
LOXIA. Curvirostra, *Crossbill.*

Fam. VII. Bucerotidæ.
BUCEROS. Rhinoceros, *Rhinoceros Hornbill.*

Order III. *SCANSORES.*
Fam. I. Rhamphastidæ.
Sub-fam. *a. Rhamphastinæ.*
RHAMPHASTOS. Toco, *Toco Toucan.*

Fam. II. Psittacidæ.
MACROCERCUS. Ararauna, *Blue and Yellow Macaw.*
PALÆORNIS. Torquatus, *Ringed Parrakeet.*
CACATUA. Sulphurea, *Great Sulphur Cockatoo.*

Fam. III. Picidæ.
Sub-fam. *c. Picinæ.*
PICUS. Major, *Great Spotted Woodpecker.*
PICUS. Viridis, *Green Woodpecker.*
Sub-fam. *g. Yuncinæ.*
YUNX. Torquilla, *Wryneck.*

Fam. IV. Cuculidæ.
Sub-fam. *e. Cuculinæ.*
CUCULUS. Canorus, *Cuckoo.*

Order IV. *COLUMBÆ.*
Fam. I. Columbidæ.
Sub-fam. *b. Columbinæ.*
COLUMBA. Palumbus, *Ring-dove.*
——— Œnas, *Stockdove.*
TURTUR. Auritus, *Turtle-dove.*
ECTOPISTES. Migratoria, *Passenger Pigeon.*

Order V. *GALLINÆ.*
Fam. III. Phasianidæ.
Sub-fam. *a. Pavoninæ.*
PAVO. Cristatus, *Peacock.*
Sub-fam. *b. Phasianinæ.*
ARGUS. Giganteus, *Argus Pheasant.*
PHASIANUS. Colchicus, *Pheasant.*
Sub-fam. *c. Gallinæ.*
GALLUS. Domesticus, *Domestic Fowl.*
Sub-fam. *d. Meleagrinæ.*
MELEAGRIS. Gallapavo, *Turkey.*
NUMIDA. Meleagris, *Guinea Fowl.*

Fam. IV. Tetraonidæ.
Sub-fam. *a. Perdicinæ.*
PERDIX. Cinerea, *Partridge.*
COTURNIX. Communis, *Quail.*
Sub-fam. *b. Tetraoninæ.*
TETRAO. Urogallus, *Capercaillie.*
——— Tetrix, *Black Grouse.*
LAGOPUS. Scoticus, *Red Grouse.*
——— Albus, *Ptarmigan.*
Fam. V. Megapodidæ.
TALEGALLUS. Lathami, *Brush Turkey.*
MEGAPODIUS. Tumulus, *Mound-making Megapode.*

Order VI. *STRUTHIONES.*
Fam. I. Struthionidæ.
Sub-fam. *a. Struthioninæ.*
STRUTHIO. Camelus, *Ostrich.*
CASUARIUS. Casoar, *Cassowary.*
DROMAIUS. Novæ-Hollandiæ, *Emu.*
Sub-fam. *b. Apteryginæ.*
APTERYX. Australis, *Apteryx.*

Sub-fam. c. *Didinæ.*
Didus. Ineptus, *Dodo.*
Sub-fam. d. *Otinæ.*
Otus. Tarda, *Bustard.*

Order VII. *GRALLÆ.*
Fam. I. Charadridæ.
Sub-fam. a. *Œdicneminæ.*
Œdicnemus. Crepitans, *Stone Curlew.*
Sub-fam. c. *Charadrinæ.*
Vanellus. Cristatus, *Lapwing.*

Fam. II. Ardeidæ.
Sub-fam. b. *Gruinæ.*
Grus. Cinerea, *Crane.*
Sub-fam. c. *Ardeinæ.*
Ardea. Cinerea, *Heron.*
Anthropoides. Virgo, *Demoiselle Crane.*
Botaurus. Stellaris, *Bittern.*
Platalea. Leucorodia, *White Spoonbill.*
Sub-fam. d. *Ciconinæ.*
Ciconia. Albo, *Stork.*
Leptoptilos. Argala, *Adjutant.*
Sub-fam. f. *Tantalinæ.*
Ibis. Religiosa, *Sacred Ibis.*

Fam. III. Scolopacidæ.
Sub-fam. a. *Limosinæ.*
Cracticornis. Arquatus, *Curlew.*
Sub-fam. b. *Totaninæ.*
Tringoides. Hypoleuca, *Sandpiper.*
Himantopus. Candidus, *Stilt Plover.*
Sub-fam. c. *Recurvirostrinæ.*
Recurvirostra. Avocetta, *Avocet.*
Sub-fam. d. *Scolopacinæ.*
Scolopax. Rusticola, *Woodcock.*
Numenius. Scolopacinus, *Snipe.*
Sub-fam. e. *Tringinæ.*
Philomachus. Pugnax, *Ruff.*

Fam. IV. Palamedeidæ.
Sub-fam. a. *Parrinæ.*
Parra. Jacana, *Jacana.*

Fam. V. Rallidæ.
Sub-fam. a. *Rallinæ.*
Orygometra. Crex, *Corncrake.*
Sub-fam. b. *Gallinulinæ.*
Gallinula. Chloropus, *Water Hen.*
Fulica. Atra, *Coot.*

Order VIII. *ANSERES.*
Fam. I. Anatidæ.
Sub-fam. a. *Phœnicopterinæ.*
Phœnicopteros. Rubra, *Flamingo.*
Sub-fam. c. *Anserinæ.*
Bernicla. Leucopsis, *Bernicle Goose.*
Sub-fam. d. *Cygninæ.*
Cygnus. Olor, *Mute Swan.*
——— Ferus, *Whistling Swan.*
Chenopis. Atrata, *Black Swan.*
Sub-fam. e. *Anatinæ.*
Anas. Boschas, *Mallard.*
Mareca. Penelope, *Widgeon.*
Querquedula. Crecca, *Teal.*
Sub-fam. f. *Fuligulinæ.*
Somateria. Mollissima, *Eider Duck.*

Fam. II. Colymbidæ.
Sub-fam. a. *Colymbinæ.*
Colymbus. Glacialis, *Great Northern Diver.*
Sub-fam. b. *Podicepinæ.*
Podiceps. Cristatus, *Crested Grebe.*
——— Minor, *Dabchick.*

Fam. III. Alcidæ.
Sub-fam. a. *Alcinæ.*
Fratercula. Arctica, *Puffin.*
Alca. Impennis, *Great Auk.*
Sub-fam. c. *Spheniscinæ.*
Spheniscus. Demersus, *Cape Penguin.*
Sub-fam. d. *Urinæ.*
Uria. Troilé, *Guillemot.*

Fam. IV. Procellaridæ.
Sub-fam. a. *Procellarinæ.*
Procellaria. Glacialis, *Fulmar Petrel.*
Thalassidroma. Pelagica, *Stormy Petrel.*
Diomedea. Exulans, *Albatros.*

Fam. V. Laridæ.
Sub-fam. b. *Larinæ.*
Larus. Marinus, *Black-backed Gull.*
Sub-fam. c. *Sterninæ.*
Sterna. Hirundo, *Tern.*

Fam. VI. Pelecanidæ.
Sub-fam. b. *Phaetoninæ.*
Phaeton. Æthereus, *Tropic Bird.*
Sub-fam. c. *Pelecaninæ.*
Sula. Bassanea, *Gannet.*

PHALACROCORAX. Carbo, *Cormorant.*
PELECANUS. Onocrotalus, *White Pelican.*
FREGATA. Aquila, *Frigate Pelican.*

Class III. REPTILIA.
Order I. *SAURA.*
Sub-order I. LEPTOGLOSSÆ.
Tribe I. CYCLOSAURA.
Fam. IV. Lacertinidæ.
ZOOTOCA. Vivipara, *Lizard.*
Tribe II. GEISSOSAURA.
Fam. XV. Scincidæ.
ANGUIS. Fragilis, *Blind-worm.*
Sub-order II. PACHYGLOSSÆ.
Tribe III. NYCTISAURA.
Fam. XXII. Geckotidæ.
GECKO. Verus, *Gecko.*
Tribe IV. STROBILOSAURA.
Fam. XXIII. Iguanidæ.
IGUANA. Tuberculata, *Iguana.*
Fam. XXIV. Agamidæ.
DRACO. Volans, *Flying Dragon.*
Tribe V. DENDROSAURA.
Fam. XXV. Chameleonidæ.
CHAMELEON. Vulgaris, *Chameleon.*
Order II. *OPHIDIA.*
Sub-order I. VIPERINA.
Fam. I. Crotalidæ.
UROPSOPHUS. Durissus, *Rattle-snake.*
Fam. II. Viperidæ.
CLOTHO. Arietans, *Puff-Adder.*
CERASTES. Hasselquistii, *Cerastes.*
PELIAS. Berus, *Viper.*
Sub-order II. COLUBRINA.
Fam. IV. Boidæ.
BOA. Constrictor, *Boa.*
Fam. V. Colubridæ.
NAJA. Tripudians, *Cobra.*
NATRIX. Torquata, *Ringed Snake.*
Order III. *CHELONIA.*
Fam. I. Testudinidæ.
TESTUDO. Græca, *Tortoise.*
Fam. V. Cheloniadæ.
CHELONIA. Viridis, *Turtle.*
Order IV. *EMYDOSAURI.*
Fam. I. Crocodilidæ.
CROCODILUS. Vulgaris, *Crocodile.*
Fam. II. Alligatoridæ.
ALLIGATOR. Mississipensis, *Alligator.*

Class IV. AMPHIBIA.
Order I. *BATRACHIA.*
Sub-order I. SALIENTIA.
RANA. Temporaria, *Frog.*
BUFO. Vulgaris, *Toad.*
Sub-order II. GRADIENTIA.
Fam. I. Salamandridæ.
TRITON. Cristatus, *Newt.*
Order V. *MEANTIA.*
Fam. I. Proteidæ.
PROTEUS. Anguinus, *Proteus.*

Class V. PISCES.
Sub-class I. PISCES OSSEI.
Order I. *ACANTHOPTERYGII.*
Sub-order I. DACTYLOPHORI.
Fam. I. Triglidæ.
TRIGLA. Cuculus, *Gurnard.*
Sub-order II. HOLODACTYLI.
Fam. IV. Percidæ.
PERCA. Fluviatilis, *Perch.*
Fam. II. Cottidæ.
COTTUS. Gobio, *Bull-head.*
Fam. XIII. Scomberidæ.
SCOMBER. Scombrus, *Mackarel.*
THYNNUS. Thynnus, *Tunny.*
XIPHIAS. Gladius, *Sword-fish.*
GASTEROSTEUS. Aculeatus, *Stickleback.*
Fam. XIV. Zeidæ.
ZEUS. Faber, *John Dory.*
Fam. XVII. Syngnathidæ.
HIPPOCAMPUS. Brevirostris, *Sea-horse.*
Fam. XXII. Echeneidæ.
ECHENEIS. Remora, *Sucking-fish.*
Fam. XXIII. Lophiidæ.
LOPHIUS. Piscatorius, *Angler.*

Order II. *MALACOPTERYGII.*
Sub-order I. ABDOMINALIA.
Fam. I. Cyprinidæ.
CYPRINUS. Carpio, *Carp.*
——— Barbus, *Barbel.*
——— Auratus, *Gold-fish.*
ABRAMIS. Brama *Bream.*
GOBIO. Fluviatilis, *Gudgeon.*
TINCA. Vulgaris, *Tench.*
LEUCISCUS. Rutilus, *Roach.*
——— Leuciscus, *Dace.*
——— Cephalus, *Chub.*
COBITIS. Barbatula, *Loach.*
Fam. II. Esocidæ.
ESOX. Lucius, *Pike.*
EXOCŒTUS. Volitans, *Flying-fish.*
Fam. IV. Salmonidæ.
SALMO. Salar, *Salmon.*
——— Fario, *Trout.*
Fam. V. Clupeidæ.
CLUPEA. Pilchardus, *Pilchard.*
——— Harengus, *Herring.*
ENGRAULIS. Enchrasicholus, *Anchovy.*
Sub-order II. SUB-BRACHIATA.
Fam. VI. Gadidæ.
MORRHUA. Callarias, *Cod.*
Fam. VII. Pleuronectidæ.
PSETTA. Maxima, *Turbot.*
SOLEA. Vulgaris, *Sole.*
Sub-order III. APODA.
Fam. IX. Murænidæ.
ANGUILLA. Acutirostris, *Sharp-nosed Eel.*
CONGER. Vulgaris, *Conger.*
Fam. X. Gymnotidæ.
GYMNOTUS. Electricus, *Electric Eel.*

Order III. *PLECTOGNATHI.*
Fam. I. Diodontidæ.
ORTHAGORISCUS. Mola, *Short Sun-fish.*

Sub-class II. PISCES CHONDROPTERYGII.
Sub-order I. ELEUTHEROPOMI.
Fam. I. Acipenseridæ.
ACIPENSER. Sturio, *Sturgeon.*
Sub-order II. TREMATOPNEI.
Sub-section I. *SQUALI.*
Fam. I. Scyllidæ.
SCYLLIUM. Canicula, *Little Spotted Dog-fish.*
Fam. II. Squalidæ.
SQUALUS. Carcharias, *White Shark.*
SPHYRNIAS. Zygæna, *Hammer-headed Shark.*
Sub-section II. *RAII.*
Fam. I. Pristidæ.
PRISTIS. Antiquorum, *Sawfish.*
Fam. II. Raidæ.
TORPEDO. Scutata, *Torpedo.*
RAIA. Clavata, *Thornback Skate.*
Sub-order III. CYCLOSTOMI.
Fam. I. Petromyzonidæ.
PETROMYZON. Marinus, *Lamprey.*
LAMPETRA. Fluviatilis, *Lampern.*
MYXINE. Glutinosa, *Myxine.*

Division II. INVERTEBRATA.
Class VI. MOLLUSCA.
Order *CEPHALOPODA.*
Fam. Octopidæ.
OCTOPUS. Vulgaris, *Cuttle-fish.*
ARGONAUTA. Argo, *Nautilus.*

Order *GASTEROPODA.*
Sub-order PULMOBRANCHIATA.
Fam. Limacidæ.
LIMAX. Ater, *Black Slug.*
Fam. Helicidæ.
HELIX. Aspersa, *Snail.*
Fam. Turbinidæ.
SCALARIA. Pretiosa, *Royal Staircase Wentletrap.*
Fam. Coniidæ.
CONUS. Generalis, *Cone.*
Fam. Cypræidæ.
ARICIA. Moneta, *Money Cowry.*
Fam. Buccinidæ.
BUCCINUM. Undatum, *Whelk.*
Fam. Muricidæ.
MUREX. Tribulus, *Thorny Woodcock.*

Order *CYCLOBRANCHIATA.*
Fam. Patellidæ.
PATELLA. Vulgata, *Limpet.*

Order *CONCHIFERA.*
Fam. Pectinidæ.
PECTEN. Jacobæus, *Scallop.*
OSTREA. Edulis, *Oyster.*

Fam. Meleagrinidæ.
MELEAGRINA. Margaratifera, *Pearl Oyster.*

Fam. Mytilidæ.
MYTILUS. Edulis, *Edible Mussel.*

Order *CIRRHOPODA.*
PENTALASMIS. Anatifera, *Bernicle.*

Class VII. CRUSTACEA.
Sub-class I. MALACOSTRACA.
Order I. *DECAPODA.*
Sub-order I. DECAPODA-BRACHYURA.
Fam. I. Canceridæ.
CANCER. Pagurus, *Crab.*

Sub-order II. DECAPODA-ANOMOURA.
Fam. III. Paguridæ.
PAGURUS. Bernhardus. *Hermit Crab.*

Sub-order III. DECAPODA-MACROURA.
Fam. V. Astacidæ.
POTAMOBIUS. Astacus, *Cray-fish.*
ASTACUS. Gammarus, *Lobster.*
Fam. VI. Crangonidæ.
CRANGON. Vulgaris, *Shrimp.*

Fam. VIII. Palæmonidæ
PALÆMON. Serratus, *Prawn.*

Class VIII. ARACHNIDA.
Order *PULMONARIA.*
Fam. Araneidæ.
MYGALE. Avicularia, *Bird Spider.*
Fam. Scorpionidæ.
SCORPIO. Europæus, *Scorpion.*

Order *TRACHEARIA.*
Fam. Acaridæ.
LEPTUS. Autumnalis, *Harvest Bug.*

Class IX. INSECTA.
Sub-class I. INSECTA MANDIBULATA.
Order I. *COLEOPTERA.*
Fam. Cicindelidæ.
CICINDELA. Campestris, *Tiger-beetle.*

Fam. Carabidæ.
CARABUS. Cancellatus, *Ground-beetle.*
Fam. Silphidæ.
NECROPHAGUS. Vespillo, *Burying-beetle.*

Fam. Lucanidæ.
LUCANUS. Cervus, *Stag-beetle.*
GEOTRUPES. Stercorarius, *Dor-beetle.*
MELOLONTHA. Vulgaris, *Cockchaffer.*

Fam. Lampyridæ.
LAMPYRIS. Noctiluca, *Glowworm.*

Fam. Ptinidæ.
ANOBIUM. Tesselatum, *Death-watch.*

Fam. Cerambycidæ.
CERAMBYX. Moschatus. *Musk-beetle.*

Fam. Staphylinidæ.
CREOPHILUS. Maxillosus, *Rove-beetle.*

Order II. *DERMAPTERA.*
FORFICULA. Forcipata, *Earwig.*

Order III. *ORTHOPTERA.*
Fam. Locustidæ.
LOCUSTA. Tartarica, *Locust.*

Fam. Achetidæ.
ACHETA. Campestris, *Field Cricket.*
——— Domesticus, *House Cricket.*
GRYLLOTALPA. Vulgaris, *Mole Cricket.*
PHYLLIA. Foliata, *Leaf Insect.*

Fam. Blattidæ.
BLATTA. Orientalis, *Cockroach.*

Order IV. *NEUROPTERA.*
Fam. Ephemeridæ.
EPHEMERA. Vulgata, *May-fly.*

Fam. Libellulidæ.
LIBELLULA. Depressa, *Dragon-fly.*

Fam. Myrmeleonidæ.
MYRMELEON. Formicarum, *Ant-lion.*

Fam. Termitidæ.
TERMES. Bellicosus, *Termite.*

Order V. *TRICHOPTERA.*

Fam. Phryganidæ.
PHRYGANEA. Grandis, *Caddis-fly.*

Order VI. *HYMENOPTERA.*

Fam. Ichneumonidæ.
PIMPLA. Manifestator, *Ichneumon-fly.*

Fam. Formicidæ.
FORMICA. Rufa, *Wood Ant.*

Fam. Vespidæ.
VESPA. Crabro, *Hornet.*
——— Vulgaris, *Wasp.*

Fam. Apidæ.
APIS. Mellifica, *Honey Bee.*

Sub-class II. INSECTA HAUSTELLATA.

Order I. *LEPIDOPTERA.*

Fam. Papilionidæ.
PAPILIO. Machaon, *Swallow-tailed Butterfly.*
ARGYNNIS. Adippe, *Silver-spotted Fritillary.*
VANESSA. Atalanta, *Red Admiral.*

Fam. Sphingidæ.
ACHERONTIA. Atropos, *Death's-head Moth.*

Fam. Sesiidæ.
MACROGLOSSA. Stellatarum, *Humming-bird Moth.*

Fam. Arctiadæ.
ARCTIA. Caja, *Tiger Moth.*

Fam. Geometridæ.
OURAPTERYX. Sambucaria, *Swallow-tailed Moth.*

Fam. Alucitidæ.
ALUCITA. Hexadactyla, *Many-plumed Moth.*

Order II. *DIPTERA.*

Fam. Culicidæ.
CULEX. Pipiens, *Gnat.*

Fam. Œstridæ.
ŒSTRUS. Bovis, *Gadfly.*

Fam. Bombylidæ.
BOMBYLIUS. Medius, *Humble-bee Fly.*

Order IV. *APHANIPTERA.*

Fam. Pulicidæ.
PULEX. Irritans, *Flea.*

NATURAL HISTORY.

Division I. VERTEBRĀTA.—(Lat. *possessing vertĕbræ.*)
Class I. . . MAMMALĬA.—(Lat. *suckling their young.*)
Order I. . . *PRIMĀTES.*—(Lat. *primus,* first.)
Family I. . Hominĭdæ.—(Lat. *homo,* a man— mankind.)
Genus I. . Homo.

BOSJESMAN AND LION.

Species I. Sapiens (Lat. *wise*), *Man.*

Man holds the foremost place in the order of creation. The perfection of his bodily form is as far superior to that of other beings as his intellect surpasses their instinct, beautiful and marvellous though it be. Between man and brutes there is an impassable barrier, over which man can never fall, or beasts hope to climb. The lower animals are but "beasts that perish," whereas man is a LIVING SOUL. It is the soul that gives consistency and force to the reason, and therefore, man, even when fallen from his high estate, and deprived of the right use of his reason, still holds his supremacy over the lower animals by the power of the still living soul, and is not subject even to the most perfect and powerful brutes.

There is but one genus of mankind, HOMO, and but one species, *Sapiens;* that is, the rational human being. Intellect, or reason, differs from instinct in its power of accommodation to circumstances; whereas instinct ever remains unchanged. The beaver, when confined in a cage, still builds dams in order to confine the stream that never visits it; the captive squirrel, when satiated with food, still conceals the remnants for a future repast, although it is regularly supplied with its daily meals; the magpie approaches a dead wasp with the same caution as if it were living; and the dog flies from a recently flayed tiger skin with no less fear than if the living tiger stood before him.* On the contrary, the power of man's reason enables him to alter his habits and actions according to the change of external circumstances. The same man can inhabit the burning sands of the tropics, or the everlasting snows of the north pole; and can defend himself from the scorching heat of the one, or set at defiance the piercing cold of the other.

The forms and habits of men are modified according to the different climates and positions in which they are placed. These modifications are in some cases so great, that many philosophers, and not a few naturalists, have imagined that there are several distinct classes of mankind, which derive their origin from different sources. There is certainly no doubt that the educated human being who peruses these pages, seated in a comfortable apartment, surrounded with luxuries brought from almost every country on the face of the earth, within sound of church bells, and clothed in garments fitted to defend him from the heat of summer or the cold of winter, is far superior to the half-naked Bosjesman, who has no conception of a God, who lives in caves, or scrapes a hole in the sand, in which he crouches until he has devoured the last putrid morsel of the prey which he has been fortunate enough to secure, and which he then abandons to the beasts of the desert, scarcely less provident than himself. Yet this superiority results entirely from the external circumstances in which each is placed. Let each be transplanted into the country of the other, and in a few generations we should find the Bosjesman civilized, and capable of reading how his former superior, now sunk into the savage state, gains a precarious subsistence by hunting, and passes his life in caves.

All men do not see the case in this light, for some theorists have ventured so far as to assert that the Negro is but an improved monkey, and that his

* In those cases where animals alter their habits to suit the changed circumstances in which they find themselves, their *reason*, not their *instinct*, acts. See this point more at large, in the introduction to "Sketches and Anecdotes of Animal Life."

reason is nothing but a partially civilized instinct. That these theorists were no anatomists is sufficiently evident, and it would not be necessary to prove the absurdity of their assertion, were it not that many have actually been deceived by their flimsy though specious arguments. Indeed, at the present time, when we find one philosopher giving what he considers satisfactory proofs that salt is the cause of all earthly misery, and the reason why the sun is at so great a distance from us; another reviving the very ancient and venerable belief, that the earth is flat like a plate; and a third pretending to read a sealed letter with the point of his toe, or to examine the interior of a friend some hundred miles distant; it is difficult to say to what extent credulity can proceed.*

We will, however, briefly examine this theory respecting the humanity of the Negro, partly by anatomy, but mostly (for which delinquency we must ask pardon of the theorists in question) by common sense. That monkey, or rather ape, whose form most resembles that of man, is the Orang-outan. Let us compare the skull of this animal with that of the Negro. Will any one venture to deny that the noble sweep of cranium, and the smooth globular surface of the human skull, demonstrating the volume of the brain within, is a proof of far superior intellect than is indicated by the heavy ridges, the irregular prominences, and the small capacity of the ape's skull? The face of the ape is an instrument for procuring food, and a weapon for attack and defence, while that of man is an ever-changing index of the workings of the mind within. We therefore find that the jaws of the ape are enormously developed, armed with formidable fangs, and marked with strong bony ridges, to which the powerful muscles which move the jaws are attached. On the other hand, as man is enabled to procure food, and to manufacture weapons by means of his hands, his jaws and teeth are reduced to the smallest size compatible with the preservation of life.

The habitually erect posture is another characteristic of mankind. Other animals are not fitted for it; since, when they attempt to assume that position, their head is thrust so far forward that its weight destroys their balance, and the bones of the leg and the pelvis are so formed as to give them a tottering gait. When the ape attempts to stand erect, it is forced to balance itself by its immensely long arms, and cannot walk without assisting itself along by the knuckles pressed on the ground. The fingers on the feet, or more properly the hinder hands, prevent the ape from planting more than the heel upon the ground. It therefore hobbles along with its body bent, and at best can only contrive to manage an uncertain and vacillating shuffle; nor does it ever walk so well or so

* I say "pretending," because, although there are sundry accounts of such sealed letters proving legible to the "clairvoyant," there still exists a crucial test in the shape of a very thick sealed envelope, containing a bank note of considerable value, which will become the property of any one who can read its number and signatures without opening the envelope. It is almost unnecessary to say that the bank note has reposed in illegible security ever since its first enclosure, and is likely so to do until the patience of its owner is exhausted.

gracefully in the erect posture as many of the performers at Astley's do on their hands, which are apparently less fitted for walking than those of the ape.

The power of the thumb is much greater in man than in the apes; it is by means of this instrument that man is able to handle large or small objects, to wield a sword or a pen, to cast a spear or thread a needle. There are also many anatomical differences which need not be described.

The intellectual power in man shows its supremacy over the instinct of the ape in many ways. We will take as our example of mankind, the most abject of the human race, the Bosjesman, as represented at the commencement of this chapter. Surely the slain lion was not destroyed by an ape. No ape or monkey was ever able to manufacture weapons for itself. It may, indeed, take up a stick or a stone and defend itself vigorously,* but it could never form a bow and arrow, much less reflect that the juices of certain plants rubbed on the points of its weapons would cause inevitable death to any person wounded by them. Yet the diminutive Bosjesman, who is far lower in intellect, and much less civilized than the calumniated Negro, boldly attacks, with perfect certainty of success, an animal before which the most intelligent ape that ever lived would fly in helpless terror.

Neither can an ape procure fire, nor even renew it. It will sit delighted by a flame which a chance traveller has left, and spread its hands over the genial blaze; but when the glowing ashes fade, it has not sufficient understanding to supply fresh fuel, but sits and moans over the expiring embers.

The Bosjesman makes a bow and arrow; he tips the arrow with a hard substance to make it penetrate; he imbues the point with substances which he has learned are fatal when mingled with the blood, and then sallies forth in search of some animal whose skin may serve as a dress, and whose flesh may furnish him a meal. When by his unerring weapons he has succeeded in destroying the terrible and ferocious lion, the swift antelope, or the wary ostrich, he constructs for himself a hut by the side of his prey, strikes fire, fetches fuel, and dresses his meat. These are actions which no beast ever performed, and no ape could ever imitate.

One point of difference between man and brutes has yet to be mentioned—LANGUAGE. This one word includes almost every distinction mentioned, as it is by the use of language that we are enabled to communicate our ideas to each other, to give the thoughts hidden in our minds an almost visible shape, to record our experience for the benefit of others; in a word, it is by language that we are civilized. The ape has no language, although there is no apparent anatomical reason why apes should not speak, and therefore, the Orang-outan in the gardens of the Zoological Society is no more refined, nor does it make a nearer approach to civilization, than its ancestors in the time of Adam.

* Even this statement is generally discredited by naturalists.

Yes, it is so. The whole of mankind has one common source, springing from one origin, from him who was formed out of the dust of the earth, and from her whose existence took its rise from his living side. All are one family, sprung from one father on earth, by the will of their one Father in Heaven.

The mind of man is much influenced by outward objects and the society by which it is surrounded. If a man be confined to one spot, or within certain bounds, his mind becomes feeble in proportion to the isolation. The rustic, whose ideas never wander from the farm on which he works, and whose travels are circumscribed by his native village, or, at most, by a casual visit to the nearest market town, exhibits a mind which has received a certain set of ideas, false as well as true, and which refuses alike to admit new notions or to give up any of the old.

So great is the influence of society on the mind, that an experienced clergyman, while examining some candidates for Confirmation, observed that the Oxford children were two years in advance of those of the same age who had been bred in the country. So with music, a town child is accustomed to hear street music, and readily catches the air, while the country child, whose notions of music are confined to the dismal hosannas and lugubrious psalmody of the village church, is usually devoid of musical ear, but is great in imitation of rooks, cows, pigs, and donkeys.

The most perfect case of isolation known, was that of the celebrated Kaspar Hauser, who had been confined for the first fourteen or fifteen years of his life in a dark cave, and was never permitted even to see his keeper. In consequence, when he at length left his dungeon, his mind was that of an infant, his body that of a man. It would have been a most interesting and important experiment to watch the gradual development of his mind, but, unfortunately for science, an unknown dagger reached his heart, and this mysterious victim of a hidden plot perished, leaving the riddle of his life unsolved and the development of his intellect unfinished. His history furnishes us with another distinction between man and beasts. When the mind of Hauser was released from its bands, it at once began to expand, and every day gave it fresh powers, while the brain of the ape is rapidly developed when young, and receives no further increase as it grows in stature, or if any change at all takes place, rather diminishes in power than develops by increasing years.

We have seen that mankind have little in common with brutes, and that the barrier between the two can be passed by neither: but there is another question to be considered.

Many theorists have boldly denied the revelation of man's origin as given in the Scriptures, and asserted that at least five distinct races of men must have been created simultaneously, each deriving its origin from a different progenitor. Let us see how this theory will hold.

It has already been stated that man is modified according to the climate and position in which he is placed. There are several of these modifications, or varieties as they are called, but authors do not agree as to their number. Some describe the human family as divided into five varieties or races: the Caucasian, the Mongolian, the Ethiopian, the Malayan, and the American; each of these being subdivided into families, as for instance, the Caucasian race subdivided into the Caucasian, the Celtic, the Germanic, the Arabian, the Libyan, the Nilotic, and the Indostanic families. The division generally received is that of Pickering, who enumerates *eleven distinct races of men*, all of whom he has seen; the Arabian, Abyssinian, Mongolian, Hottentot, Malay, Papuan, Negrillo, Telingan, Ethiopian, Australian, and Negro. He differs from Prichard in several points, but especially in referring the population of America to the Mongolian race, whereas Prichard considers it as entirely separate.

The characteristics and distribution of each race are briefly these. The *Arabian* race extends over the whole of Europe, excepting Lapland, about half of Asia, including the greater part of India, and most of the northern third of Africa. The complexion is light, the lips are thin, the nose is prominent, and the beard thick. Number, about 350,000,000.

The *Abyssinian* race occupies a small tract towards the east of Africa, including part of Abyssinia, and part of Nubia. The features are like those of Europeans, the complexion is light, the hair is crisp, and the beard moderate. Number about 3,000,000.

The *Mongolian* race is remarkable for a feminine aspect in both sexes, so that a stranger is often perplexed to distinguish a man from a woman at a short distance; the hair is straight, and the beard is wanting. It extends over the eastern half of Asia, except Corea, over Lapland, and the whole of America, except the western coast by California, and the upper part of South America. Number 300,000,000.

The *Hottentot* race occupies the southern extremity of Africa. The complexion is not so dark as that of the Negro, the hair is woolly, and frequently grows in irregular patches, leaving a bald spot in the centre of each patch. This race includes the Bechuanas and the Bosjesmans. The complexion of the Bosjesmans, or Bushmen, is very light, and strongly resembles that of an European, with a few sooty patches irregularly placed. Number about 500,000.

The *Malay* race is almost amphibious, and is never found far inland. It is widely spread, and inhabits the centre of Madagascar, the whole of the islands in the Pacific Ocean, except the Fiji, New Hebrides, Solomon's Isles, Papua, and parts of the Philippines. The parts of America not populated by the Mongolians, are also inhabited by this race. The complexion is a dark copper, the hair straight, when cut it stands erect, and the beard is thin. Number 120,000,000.

The *Papuan* race inhabits about two-thirds of Papua, and the Fiji Islands, where Pickering saw the only individuals of this race who came under his notice. The complexion is dark, the hair bushy, the beard copious. The most remarkable point in this race is the skin, which is astonishingly rough and harsh. Number 3,000,000.

The *Negrillo* race is like the Papuan in colour, but the hair is more woolly, the stature is small, and the beard absent. The Negrillos inhabit part of Papua, Solomon's Isles, the northern extremities of Luzon and Sumatra, and the New Hebrides. Number 3,000,000.

The *Telingan*, or Indian race, inhabits the eastern parts of India, especially about Calcutta, several isolated spots in other parts of India, and the east coast of Madagascar. The complexion is dark, (best imitated by a mixture of red and black,) the skin is soft, the features are like those of Europeans, hair straight and fine, and the beard copious. Number 60,000,000.

The *Ethiopian* race is darker than the Telingan, the hair is crisp and fine, skin soft, and the features are more like European features than those of the Negro. This race inhabits the north-eastern portion of Africa, including Southern Egypt, part of Nubia, and part of Abyssinia; a few detached spots toward the north-west, and a large tract of country by Senegambia. Number 5,000,000.

The *Australian* race inhabits Australia alone. The complexion is like that of the Negro, but the hair is not woolly like that of the Negro. Number 500,000.

The *Negro* race inhabits the central parts of Africa, from the north of Ashanti to a little southward of Zanzibar. The complexion is black, the lips are immensely thick, the nose is flat, and the hair is close and curly, strongly resembling wool. Number 55,000,000. The numbers given in this distribution are of course in many cases only conjectural.

In the distribution of races, it is most interesting to observe the influence of climate and vegetation on the character of man. The vast tract of desert extending from the north-west of Africa, through Arabia, part of India and Tartary, as far as Mongolia, is inhabited by nomadic, or wandering, tribes, who depend principally on the milk of their domesticated animals for subsistence.

The interminable and trackless woods of North America develop tribes whose faculties are moulded to the exigencies of their position. To their practised senses the tangled forests are as clear as the highway; the moss on the trees, the sun by day, the stars by night, the rushing of the wind, or the sounds of animal life, are as broad roads and legible signs to them, although we could discover no means to escape from the wilderness of trees. Dependent in a great measure on hunting for their subsistence, their keen eye marks the slightest trace of the expected prey; a drooping leaf, a twisted blade of grass, a bent twig, a ripple in the stream, are all noticed and all understood. Ever eagerly bent on the destruction of inimical tribes, and deeming the number of "scalps" attached to their dress, each designating a slain enemy, as the best mark of nobility, they learn to track an enemy by his footsteps with unexampled patience and untiring assiduity. No bloodhound ever followed his prey with more certainty than the American Indian when on his "war-path" tracks his retiring enemies, and when near them his approach is silent as the gliding of the serpent, his blow as deadly as its fangs.

The Malay race, whose lot is thrown amid islands and coasts, are as crafty and fierce on the waters as the American Indians in their woods. Accustomed to the water from their earliest infancy, able to swim before they can walk, using as their toys waves that would dash an ordinary swimmer to pieces against the rocks, their existence is almost entirely passed on the water. As the American Indians are slayers and robbers by land, so are the Malays murderers and pirates by sea. They have been known to capture a ship in the midst of a storm by swimming to it and climbing up the cable, and many instances of their crafty exploits in ship-taking are on record. For a full account of their ferocity, cunning, and endurance, the reader is referred to Sir James Brooke's reports on the Borneo pirates.

The Esquimaux, situated among ice and snow, where mercury freezes in the open air and water becomes ice within a yard of a blazing fire, pass a comparatively inactive life. They actually form the ice and snow into warm and comfortable houses; wrapped up in enormous fur garments that almost disguise the human form, they defy the intensity of the frost, and place their highest happiness in the chance possession of a whale, which will furnish them with food, clothing, and light, through their long winter.

All these races, although they differ in habits and external appearance, are not different genera, or even different species, but only *varieties* of *one* species. There is not so marked a distinction between the European and Negro, as between the light and active racer and the heavy brewer's horse; yet no one attempts to deny that these two animals belong to one species. The varieties in man are permanent; that is, the child of Negro parents will be a Negro, and the child of Malay parents will be a Malay, but that is no proof of a distinct species, as precisely the same argument may be used with regard to the horse. The *soul* is the important part of man, not the body; and though the outward bodies of men differ, the soul is the same in all, and therefore in all the mind is capable of improvement and cultivation.

It were an easy task to prove the unity of mankind by Scriptural proofs, but I have thought it better to use rational arguments, as so-called reason was the weapon used to disprove the facts which the Scriptures asserted. Sufficient, I trust, has been said to show that man "has dominion over the fish of the sea, and over the fowl of the air, and over every living thing that moveth upon the earth;" and also that the whole of mankind forms one great family, precisely according to the Scriptural assertion, that Eve was "the mother of all living."

The distribution of the different nations over the face of the earth, and especially the presence of man in certain situations, forms one of the principal arguments of those who deny the unity of mankind. Among such lands may be named America, Australia, New Zealand, and several other countries, whose inhabitants were supposed to have been destitute of artificial means of crossing the expanse of waters that divides their lands from others.

So to obviate this difficulty the perplexed philosophers invented a theory that each race was separately and simultaneously created,—a theory which has but one disadvantage, that of being entirely false. There are many other theories on the same subject, all differing from each other in essentials, but all remarkable for their ingenuity and folly.

As, however, such theories have been promulgated, it cannot but be interesting to one who holds in its fullest sense the doctrine of the unity of mankind as given in the Scriptures, to search after the means by which men were enabled to pass from one common centre to all parts of the earth.

Many portions of the globe, such as islands, could not be reached without some artificial means to enable men to cross the waters. This implies some degree of civilization, as boats or rafts are the result of much thought and some skill. Pickering has published a map containing the probable route of mankind through the earth. He appears to think that the most perplexing question of all, namely the problem of the population of America, is not very difficult of solution, as the Aleutian Isles form a chain of spots easily traversed by the skin-covered canoes which are still in use among those islands.*

All nations which have preserved traditions of past events agree in many points in a very remarkable manner. All have some traditions of a creation, not always of a world, but of that particular part in which they reside. The Fiji islanders believe that one of their gods fished up Fiji from the bottom of the sea, by entangling his fish-hook in a rock, and that the island would have been higher had not the line broken. The fish-hook is still preserved as a proof, but they do not state where the god stood while fishing. A traveller asked one of the priests why the hook, an ordinary tortoishell one, did not break? "Oh! it was a god's hook, and could not break." But why then did the line break? was the traveller's very natural response. Whereupon the man, according to the prevailing system of argument in those countries, and perhaps in a few others, threatened to knock him down if he abused the gods any more. Most nations have dim notions of a deluge which overwhelmed the whole world, and from which only a few individuals escaped, by whom the earth was repeopled. Nearly all believe in a good and an evil power continually at warfare, and that the good will finally subdue the evil. Many savage nations, in consequence, seek to propitiate the evil power with prayers and offerings, feeling sure that the good one will not injure them.

* Pickering's Races of Man. Hall's Edition, p. 296.

All nations, (except one or two, such as the abject Bosjesman, who can form no idea of what he cannot see, and whose answer when told of a God is, "Let me see him,") believe in a future state. Their belief is invariably modified according to their habits. Some of the debased dark races believe that after death they become white men and have plenty of money; the Mahometan considers his paradise as an abode of everlasting sensual indulgence; the savage believes that when he leaves this world he will pass to boundless hunting-fields, where shall be no want of game, and where his arrows shall never miss their mark; while the Christian knows his heaven to be a place of unspeakable and everlasting happiness, where the power of sin shall have ceased for ever.

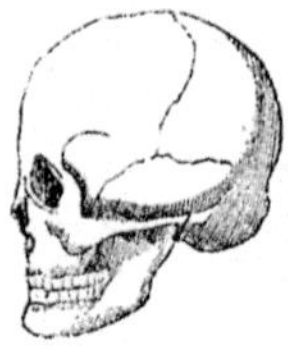

SKULL OF MAN.

THE section Quadrumana includes the apes, baboons, and monkeys. The name of Quadrumana is given to these animals because, in addition to two hands like those of man, their feet are also formed like hands, and are capable of grasping the branches among which most monkeys pass their lives.

Apes are placed at the head of the Quadrumana because their instinct is mostly superior to that of the baboons and monkeys, of whom the former are usually sullen and ferocious, when arrived at their full growth, and the latter volatile and mischievous.

The Chimpansee and the Orang-outan have been confounded together by the older naturalists, whose pardonable error has been unpardonably repeated even in books professing to instruct the young in the present state of natural science. That they are really distinct animals a glance at the skull of each will at once prove. The Chimpansee is a native of Western Africa, and is tolerably common on the banks of the Gambia and in Congo.

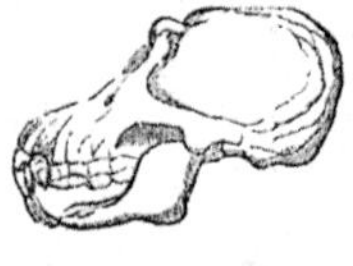

SKULL OF CHIMPANSEE.

Large bands of these formidable apes congregate together and unite in repelling an invader, which they do with such fury and courage that even the dreaded elephant and lion are driven from their haunts by their united efforts. They live principally on the ground, and, as their name imports, spend much of their time in caves or under rocks. Their height is from four to five feet, but they are said not to reach this growth until nine or ten years of age.

Several young chimpansees have been recently imported into this country, and have shown themselves very docile and gentle; but, had they lived, they would probably in a few years have become fierce and obstinate, as apes almost invariably are when they reach their full growth.

I lately saw one of these animals at the Zoological Gardens in the Regent's Park. It was very gentle, and seemed to be above that restless spirit of curiosity that is so characteristic of the monkey tribe. Its countenance appeared expressive of the deepest distress, and it moved about with a gravity that contrasted curiously with the lively movements of the body of monkeys living in the large cage.*

Family II. Simiădæ.—(Lat. *Sĭmĭa*, an ape—Ape kind.)

Troglodȳtes.—(Gr. τρώγλη, a hole ; δύω, to creep.)

Niger (Lat. *black*), *the Chimpansee.*

When I saw it an unmistakeable hacking cough proved that the scourge of these animals had fallen upon it. When monkeys are brought to England they are generally carried off by consumption within a few years. This disease is not so prevalent now in the Zoological Gardens as it was when the keepers thought that the animals ought to be shut up close in a warm atmosphere. This treatment predisposed them to taking cold on the slightest occasion, and thus paved the way for their deadly

* There is a remarkably fine specimen of the Chimpansee in the Jardin des Plantes at Paris. It is very lively, and is fond of swinging on the ropes that are so plentifully hung from the roof of the great monkey cage at that establishment. The Chimpansee appears to be entirely free from disease of the lungs.

foe. Since the monkeys have been permitted to range freely in the open air, they have proved much more healthy than under the old system of confinement.

The ORANG-OUTAN inhabits Borneo and Sumatra. In Borneo there are certainly two species of orang, called by the natives the Mias-kassar and the Mias-pappan. Some naturalists suppose that the Sumatran orang is also a distinct species.

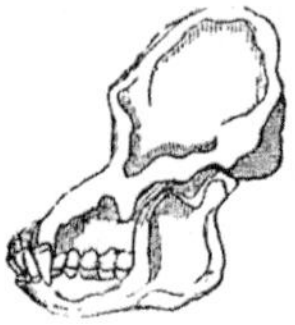

SKULL OF ORANG-OUTAN.

This is the largest of all the apes, as it is said that orangs have been obtained from Borneo considerably above five feet in height. The strength of this animal is tremendous; a female snapped a strong spear asunder after having received many severe wounds. Its arms are of extraordinary length, the hands reaching the ground when it stands erect. This length of arm is admirably adapted for climbing trees, on which it principally resides. Mr. Brooke, the Rajah of Sarāwak, gives the following account of the orangs of Borneo. There appears also to be a third species, the Mias-rombi:—

"On the habits of the orangs, as far as I have been able to observe them, I may remark that they are as dull and as slothful as can well be conceived, and on no occasion, when pursuing them, did they move so fast as to preclude my keeping pace with them easily through a moderately clear forest; and even when obstructions below (such as wading up to the neck) allowed them to get away some distance, they were sure to stop and allow us to come up. I never observed the slightest attempt at defence; and the wood, which sometimes rattled about our ears, was broken by their weight, and not thrown, as some persons represent. If pushed to extremity, however, the pappan could not be otherwise than formidable; and one unfortunate man, who with a party was trying to catch one alive, lost two of his fingers, besides being severely bitten on the face, whilst the animal finally beat off his pursuers and escaped. When they wish to catch an adult, they cut down a circle of trees round the one on which he is seated, and then fell that also, and close before he can recover himself, and endeavour to bind him.

"The rude hut which they are stated to build in the trees would be more properly called a seat, or nest, for it has no roof or cover of any sort. The facility with which they form this seat is curious; and I had an opportunity of seeing a wounded female weave the branches together, and seat herself in a minute. She afterwards received our fire without moving, and expired in her lofty abode, whence it cost us much trouble to dislodge her.

"The pappan is justly named Satyrus, from the ugly face and disgusting callosities. The adult male I killed was seated lazily on a tree; and when approached only took the trouble to interpose the trunk between us, peeping at me and dodging as I dodged. I hit him on the wrist, and he was afterwards dispatched. I send you his proportions, enormous relative to his height; and until I came to actual measurement my impression was that he was nearly six feet in stature.

SIMĬA.—(Lat. *an Ape.*)

Satȳrus (Gr. Σάτυρος, a satyr), *the Orang-outan.*

"The great difference between the kassar and the pappan in size would prove at once the distinction of the two species; the kassar being a small slight animal, by no means formidable in his appearance, with hands and feet proportioned to the body, and they do not approach the gigantic extremities of the pappan either in size or power; and, in short, a moderately strong man would readily overpower one, when he would not stand a shadow of a chance with the pappan."

I saw a young Orang-outan not long since. It was rather spidery in its development, having a very small and very rotund body, to which were affixed very long and slender limbs. Its face was like that of a very misanthropical old miser, thoroughly wearied of life, and contemplating surrounding objects with a calm but derisive pity. The whole form of the creature reminded me strongly of the Goblin king that appeared to Gabriel Grubb on Christmas Eve, and whose *degagé* ease of posture upon a tall tombstone has been so admirably depicted by "Phiz."

It possessed in a high degree the expressive mobile character of the lips, which appeared to express its feelings much in the same manner as do the ears of a horse. When it was alarmed or astonished at any object it was accustomed to shoot out both its lips, and to form its mouth into a trumpet kind of shape. A snail was very effectual in producing this contortion of countenance.

The creature was very tame, and delighted in walking about the garden leaning on the arm of its keeper, and if any lady would venture to be its guide, it appeared as happy as any such misanthropical being could be.

When young the Orang-outan is very docile, and has been taught to make its own bed, and to handle a cup and saucer, or a spoon, with tolerable propriety. For the former occupation it proved itself particularly apt, as it not only laid its own bed-clothes smooth and comfortable, but exhibited much ingenuity in stealing blankets from other beds, which it added to its own. The young Orang in the collection of the Zoological Society evinced extreme horror at the sight of a small tortoise, and, when the reptile was introduced into its den, stood aghast in a most ludicrously terrified attitude, with its eyes intently fixed on the frightful object.

HYLOBĂTES.—(Gr. ὕλη, a wood; βαίνω, to traverse.)

Agĭlis (Lat. *active*), *the Agile Gibbon*, or *Oungka*.

The AGILE GIBBON is a native of Sumatra. It derives its name of Agile, from the wonderful activity it displays in launching itself through the air from branch to branch. One of these creatures that was exhibited in London some time since, sprang with the greatest ease through distances of twelve and eighteen feet; and when apples or nuts were thrown to her while in the air, she would catch them without discontinuing her course. She kept up a succession of springs, hardly touching the branches in her progress, continually uttering a musical but almost deafening cry. She was very tame and gentle, and would permit herself to be touched or caressed. The height of the Gibbon is about three feet, and the reach of the extended arms about six feet. The young Gibbon is usually of a paler colour than its parent. There are several species of Gibbon, amongst which some naturalists include the Siamang, a monkey chiefly celebrated for the pains it takes to wash the faces of its young, a duty which it conscientiously performs in spite of the struggles and screams of its aggrieved offspring.

The KAHAU is a native of Borneo. It derives its name from the cry it utters, which is a repetition of the word "Kahau." It is remarkable for the extraordinary size and shape of its nose, and the natives relate that while leaping it holds that organ with its paws, apparently to guard it against the branches. As may be seen from the engraving, it is not an animal of very captivating appearance; but when it has been macerated in spirits of wine for a few months, its ugliness is quite supernatural. Naturalists formerly supposed that there were two species of this animal,—the nose of one being aquiline like that of the monkey in the accompanying cut; and that of the other being slightly retroussée. It was discovered, however, that the latter animal was only the young Kahau, whose nose had not reached its full beauty.

PRESBYTES.—(Gr. πρεσβύτης, an old man.)

Larvātus (Lat. *masked*) Kahau, or *Proboscis Monkey.*

The length of the animal from the head to the tip of the tail is about four feet four inches; and its general colour is a sandy red, relieved by yellow cheeks and a yellow stripe over the shoulders.

The ENTELLUS, or Hoonuman, is a native of India. It is astonishingly active in the capture of serpents. It steals upon the snake when asleep, seizes it by the neck, runs to the nearest stone, and deliberately grinds down the reptile's head until the poisonous fangs are destroyed, frequently inspecting its work and grinning at the impotent struggles of the tortured reptile. When the snake is rendered harmless the monkey casts it to its young, who, after tossing about and exulting over their fallen enemy for some time, finally destroy it. The length of its head and body is about two feet two inches.

PRESBYTES.

Entellus (Lat. *A proper name*), *the Entellus Monkey.*

Seven genera are omitted on account of want of space.

CYNOCEPHALUS.—(Gr. κύων, a dog; κεφαλή, a head.)

Mormon (Gr. Μορμών, a bogie), *the Mandrill.*

We now arrive at the BABOONS. This tribe is principally distinguished from the apes by their short and insignificant looking tails. The baboons are the only mammalia which exhibit brilliant colours; on some parts of these, however, nature has bestowed vivid tints hardly to be surpassed even by the gorgeous plumage of the tropical birds.

The Mandrill, which is the most conspicuous of the

baboon tribe, is a native of Guinea and Western Africa, and is chiefly remarkable for the vivid colours with which it is adorned. Its cheeks are of a brilliant blue, its muzzle of a bright scarlet, and a stripe of crimson runs along the centre of its nose. These colours are agreeably contrasted by the purple hues of the hinder quarters. It lives principally in forests filled with brushwood, from which it makes incursions into the nearest villages, plundering them with impunity. On this account it is much dreaded by the natives, who feel themselves incapable of resisting its attacks. It is excessively ferocious, and easily excited to anger; and when enraged, so boundless is its rage, that Cuvier relates that he has seen several of these animals actually expire from the violence of their fury.

The greenish brown colour of the hair of this and other monkeys is caused by alternate bands of yellow and black, which exist on each hair. The brilliant colours referred to above belong to the skin, and fade away entirely after death, becoming paler when the animal is not in perfect health.

Family III. *Cebĭdæ.—(Gr. *κῆβος*, a monkey Monkey kind.)

Atĕles.—(Gr. *ἀτελής*, imperfect.)

Paniscus (Gr. *Πανίσκος*, dim. of *πᾶν*, a little Pan), *the Coaita Spider Monkey.*

The **American Monkeys**, or Cebidæ, are found exclusively in South America, and are never seen north of Panama. Their tails are invariably long, and in some genera, prehensile.

The Coaita is one of the Spider Monkeys, so called from their long slender limbs, and their method of progressing among the branches. The tail seems to answer the purpose of a fifth hand, as it is capable of being used for every purpose to which the hand could be applied; indeed, the Spider Monkeys are said to use this member for hooking out objects where a hand could not be inserted. In this manner they often rob nests of birds, who thought that they had laid their eggs safe from all danger. The tail is also of considerable use in climbing among the branches of trees; they coil it round the boughs to lower or raise themselves, and often will suspend themselves entirely

* Pronounce *Kebidæ.*

by it, and then by a powerful impetus swing off to some distant branch. The habits of all the Spider Monkeys are very similar. They are extremely sensitive to cold, and when chilly are in the habit of wrapping their tail about them, so that this useful organ answers the purpose of a boa as well as a hand. They will also, when shot, fasten their tail so firmly on the branches, that they remain suspended after death. The great length of their tail enables them to walk in the erect attitude better than most monkeys. In walking, they cast their tails upwards as high as the shoulders, and then bend it over so as to form a counterbalance against the weight of the body, which is thrown very much forward in that and most other animals. The genus is called Ateles, or imperfect, because in most of the species the thumb is wanting. The Coaita inhabits Surinam and Guinea.

Several genera are omitted.

MYCĒTES.—(Gr. μυκητής, a howler.)

Ursīnus (Lat. *Ursa*, a bear—Bearlike), *the Ursine Howler*.

The HOWLING MONKEYS are larger and not so agile as the Spider Monkeys, and are chiefly remarkable for the peculiarity from which they derive their name. These animals possess an enlargement in the throat, composed of several valvular pouches, which apparatus renders their cry exceedingly loud and mournful. An arrangement somewhat similar may be seen in the throat of several loud-voiced birds.

They howl in concert, principally at the rising and setting of the sun; one monkey begins the cry, which is gradually taken up by the rest, precisely as may be observed in a colony of rooks. They are in great request among the natives as articles of food, their slow habits rendering them an easy prey.

The Ursine Howler, or Araguato, is common in Brazil, where forty or fifty have been observed on one tree. They generally travel in files, an old monkey taking the lead, and the others following in due order. They feed principally on leaves and fruit; the tail is prehensile like that of the Spider Monkeys.

The genus Cebus is omitted.

CALLĬTHRIX.—(Gr. καλός, beautiful; θρίξ, hair.)

Torquătus (Lat. *torquis*, a necklace—Collared), *the Collared Tee Tee.*

The beautiful little animals here represented belong to the Squirrel Monkeys, so called on account of their large bushy tails.

The COLLARED TEE TEE, or White-throated Squirrel Monkey, is found to the east of the Orinoco. It lives on small birds, insects and fruits. Its habits are, apparently, mild and inoffensive, but its acts belie its looks, for when a small bird is presented to it, it springs upon its prey like a cat and speedily devours it.

The MARMOSET is a most interesting little creature. It is exceedingly sensitive to cold, and when in England is usually occupied in nestling among the materials for its bed, which it heaps up in one corner, and out of which it seldom emerges entirely. It will eat almost any article of food, but is especially fond of insects, which it dispatches in a very adroit manner. It will also eat fruits, especially those of its native country. Its fondness for insects is carried so far, that it has been known to pinch out the figures of beetles in an entomological work, and swallow them.

JACCHUS.—(Gr. Ἴακχος, Bacchus.)

Vulgăris (Lat. *common*), *the Marmoset.*

A beautiful little marmoset in the Zoological Gardens ate a great number of flies which I caught and presented to it. Its little eyes sparkled with eagerness each time that it saw my hand moving towards a fly settled out of its reach, and it even ventured from its warm woolly nest, and climbed up the wires of its cage as it saw the fly approaching. It was also rather expert at catching for itself the flies that settled on the bars of the cage. A blue-bottle fly was evidently considered a great prize.

This pretty little monkey is also called the Ouistiti, from its peculiar whistling cry when alarmed or provoked.

Several genera are omitted between Callithrix and Jacchus.

Family IV. Lemurĭdæ.—(Lat. *lemŭres*, ghosts—Ghostlike.)

LEMUR.

Macáco (Native name), ***the Ruffled Lemur.***

The Lemurs derive their name from their nocturnal habits, and their noiseless movements. The Ruffled Lemur is a native of Madagascar. It lives in the depths of the forests, and only moves by night, the entire day being spent in sleep. Its food consists of fruits, insects and small birds, which latter it takes while they are sleeping. This is the largest of the Lemurs, being rather larger than a cat.

The Slender Loris is a native of India, Ceylon, &c. It, like the Lemur, seldom moves by day, but prowls about at night in search of food. No sooner does it espy a sleeping bird, than it slowly advances until within reach; then putting forward its paw with a motion slow and imperceptible as the movement of the shadow on the dial, it gradually places its fingers over the devoted bird; then, with a movement swifter than the eye can follow, it seizes its startled prey.

LORIS. (Native name.)

Gracĭlis (Lat. *slender*), *the Slender Loris.*

Several specimens of this animal have been kept in captivity and have proved very interesting in their habits. Their sleep appeared to be almost torpidity, as they would suffer their cage to be cleaned at any time between six in the morning, and the dusk of the evening. One of them when quietly awakened by thrusts of a stick, would utter a low plaintive cry, and walk once or twice slowly round its cage, and then sleep again.

Two genera are omitted between Lemur and Loris, and several more between Loris and the Vespertilionidæ.

Family V. . . Vespertilionĭdæ.—(Lat. *vespertilio*, a bat. Bat kind.)
Sub-family *a. Phyllostomīna.*—(Gr. *φύλλον*, a leaf; *στόμα*, a mouth.)

VAMPĪRUS (*"said by Adelung to be of Servian origin"*).

Spectrum (Lat. *a spectre*), *the Vampire Bat.*

We now arrive at the BATS, or Cheiroptĕra. This name is derived from the singular manner in which their fore-paws, or hands, are developed into wings. If the fingers of a man were to be drawn out like wire to about four feet in length, a thin membrane to extend from finger to finger, and another membrane to fall from the little finger to the ancles, he would make a very tolerable imitation of a Bat.

The usual food of Bats is insects, which they mostly capture on the wing, but some, as the Vampires, suck blood from other animals, and a few, as the Kalongo, or Flying Fox, live upon fruits, and so devour the mangoes, that the natives are forced to cover them with bamboo baskets to preserve them from the ravages of these animals, who would soon strip the fruit-trees without these precautions. Even the cocoa nut is not secure from their depredations.

The membrane of the Bat's wing is plentifully supplied with nerves, and is extremely sensitive, almost appearing to supply a sense independent of sight. Spallanzani cruelly deprived several Bats of their eyes, and then let them fly loose in his room, across which he had stretched strings in various places. The unfortunate Bats, however, did not strike against the strings or any other obstacles, but threaded their way among them with a degree of accuracy perfectly wonderful. Many Bats possess a similar membrane on the nose, which is possibly used for the same purpose.

The long and muscular structure of these animals seems principally directed towards their organs of flight. In order to give support to the powerful muscles that move the wings, the breast-bone is developed in front with a strong ridge, like that of a bird.* The shoulder blades and collar bones are also exceedingly strong for the same reason, and as the power of rotating the bones of the forearm would not only be useless but even hurtful, these bones can only move backwards and forwards, while the shoulder has considerable mobility.

The object of the extension of the finger joints is to give the animal the power of extending the wing membrane or folding it at pleasure. When the bat wishes to walk, it half folds the membrane and assumes an attitude admirably represented in the cut of the Long-eared Bat.

The thumb joint has no part of the wing attached to it, but is left free, and is armed with a hook at the extremity, by means of which it is enabled to drag itself along in that singular vacillating hobble which constitutes a Bat's walk.

There are five tribes, or sub-families, of Bats, according to Gray, each tribe including many genera. The British Museum alone possesses seventy-seven genera.

The Vampire Bat is a native of South America, where it is very common, and held in some dread. It lives on the blood of animals, and sucks usually while its victim sleeps. The extremities where the blood flows freely, as the toe of a man, the ears of a horse, or the combs and wattles of fowls, are its favourite spots. When it has selected a subject, on which it intends to feed, it watches until the animal is fairly asleep. It then carefully fans its victim with its wings while it bites a little hole

* See the cut of the breast-bone of a bird at the commencement of the division AVES.

in the ear or shoulder, and through this small aperture, into which a pin's head would scarcely pass, it contrives to abstract sufficient blood to make a very ample meal. The wound is so small, and the Bat manages so adroitly, that the victim does not discover that anything has happened until the morning, when a pool of blood betrays the visit of the Vampire. Darwin relates, that while travelling in Chili, "We were bivouacking late one evening near Coquimbo, when my servant, noticing that one of the horses was very restive, went to see what was the matter, and fancying he could distinguish something, suddenly put his hand on the beast's withers, and secured a Vampire. In the morning the spot where the bite had been inflicted was easily distinguished, from being slightly swollen and bloody."

SKULL OF COMMON BAT.

The wound made by the bat's teeth is no larger than that made by a needle, and hardly penetrates the skin, so that the blood must be extracted by suction. There have been very different accounts of the Vampires from travellers, some denying that they suck blood at all, and others narrating circumstantially the injuries inflicted upon their own persons. The cause for these discrepancies is probably owing to the constitution of the narrators, there being some persons whom a Vampire will not touch, while others are constantly victimised.

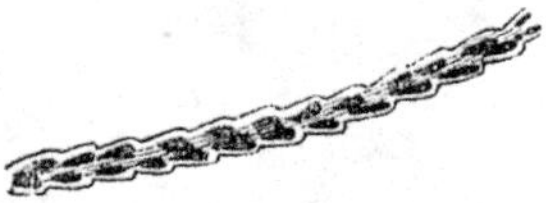

HAIR OF LONG-EARED BAT.*

This Bat is placed among the Phyllostomina, because the membrane on its nose resembles a leaf. The length of its body is about six inches.

THE LONG-EARED BAT.

The LONG-EARED BAT is found in most parts of Europe, and is common in England. It may be seen any warm evening flying about in search of insects, and uttering its peculiar shrill cry. It is very common on Hampstead Heath. The ears are about an inch and a half in length, and have a fold in them reaching almost to the lips, from which peculiarity the genus is called Plecotus.

This Bat is very easily tamed, and will take flies and other insects from the hand. One that I had in my own possession used to hang by the wing-hooks during the whole of the day, and could hardly be persuaded to move, or even to eat; but when the evening came on it became very brisk indeed, and after carefully combing itself with its hind feet, it would eagerly seize a fly or beetle and devour it, always rejecting the head, legs, and wings. It was then very impatient to be released from

* Magnified about 200 diameters.

the cage, and would show its uneasiness by climbing about the cage and fluttering its wings. It unfortunately died before further investigations could be made, but during the short time that it survived, it seemed very gentle, and only bit me once, although I used frequently to handle it.

Sub-family c. *Vespertilionīna.*
(Lat. *Vespertilio,* a Bat.)

PLECŌTUS.—(Gr. πλέκω, I fold; οὖς, an ear.)

Aurītus (Lat. *auris,* an ear—Eared), *the Long-eared Bat.*

The singular appearance of the hair of the Bat as seen through a microscope is caused by a number of scales adhering to the exterior of the hair. These scales can be rubbed off, and in consequence of this property, the bat's hair often assumes very singular forms. The hair that is figured was drawn by means of the Camera Lucida, from a specimen seen by transmitted light, but if it had been seen by reflected light, it would have presented quite a different appearance, not very unlike the plant called Mare's tail before its leaves are grown.

The details of the hair are different in the differing species of Bats, but there is always a character about them which is not to be mistaken.

When the Long-eared Bat is suspended by its hinder claws it assumes a most singular aspect. The beautiful long ears are tucked under its wings, which envelop great part of its body. The tragus, that pointed membrane visible inside the ear, is then exposed, and appears to be the actual ear itself, giving the creature a totally different cast of character.

QUADRUPEDS.

THE former sections have been characterised by the number and properties of the *hands.* In the section that we are about to consider, the hands have been modified into feet. At the head of the quadrupeds, or four-footed animals, are placed the carnivora, or flesh-eaters, and at the head of the carnivora, the Felidæ, or Cat kind are placed, as being the most perfect and beautiful in that section. The Felidæ all take their prey by creeping as near as they can without observation, and then springing upon their unfortunate victim, which seldom succeeds in making its escape, as the powerful claws and teeth of its enemy usually dash it insensible to the ground. The jaws and teeth of the Felidæ are very different from those of the animals already described; their jaws are more powerful, and their teeth longer and sharper. Their claws, too, are necessarily very long, curved and sharp, and to prevent them from being injured by coming into contact with the ground, they are concealed, when not in use, in a sheath, which effectually guards them and keeps them sharp. There are five claws on the fore feet, and four on the hinder feet. The tongue of the Felidæ is very rough, as may be proved by feeling the tongue of a cat. This roughness is occasioned by innumerable little hooks which cover the tongue, point backwards, and are used for the purpose of licking the flesh off the bones of their prey. The bristles of the mouth, or whiskers, are each connected with a large nerve, and are exceedingly useful in indicating an obstacle when the animal prowls by night. Their eyes are adapted for nocturnal vision by the dilating power of the pupil, which expands so as to take in every ray of light.

THE LION.

The LION stands at the head of the wild beasts. His noble and dignified bearing, the terrific power compressed into his comparatively small frame, and the deep majesty of his voice, have gained for him the name of "king of beasts." The Lion inhabits Africa and certain parts of Asia, such as portions of Arabia and Persia, and some parts of India. It varies in external appearance according to the locality, but there is little doubt that there is but one species. We are indebted to Mr. Cumming for many interesting notices of this noble animal, observed during his residence in Southern Africa, and from his book many extracts will be given in the course of this work, as by his cool and daring courage he has been enabled to watch the habits and actions of the most ferocious beasts in the depths of their own haunts.

The Lion is barely four feet high, and eight in length, yet he can, with little difficulty, dash the giraffe to the earth, or overcome the powerful buffalo. He has been known to carry off a heifer in his mouth, and although encumbered with such a burden, to leap a broad dyke, apparently with the greatest ease. No animal willingly molests the Lion, and there are but very few which he cannot overcome. The rhinoceros and elephant are almost the only quadrupeds he dare not meddle with, but he does not seem to stand in much fear of them. Gnoos, zebras, and antelopes, seem to be his favourite prey, although one of the antelopes, the oryx, or gemsbok, not unfrequently avenges its own death by the destruction of its pursuer, its long straight horns impaling the Lion from side to side. The two skeletons have been seen lying together. The roar of the Lion is one of its chief peculiarities; the best description of it is in Cumming's Adventures:—

"One of the most striking things connected with the Lion is his voice, which is extremely grand and peculiarly striking. It consists, at times, of a low deep moaning, repeated five or six times, ending in faintly audible sighs; at other times he startles the forest with loud, deep-toned, solemn roars, repeated five or six times in quick succession, each increasing in loudness to the third or fourth, when his voice dies away in five or six low, muffled sounds, very much resembling distant thunder. At times, and not unfrequently, a troop may be heard roaring in concert, one assuming the lead, and two, three, or four more regularly taking up their parts like persons singing a catch."

Order II. . . . *FERÆ.*—(Lat. *ferus*, wild. Wild beasts.)
Family I. . . . Felĭdæ.—(Lat. *felis*, a cat. Cat kind.)
Sub-family *a*. *Felīna*.

LEO.—(Lat. *a Lion.*)

Barbărus (Lat. *fierce*), *the Lion.*

"As a general rule lions roar during the night, their sighing moans commencing as the shades of evening envelope the forest, and continuing at intervals throughout the night. In distant and secluded regions, however, I have constantly heard them roaring loudly as late as nine and ten o'clock on a bright sunny morning. In hazy and rainy weather they are to be heard at every hour in the day, but their roar is subdued."

It is well known that the power of the human voice is very efficacious in alarming even the most savage wild beasts. But I never remember seeing it mentioned, that it is not so much the sound of the *voice*, as the sound of the *words* that alarms these animals. Any one may test the fact for themselves, for whereas they may shout inarticulately without scaring away the wild birds and beasts of our own country, a few words spoken in comparatively a low key, sends them off at once.

Indeed we involuntarily use words on such an occasion, as for example, if we are attacked by a dog or a bull, we do not content ourselves with shouting, but speak words to it. The dumb brute seems to be cowed by the "winged words" of human reason.

There is a remarkable instance of this faculty in the work of the author whom I have before quoted. A lioness whom he had wounded was about to spring upon him, but as he stood quite still, and recommended her in a commanding tone to "take it easy," she halted and permitted her assailant to retreat, which he did very slowly, still continuing to talk to the lioness until he had made his escape.

The opinion that lions will not touch a dead animal is erroneous; as they were frequently shot by Mr. Cumming while devouring gnoos, &c. that had fallen by his rifle. Those lions who have once tasted human flesh are generally the most to be dreaded, as they will even venture to spring in among a company of men, and seize their victim. These lions are called Man-eaters. During the latter part of Cumming's residence in South Africa a dreadful instance of their ferocity occurred.—While the hunting party was encamped for the night in the territory of the Balakahari, a lion, taking advantage of the stormy night, suddenly sprang upon two men, Hendrick, the driver, and Ruyter, the Bosjesman tracker, who were wrapped in the same blanket, by the fire. It seized Hendrick by the neck, and dragged him into the bushes, in spite of the blows which another man gave it with a burning brand, leaving Ruyter unhurt except by a few scratches with its claws. Next morning it was shot by Mr. Cumming, who placed its skin in his magnificent collection, where Ruyter points it out with great glee.

The Lioness is much smaller than the Lion, and is destitute of the magnificent mane which is so great an ornament to her mate. As a general rule she is more fierce and active than the male, especially before she has had cubs, or while she is suckling them. She has usually from two to four cubs at a time. They are beautiful playful little things, and are slightly striped. They have no mane until about two years old. While her cubs are small the Lioness knows no fear, and will attack a company of men or a herd of oxen if they come too near her den. Her mate also ably seconds her endeavours, and has been known to keep the hunters at bay until she has withdrawn her cubs to a place of safety, after which he bounds off in the direction which she has taken.

The cubs are remarkably heavy for their age. Many years ago, I had a pair of young lion cubs in my hands. They were about the size of very large cats, but weighed considerably more than their size led me to believe.

They were playful little animals, but struck rather too hard to be agreeable.

The Lion when young is easily tamed, and shows a strong attachment to its keeper. Those who have seen Van Amburgh will know what influence man may obtain over this powerful creature. Many anecdotes have been told of the celebrated lion "Nero," who would suffer even strangers to caress him, and carry children on his back with the greatest good-nature.

Many naturalists, of whom Buffon is the chief, have fallen into errors concerning the contradictory dispositions of the lion and tiger. "The lion unites with a high degree of fierceness, courage, and strength, the more admirable qualities of nobleness, clemency, and magnanimity. Walking with a gentle step, he does not deign to attack man unless provoked to the combat. He neither quickens his step, nor flies, and never pursues the inferior animals except when urged by hunger;" while the tiger "presents a compound of meanness and ferocity; he seems always thirsty for blood," &c. &c. Now nothing can be more erroneous than these sentences. The tiger is as tameable as the lion, the tiger and lion seize their prey with equal ferocity, and neither will attack a man or any other animal when satisfied with food.

There is one remarkable difference in the characters of the feline and canine tribes. If a man is overcome by a wolf or a dog, the animal ceases not to mangle its vanquished foe until life is quite extinct. A dog killing a rat is a good instance of this trait of character. But if a lion or any other feline animal vanquishes a man, it contents itself with the victory for some time without making any attempt to injure him, unless he tries to escape, in which case he is again dashed to the earth, and probably considerably bitten as a warning. A cat treats a mouse just as a lion treats a man.

This propensity in the lion has been the cause of saving several lives, the men having been able either to destroy their foe by cautiously getting out a weapon, or by lying still until they were succoured.

At the extremity of the lion's tail there is a small hook or claw, which has been represented as the means by which the animal lashes itself into fury, using it as a spur. This is impossible, as the claw or prickle is very small, not fixed to the bone as the claws of the feet are, but merely attached to the skin, and falls off if roughly handled. It is nct present in all lions, as Mr. Wood only discovered it once out of numerous specimens which he examined.

As an example of the accuracy of the Nineveh sculptures we may notice that the little horny claw is faithfully represented. Indeed, the sculptures of hunting scenes are executed with wonderful spirit and truth, even delineating the different attitude in which a bull and a lion fall in death. There is one group representing a combat between a lion and a bull admirably represented, the whole action of the lion being wonderfully correct, although slightly exaggerated.

Tigris.—(Lat. *a Tiger.*)

Regālis (Lat. *royal*), *the Tiger.*

This magnificent animal is found only in Asia, Hindostan being the part most infested by it. In size it is almost equal to the lion, its height being from three to four feet, and its length rather more than eight feet. It has no mane, but to compensate for this deficiency it is decorated with black stripes, upon a ground of reddish yellow fur, which becomes almost white on the under parts of the body. The chase of the Tiger is among the most exciting and favourite sports in India. A number of hunters assemble, mounted on elephants trained to the sport, and carry with them a supply of loaded rifles in their howdahs, or carriages mounted on the elephants' backs. Thus armed, they proceed to the spot where a tiger has been seen. The animal is usually found hidden in the long grass or jungle, which is frequently eight or more feet in height, and when roused, it endeavours to creep away under the grass. The movement of the leaves betrays him, and he is checked by a rifle ball aimed at him through the jungle. Finding that he cannot escape without being seen, he turns round, and springs at the nearest elephant, endeavouring to clamber up it, and attack the party in the howdah. This is the most dangerous part of the proceedings, as many elephants will turn round and run away, regardless of the efforts of their drivers to make them face the tiger. Should, however, the elephant stand firm, a well-directed ball checks the tiger in his spring, and he then endeavours again to escape, but a volley of rifle balls from the backs of the other elephants, who by this time have come up, lays the savage animal prostrate, and in a very short time his skin decorates the successful marksman's howdah. These hunts are not carried on without considerable danger, as in some cases the tiger has succeeded in reaching the howdah, and more than one hunter has been known to overbalance himself in his anxiety to get a shot at his game, and has fallen into the very claws of the enraged brute. Once a wounded tiger sprang at a badly trained elephant, who immediately turned round and made off. The tiger succeeded in reaching the elephant's tail, which it mangled dreadfully, but could climb no higher, partly on account of its wounds, and partly through the exertions of a native, who kept it back with a spear. The tiger hung in this way for the greater part of a mile, when another hunter succeeded in overtaking the terrified elephant, and with a single ball freed the poor animal from its tormentor.

Of late years tiger hunting has become less dangerous, principally on account of the innate fear that all wild beasts seem to have of the power of fire-arms. When mankind first waged war against the tigers, they did not heed the fire-arms, but experience has taught them a fear of those terrible weapons, which appears to have been communicated to their posterity, just as the puppy of a retriever dog will plunge into the water and fetch a stick without being taught.

Tigers are usually taken by the natives in pitfalls, at the bottom of which is planted a bamboo stake, the top of which is sharpened into a point. The animal falls on the point and is impaled.

The general notion that tigers cannot be tamed is erroneous. They can be tamed as easily as the lion; but great caution must be used with all wild animals, as in a moment of irritation, their savage nature breaks out, and the consequences have more than once proved fatal. The melancholy death of the "Lion Queen," in Wombwell's Menagerie, is a recent example of this propensity.

In the British Museum are three cubs bred between a lion and a tigress. They are not unlike lion cubs, but the stripes are much darker, and the colour of the fur is brighter.

The colouring of the tiger is a good instance of the manner in which animals are protected by the similarity of their external appearance to the particular locality in which they reside. The stripes on the tiger's skin so exactly assimilate with the long jungle grass amongst which it lives, that it is impossible for unpractised eyes to discern the animal at all, even when a considerable portion of its body is exposed.

LEOPARDUS.—(Lat. *leo*, a lion; *pardus*, a panther.)

Varius (Lat. *varied*), *the Leopard*, or *Panther*.

The LEOPARD is an inhabitant of Africa, India, and the Indian Islands. A black variety inhabits Java, and is not uncommon there. Its height is about two feet. This and the following Felidæ are accustomed to live much on trees, and are on that account called Tree-tigers by the natives. Nothing can be more beautiful than the elegant and active manner in which the leopards sport among the branches of the trees: at one time they will bound from branch to branch with such rapidity that the eye can scarcely follow them; then as if tired, they will suddenly stretch themselves along a branch so as to be hardly distinguishable from the bark, but start up again on the slightest provocation, and again resume their graceful antics. It is easily tamed, and expresses great fondness for its keeper, and will play with him like a cat. A remarkably beautiful specimen in Wombwell's Menagerie was exceedingly fond of playing with the tuft at the extremity of a lion's tail, and from the familiar manner in which he patted and bit it, he evidently considered it as manufactured for his own particular entertainment.

This animal is exceedingly fond of some scents, especially preferring lavender water, by means of which predilection, it has been taught to perform several tricks.

The Leopard and Panther are considered as the same animal, on the authority of Mr. Gray.

LEOPARDUS.

Uncia (Lat. *uncia*, an ounce), *the Ounce.*

The OUNCE is a native of India, and has been often confounded with the Leopard. Its fur is much more rough than that of the leopard, and the tail is almost bushy, especially towards the extremity. Its body is marked with irregular wavy stripes, and the head is adorned with black spots. The general colour is a yellowish grey.

It is easy to distinguish the Ounce from the Leopard, by the indistinctness of the markings, and also by the roughness of the fur, which latter distinction, in the opinion of some naturalists, shows that it lives in mountainous regions. The habits and history of this animal are but little known.

The JAGUAR inhabits America. It is larger and more powerful than the leopard, which it resembles in colour, but has a black streak across the chest, and a black spot in the centre of the rosettes. It is fond of climbing trees, and finds little difficulty in ascending, even when the trunk is smooth and destitute of branches. It chases monkeys successfully, and is said to watch for turtles on the beach, and to scoop out their flesh by turning them on their backs and inserting its paws between the shells. Nor does it confine its attention to the turtles themselves, for it watches them lay their eggs, and then scoops them out of the sand with its claws. It often makes fearful havoc among the sheepfolds, and is said to depart so far from the usual habits of the Felidæ, as to enter the water after fish, and to capture them in the shallows by striking them out of the water with a blow of its paw. There have been instances of the domestic cat acting in the same manner.

When it captures one of the larger animals it destroys it by leaping upon its back, and twisting the head of its prey round, until the neck is dislocated.

LEOPARDUS.

Onca (Gr. Ὄγκα, a proper name), *the Jaguar.*

The PUMA inhabits the whole of America, where it is held in much dread by the natives. Its colour is an uniform grey, fading into white on the under parts of its body, and this similarity of colour is the reason that the name "concolor" has been given to it. It lives much on trees, and usually lies along the branches, where its uniform dusky fur renders it so like the bark that it can scarcely be distinguished from the branch. This habit it preserves when in captivity, and many persons pass its den in the Zoological Gardens, fancying it empty, while the puma is lying along its shelf unobserved.

LEOPARDUS.

Concŏlor (Lat. *of the same colour*), *the Puma.*

Mr. Eaton Stone, the celebrated equestrian who has travelled for many years in the wilder parts of America, told me that the puma is accustomed to follow men by scent, and to track them on their journey, waiting for an opportunity to spring upon them unobserved. If the traveller keeps his eye on the animal it is perfectly harmless, but it will wait for the moment when his eye is withdrawn to spring upon him.

The Americans always speak of this animal as the panther, or "painter," as it is more familiarly pronounced; and many authors still term it the *cougar*, a word contracted from the original elongated unpronounceable Mexican name, "Gouazouara."

LEOPARDUS.

Pardălis (Gr. πάρδαλις, a pard), *the Ocelot.*

The Ocelot, one of the Tiger-cats, is a native of Mexico and Peru. Its height is about eighteen inches, and its length about three feet. It is a most beautiful animal, and is easily tamed. When in a wild state it lives principally on monkeys, which it takes by stratagem.

The domestic Cat was formerly supposed to be the same animal as the wild Cat, but it is now proved to be a distinct species, and the difference is seen at once by the form of the tail. That of the domestic cat is long and taper, while that of the wild cat is bushy and short. To make the point clearer, it may be observed that domestic cats, which have made their escape into the woods and become wild, have retained their slenderness of tail for several generations, while the wild cat on being domesticated never loses its characteristic roundness and shortness of tail. See the cut on page 36.

The cat is an animal which, whether lying curled up on the hearth-rug fast asleep and immersed in dreams of shadowy fat mice, or leisurely pacing the room, and complacently muttering its self-satisfied purr as it brushes softly against the legs of the table or chair, certainly succeeds in giving a great air of comfort to a room. On this account it is a general favourite, especially in houses where there are no children. Pussy, however, is not only ornamental, but useful also, as she is eager and successful in the pursuit of rats and mice. So strong, indeed, is the passion for hunting in the breast of the Cat, that she sometimes disdains mice "and such small deer," and trespasses on warrens or preserves. A large tabby cat, residing at no great distance from White Horse Vale, was accustomed to go out poaching in the preserves of a neighbouring nobleman, and so expert was she at this illegal sport that she constantly returned bearing in her mouth a leveret or a partridge, which she insisted on presenting to her mistress, who in vain endeavoured to check her marauding propensities. These exploits, however, brought their own punishment; for one day, when in the act of seizing a leveret, she found herself caught in a vermin trap, which deprived her of one of her hind legs. This misfortune did not damp her enthusiasm for hunting, as although the loss of a leg prevented her from chasing hares, and such like animals, she would still bring in an occasional rat.

This instinctive desire of hunting seems to be implanted in cats at a very early age. I have seen kittens but just able to see, bristle up at the touch of a mouse, and growl in a terrific manner if disturbed. Weissenborn, in his Magazine of Natural History, gives the following interesting account of the propensity of the cat to hunt, and of the

FELIS.

Domestĭca (Lat. *domestic*), *the Cat.*

mice to escape, both being at an age rendering it impossible that any instruction could have been given them by their parents.

"That instinct is an inherent or innate quality of animals is clearly proved by experience. The cat possesses the instinct of catching and eating mice, and the mouse that of shunning the cat as its most dangerous enemy. Once, in Rome, I happened to open a drawer which I seldom had occasion to use, when I saw a mouse jumping out of it, and found among the papers a nest with five young mice, naked and blind, and of a pale flesh-colour. I placed them on a table, handled them, &c., and they evinced no symptoms of fright, nor any inclination to get away, but only appeared eager to approach each other for the sake of warmth. There happened to be in the house a very young cat who had never tasted anything but milk. I placed it near the little mice by way of experiment, but to my astonishment it did not even look at them, nor perceive them, even when I turned its eyes in the proper direction, until at last, when I had repeatedly approached its nose to the mice, it suddenly caught a scent which made it tremble with desire. The propensity became more and more violent, and the cat smelled at the mice, touching them with its nose, when all at once the pale-coloured creatures became suffused with blood, and began to make great exertions to get out of the way of imminent danger, whilst the cat as eagerly followed them."

The Cat displays a great affection for her kittens, and her pride when they first run about is quite amusing. While I was an undergraduate at College, a cat belonging to the baker's department formed a great friendship for me, and used to come every morning and evening to obtain her share of breakfast and tea. She continued her attentions for some time, but one morning she was absent from her accustomed corner, nor did she return until nearly a week had passed, when she came again, but always seemed uneasy unless the door were open. A few days afterwards she came up as usual, and jumped on my knee, at the same time putting a little kitten into my hand. She refused to take it back again, so I restored it to its brothers and sisters myself. A few hours afterwards, on going into my bedroom, I found another black kitten fast asleep on the bed.

CATS' TAILS.*

Cats are very fond of aromatic plants and several powerful scents. My own cat has just been discovered in the act of eating the green tops of a musk plant that was standing in the window. Valerian appears to be the great attraction for cats; and any one who is disposed to place a plant of valerian in his garden must beware of the cats, for they wil. come in numbers, roll over it, and scratch up the plant until there is not a vestige of it left. Moreover they will fight for the fragments in various parts of the garden, and cause great confusion among the seeds.

CARĂCAL.—(Turk. *black ears.*)

Melanōtis (Gr. μέλας, black · οὖς, an ear), *the Caracal.*

There are several varieties of the domestic cat, among which the Angora cats, with their beautiful long fur, and the Manx cats of the Chartreuse breed, which have no tails, are the most conspicuous.

The CARACAL is found in most parts of Asia and Africa. It derives its name from the black tips of its ears, which render it a very conspicuous

* 1. Tail of Domestic Cat; 2. Tail of Wild Cat.

animal. It is one of the group of the Lynxes, and is generally supposed to be the animal referred to by several ancient authors under the name of Lynx. It lives on the smaller quadrupeds and birds, which it pursues even to the tops of the trees. There are no records of its being tamed, as in every instance when confined it snarls at those who approach its cage. The length of its body is about two feet, and its height about fourteen inches.

The Canada Lynx is a native of North America, and is remarkable for its gait. Its method of progression is by bounds from all four feet at once, with the back arched. It feeds principally on the American hare, as it is not courageous enough to attack the larger quadrupeds. Its length is about three feet. The natives sometimes eat its flesh, which is white and firm, and not unlike that of the American hare itself. Its skin forms an important article in commerce, and between seven and nine thousand are imported yearly by the Hudson's Bay Company, by whom the grey specimen in the British Museum was presented.

Lyncus.—(Gr. λύγξ, a Lynx.)

Canadensis (Lat. *of Canada*), *the Canada Lynx.*

The Chetah, or Hunting Leopard, as it is sometimes called, is one of the most elegant and graceful animals known. It is a native both of Africa and India, but it is only in the latter country that it is used for hunting game, as the Africans appear not to possess sufficient ingenuity to train the animal. The method of employing it is usually as follows —The Chetah is either led blind-folded in a chain, or placed upon a hackery, or native cart, and taken as near as possible to the place where antelopes or deer are feeding. When close enough, the hunter takes the band from its eyes, and directs its head towards the game. Directly the Chetah sees the deer, it creeps off the cart, and makes towards them as rapidly and silently as it can, carefully availing itself of the accidental cover of a bush, or stone, precisely as a cat does when stealing after a bird. When it has succeeded in unobservedly approaching the unsuspecting herd, it makes two or three tremendous springs, and fastens on the back of one unfortunate deer, brings it to the ground, and waits until its keeper comes up, who induces it to leave its prey by a ladle-full of the blood, which he takes care to have ready. The Chetah is then hooded and led back to his cart. It is so easily tameable and so gentle that it is frequently led about the streets by a string for sale.

GUEPARDA.

Jubāta (Lat. *crested*), *the Chetah.*

It is rather larger than the leopard, and differs from it in the length of its paws, its inability to climb trees, and the crispness of its fur. It is therefore placed in a different genus from the leopard. It derives its name of "jubata," from a thin mane running down the neck.

The HYENINA, or HYÆNAS, are remarkable for their predatory, ferocious, and withal, cowardly habits. There are several Hyænas, the striped, the spotted, and the villose, but as the habits of all are very similar, only one will be mentioned. The Hyænas, although very repulsive in appearance, are yet very useful, as they prowl in search of dead animals, especially of the larger kinds, and will devour them even when putrid, so that they act the same part among beasts, that the vultures do among birds, and are equally uninviting in aspect. They not unfrequently dig up recently interred corpses, and in Abyssinia, according to Bruce, they even flock in numbers into the village streets, where they prey on slaughtered men who are thrown out unburied. One of these animals attacked Bruce in his tent, and was only destroyed after a severe battle. Their jaws and teeth are exceedingly powerful, as they can crush the thigh-bone of an ox with apparently little effort; and so great is the strain upon the bones by the exertion of these muscles, that the vertebræ of the neck become anchylosed, as it is called, that is, become united together, and the animal has a perpetual stiff neck in consequence. Before the anatomy of the hyæna was better known, people thought that it had only one bone in its neck. The skull too is very strong, and furnished with heavy ridges for the support of the muscles which move the jaw.

Sub-family *b. Hyenīna.*
HYÆNA.—(Gr. Ὕαινα.)

Striāta (Lat. *striped*), *the Striped Hyæna.*

The hinder parts of the Hyæna are very small, and give it a strange shambling appearance when walking. The Hyæna is easily tamed, and even domesticated, so that the tales of its untameable disposition are entirely erroneous.

The striped Hyæna is found in many parts of Asia and Africa, where it is both a benefit and a pest, for when dead animals fail it, the flocks and herds are ravaged, and even man does not always escape.

The VIVERRINA, or CIVETS, are active little animals, averaging about two feet in length. The whole group is celebrated for the perfume which is secreted in a grandular pouch near the tail, and is of some importance in commerce. If the Civet is kept alive, the perfume is obtained by enclosing it in a long narrow box so that it cannot turn round, and then scraping the secretion from the pouch with a spoon. If the creature is killed, the entire pouch is usually cut off, and sells for a higher sum than when the perfume is sold separately, as it is not so liable to adulteration.

Sub-family *c. Viverrīna.*
VIVERRA.—(Lat. *a Ferret.*)

Civetta (Arabic *Zibetta*, scent), *the Civet Cat.*

The Civet is only found in North Africa, especially in Abyssinia, where it takes up its abode on uncultivated and barren hills. It feeds upon birds and the smaller quadrupeds, which it takes by surprise. As it pursues its prey by night only, its eyes are formed for seeing in the dark, and gleam as do those of a cat.

GENETTA.—(Fr. *Genette.*)

Vulgāris (Lat. *common*), *the Genet.*

The GENET slightly resembles the cat, particularly in its spots, and the power of climbing trees. It inhabits Africa, and is not unfrequently found in the south of France. At Constantinople it is domesticated, and keeps the houses free from rats and mice, which are said to be unable to endure its scent, but it is much more probable that it frees the houses from mice by devouring them.

The ICHNEUMONS, or MANGOUSTS, well deserve their name of Creepers, for with their long bodies and snouts, their short limbs and slender tails, they insinuate themselves into every crevice in their way in search of their expected food. Few animals are more useful than the Ichneumons. Snakes, lizards, crocodiles' eggs, or even young crocodiles themselves, form their principal food, and their activity is so great that when these sources fail, they are able to secure birds, and even seize upon the swift and wary lizards, which, when alarmed, dart off like a streak of green light glancing through the bushes.

HERPESTES.—(Gr. ἑρπηστής, a creeper.)

Ichneumon (Gr. ἰχνεύμων, a tracker), *the Egyptian Ichneumon.*

The Egyptian Ichneumon, or Pharaoh's Rat, as it is sometimes called, is a native of North Africa, and is often domesticated for the purpose of destroying the various snakes, and other reptile annoyances, which are such a pest in the houses of hot countries. It principally seeks its prey by night, creeping along with such noiseless and snake-like progress, that not a sound warns the unsuspecting victim of its danger. Its slender snout enables it to suck out the contents of eggs with ease, and it destroys serpents by creeping behind them, and then suddenly leaping on their heads, which it instantly crushes between its sharp teeth. Its length without the tail is about eighteen inches.

About twelve genera are omitted.

THE DOG.

We now arrive at the DOG FAMILY, which includes the Dogs, Wolves, Jackals, and Foxes. The first of the Dogs is the Kolsun, or Dhale, which inhabits Bombay and Nepaul. It hunts in packs, as most of the dogs do even in a wild state, and has been known to destroy tigers and chetahs. Let us pass to a more interesting animal, the Newfoundland Dog. This magnificent creature was originally brought from Newfoundland. It is often confounded with the Labrador Dog, a larger and more powerful animal. Both these dogs are trained by their native masters to draw sledges and little carriages, and on that account are highly esteemed. The Newfoundland dog is well known as a most faithful guardian of its

Sub-family *d. Canīna.*—(Lat. *Canis*, a Dog.)
CANIS.

Familiāris (Lat. *familiar*), *the Newfoundland Dog.*

master's property. It is remarkably fond of the water, and will fetch out any article which its master indicates, and lay it at his feet. Many instances are known of this noble animal saving the lives of people that have fallen into the water, and must have perished but for its timely aid. There is an anecdote related of one of these dogs leaping over the parapet of a bridge, and rescuing a baby who had sprung from its nurse's arms into the river. A gentleman who just came up, and was caressing the dog after its exploit, discovered, on seeing the child, that it was his own. He offered a large sum for the noble creature, but his master refused to part with him on any terms.

This is one of the largest of the dogs, as it stands nearly two feet two inches in height.

THE WATER SPANIEL.

The WATER SPANIEL, as its name denotes, delights in plunging into water, especially if any game is to be found among the rushes that fringe the rivers. It is a most useful assistant when shooting wild ducks, or water hens, as, when wounded, they conceal themselves so effectually, that, without a dog, discovery is almost impossible. It can also dive to some depth, and bring up in its mouth any small object from the bottom.

KING CHARLES'S DOG.

The KING CHARLES'S DOG is a diminutive breed of spaniels, first brought into notice by Charles the Second, who delighted in being accompanied by them in his walks, and was accustomed to admit them into his bedchamber, and even permitted them to lie on his bed.

The BLOODHOUND.—There are several varieties of this animal, inhabiting Cuba, Africa, and England. They all are endowed with a wonderfully acute sense of smell, and can trace a man or animal with almost unerring certainty. The Cuban Bloodhound was formerly employed by the Spaniards to hunt down the natives while endeavouring to escape from their invasions. A few years since, one of these dogs saved the life of its master, an American hunter, by boldly attacking a puma which had sprung on him in the darkness, and was lacerating him in a dreadful manner. The sagacious animal had been tied up at home, but apparently knowing the dangers of the forests through which his master was about to pass, he broke his chain, and arrived barely in time to save the hunter from a horrible death.

The English Bloodhound is frequently mentioned by the older historians. Bruce was repeatedly chased by bloodhounds, and at one time he was so closely pressed that he barely escaped by leaping into a brook, and wading a considerable distance up the stream, knowing that running water would not retain the scent. The bloodhounds led his pursuers as far as the place where he entered the water, but the stratagem of Bruce baffled them, and the pursuit was abandoned. The voice of these dogs is peculiarly deep, and may be heard at considerable distance. Not very long since, a sheepstealer was detected by a bloodhound when every other means had failed. The dog, on being shown the footsteps of the thief, at once set off on the track, and dashed into a cottage, where the unsuspecting robber was busily employed in skinning the sheep which he had stolen. The height of this splendid animal is about two feet four inches, and its colour a reddish tan, becoming almost black along the back.

THE BLOODHOUND.

The Foxhound and Beagle are not very dissimilar in form or habits. They both follow game by the scent, and are used in hunting. The Foxhound, as its name implies, is used for hunting the fox, and enters into the sport with extraordinary eagerness. These dogs are trained with great care: whole books have been written on their education, and men are engaged at high salaries to train them to the sport. England possesses the finest breed of foxhounds in the world, and certainly no expense is spared to improve them, as one kennel is said to have cost nearly twenty thousand pounds. The height of the foxhound is about twenty-two inches.

The BEAGLE is used principally for hare hunting. It is much smaller than the foxhound, and not nearly so swift, but its scent is so perfect that it follows every track of the flying hare, unravels all her windings, and seldom fails to secure her at last. Sportsmen usually prefer the smallest beagles obtainable. The most valuable pack of these dogs known, used to be carried to and from the field in a pair of panniers slung across a horse's back. Unfortunately, this pack was so well known, that numerous were the attempts to gain possession of it. One ill-fated evening, as the dogs were returning in their panniers after the day's sport, the keeper was decoyed away by some stratagem, and when he returned, his dismay was great to find that the dogs, panniers, and horse were all missing. No traces of them were discovered, and it was conjectured that they must have been sold on the Continent. It is a common custom in the military schools and sometimes at the universities, to follow the beagle on foot. There has been for several years a society at Oxford, who thus hunt on foot. As too much time would be lost in looking for a living hare, a dead rabbit is trailed along the ground, and as its fur has been rubbed with aniseed, the dogs can follow it easily.

THE FOXHOUND.

THE BEAGLE.

The POINTER is used by sportsmen to point out the spot where the game lies. It ranges the fields until it scents the hare or partridge lying close on the ground. It then remains still as if carved in stone, every limb fixed, and the tail pointing straight behind it. In this attitude it remains until the gun is discharged, reloaded, and the sportsman has reached the place where the bird sprung. It then eagerly searches for the game, and brings the bird in its mouth.* There are many anecdotes of its intelligence, among which the following is not the least interesting.

In 1829, Mr. J. Webster was out on a shooting party near Dundee, when a female pointer, having traversed the field which the sportsmen were then in, proceeded to a wall, and, just as she made the leap, got the scent of some partridges on the opposite side of the wall. She hung by her fore-feet until the sportsmen came up; in which situation, while they were at some distance, it appeared to them that she had got her leg fastened among the stones of the wall, and was unable to extricate herself. But, on coming up to her, they found that this singular circumstance proceeded from her caution, lest she should flush the birds, and that she had thus purposely suspended herself in place of completing her leap.

THE POINTER.

When badly trained, this dog is apt to make very absurd mistakes. A young pointer belonging to a friend disappointed him by most perversely pointing at a pig; and on another occasion was discovered feasting on a dead sheep instead of attending to its business.

The group of the Mastiff dogs is distinguished by the shortness of the nose and the breadth of the head. This group includes the mastiff, the bull-dog, and the almost obsolete absurd little pug-dog. The breadth of their heads is caused by the large muscles which move the jaw.

The English Mastiff is generally employed as a house-dog, as its powerful frame and deep voice are well fitted to scare away marauders, or to repel them if they approach too near. It is by far the most sagacious of the whole group, and exhibits much more attachment to its master than the others. This animal has been called by several names,

* Many dog-trainers do not permit the dog even to touch the bird.

THE MASTIFF.

of which "Ban-dog" is the best known. Bewick thinks that the ban-dog is a separate species, of a lighter make than the ordinary English mastiff.

The BULL-DOG is proverbial for courage and endurance. Unfortunately its social qualities are by no means pleasing, as, although it has some attachment to its master, yet it is not always safe even for him to disturb it. This dog was extensively used in the cruel sport of bull-baiting, a recreation now extinct. When opposed to the bull, the dog would fly at its nose, and there hang in spite of all the infuriated animal's struggles. So firm is its hold, that the owner of a bull-dog once laid a wager that when his dog had seized a bull he would cut off all his feet in succession without inducing the poor beast to loose his hold. The experiment was made, and the cruel master, who deserved a similar fate himself, won his wager.

THE BULL-DOG.

The PUG-DOG looks like a bull-dog in miniature. It was formerly in great request as a pet, but is now seldom seen. Its tail is curled over its back so tightly, that it is not very difficult to believe the story of a pug-dog being lifted off his hind-legs by the curliness of his tail.

The TERRIERS never grow to any considerable size. There are several breeds of terriers, among which the English and Scotch are most conspicuous. These dogs are principally used for destroying rats or other vermin, and are so courageous that they do not hesitate to unearth the fox or the badger. Otters are also hunted by them, but prove by no means an easy prey, as their snake-like body, sharp teeth, and amphibious habits, render them very difficult to seize, and their tenacity of life will frequently enable them to escape when the dog considers them dead. The Scotch Terrier is a rough, wiry little dog, with hair hanging over its eyes, so that those organs are hardly visible, and when it is in the water its wetted hair quite obscures its vision. There is a smaller breed of these dogs called the "Skye Terrier," whose principal beauty seems to consist in their ugliness.

THE ENGLISH TERRIER.

Terriers are extremely attached to their master, and are capable of learning many amusing tricks. I had a terrier, said to be of Irish breed, who had imbibed many of the eccentricities of the Irish character. He was particularly fond of terrifying lapdogs, a species of animal which he held in supreme contempt. On one occasion, he met a very fat lapdog, the property of an equally fat old lady, waddling along the street. Rory looked at it for a short time, and

THE SCOTCH TERRIER.

then gave it a pat which rolled it over on its back. Its mistress immediately snatched it up, and put it on her muff, whereupon Rory erected himself on his hind-legs, an art which he possessed in great perfection, and walked along by her side, making occasional snatches at the lapdog. The terrified old lady struck at him with her boa, which Rory immediately caught in his mouth, and carried off down the street in an ecstasy of delight, ever and anon tripping over it and rolling head over heels. He had learned to shut the door, ring the bell, bring the slippers, or put the cat down stairs, which he accomplished by pushing her with his nose down each successive stair. During his residence at College he was accustomed to sit, dressed in a cap and gown, at the breakfast table, where his deportment was always most exemplary, and afforded a good example to many of the guests.

Poor Rory is dead now, but there is a record of his life in the "Sketches and Anecdotes."

The SHEPHERD'S DOG is a rough, shaggy animal, with sharp pointed ears and nose. It is an invaluable assistant to the shepherd, as it knows all its master's sheep, never suffers them to stray, and when two flocks have mixed, it will separate its own charge with the greatest certainty. It understands every look and gesture of its beloved master, and drives the flock to any place which he points out. This is the dog alluded to by Burns in the following beautiful passage:—"Man," said he, "is the god of the dog; he knows no other; he can understand no other. And see how he worships him! with what reverence he crouches at his feet! with what love he fawns upon him! with what dependence he looks up to him! and with what cheerful alacrity he obeys him! His whole soul is wrapt up in his god! all the powers and faculties of his nature are devoted to his service! and these powers and faculties are ennobled by the intercourse. Divines tell us that it ought just to be so with the Christian but the dog puts the Christian to shame."

THE SHEPHERD'S DOG.

The Greyhound is the swiftest of all the dogs, and is principally used in the pursuit of the hare, which amusement is termed coursing. It has but little delicacy of scent, and hunts almost entirely by sight. The hare endeavours to baffle it by making sharp turns, which the dog cannot do on account of its superior size, and has therefore to take a circuit, during which the hare makes off in another direction. The hare also has the property of stopping almost instantaneously when at full speed. It puts this manœuvre into force, when it is nearing its favourite hiding place. It induces the dog to spring upon it, and then suddenly checks itself. The dog is carried twenty or thirty yards forward by its own momentum, and the hare springs off to her place of refuge.

At Ashborne, in Derbyshire, there is a public-house sign representing a black and white greyhound chasing a hare. One greyhound was a little in advance of the other, and struck the game so forcibly with its nose that the hare was thrown over its back into the jaws of the other greyhound. This animal has been known to exert rather an unexpected talent, viz. retracing a journey during which it had been a close prisoner.

THE GREYHOUND.

"The celebrated greyhound, Black-eyed Susan, was brought to Edinburgh from Glasgow in the boot of a coach, on the night of Wednesday, the 13th May, 1835. On the following Sunday evening she made her escape, and in forty-eight hours reached her kennel, eight miles beyond Glasgow, being fifty-two miles in all. The road between Glasgow and Edinburgh she had never travelled on foot, and from the time taken she cannot have come direct; but by what route or process this animal made her point good it is in vain to conjecture."

I have lately made acquaintance with a very unique dog called "Quiz." I could not class him with any of the before-mentioned dogs, as no one has been bold enough to say exactly what kind of dog he is. Some say he is a poodle, some say he is a shock, while some give their opinion that he is a very small water spaniel; but no one has as yet decided on the precise breed to which he belongs. At first sight, a stranger would say that there was no difficulty about the matter, as the object lying on the floor can be nothing more or less than two mops without handles. But the thrums appear too long even for a mop, and if the visitor looks close he may see two glistening specks buried beneath the bundle or woollen cords. These are the animal's eyes, and if the stranger speaks kindly to him, a slight agitation of the opposite extremity proclaims that a tail is situated in that vicinity. Without these tests it would be impossible to judge where his head or his tail were, or whether he possessed such organs at all. Indeed one visitor did deliberately mistake his tail for his head.

He is a wonderfully clever dog, and perfectly capable of appreciating his own abilities. When told, he will play the piano and sing, the former feat being performed by beating the keys with his forefeet, while the latter accomplishment is represented by a melodious howl, during the performance of which he throws back his head and looks at the ceiling in the most approved style. He taught himself many of his tricks, and although my space will not permit me to give many lines to one animal, one of his eccentric accomplishments must be narrated. He is extremely fond of being noticed, and will play off all kinds of antics to draw attention. Once his mistress was unwell, and sent for her medical attendant, who according to the received routine felt her pulse and inspected her tongue. This took place for several days in succession, and Quiz felt himself aggrieved that no notice should be taken of him. A few days afterwards Quiz was observed sitting on his hind legs, holding out his paw and putting out his tongue as he had seen his mistress do. I have seen him perform this feat several times. He has not patience to keep his tongue out, but continues rapidly putting it out and then drawing it back again.

Although so fond of attention he is a very modest dog in his way, and cannot endure that his face should be seen. If any one pushes aside the heavy mass of woolly ringlets which entirely obscure his head, he gives himself an impatient shake, and effectually conceals every particle of his countenance from the public gaze. I think that he must be ashamed of his nose because it is white, and not black like that of other dogs.

It is amusing to note the comments which accompany his journeys through the streets. Some spectators express their opinion that he is a young lion, while others are as strongly persuaded that he is a bear. One gentleman, seeing the dog in the arms of its mistress, took it for a muff of a novel and eccentric character. It is impossible to give any idea of his appearance when walking, except by returning to my first simile of a mop. The only image by which the appearance of this remarkable dog can be expressed, is by requesting the reader to imagine three or four double handfuls of mop thrums twice the ordinary length, trundling along the ground by some invisible power, and attaching to themselves all the bits of stick, dried bones, &c. that come in the path of the bundle. I might fill pages with descriptions of this funny animal and other dogs with whom I am acquainted, but my completed space forces me to postpone such accounts to another occasion.

The WOLF.—Ferocity, craft, and cowardice, are the well-known traits of the Wolf. Although one of the dog tribe, it is held in utter abhorrence by the domesticated dogs. The stronger pursue and destroy it, the weaker fly from it in terror. In the earlier part of English history it is frequently mentioned as a common and dreaded pest. It was finally extirpated in England about 1350, in Scotland about 1600, and was not entirely destroyed in Ireland until the beginning of 1700. It is still found in parts of France, Russia, and the whole of Western Asia.

These formidable creatures almost invariably hunt in bands, and display very great cunning in waylaying and pursuing their prey. Winter is the time of year most dreaded by those who live in countries where wolves exist, as at that season hunger renders them exceedingly ferocious and daring. They will then attack sledges, or carriages, even when guarded by armed men. They are very wary, and dislike approaching anything at all resembling a trap. A traveller, aware of this habit, saved his life by trailing a cord from his carriage window. The wolves thought that the cord looked suspicious, and before they had quite made up their minds about it, the traveller reached a station where he was in safety.

CANIS.—(Lat. *a Dog.*)

Lupus (Lat. *a Wolf*), *the Wolf.*

The bite of the wolf is extremely dangerous, as its jaws are immensely strong, and it generally brings away the part it seizes. It does not bite like any other animal, but makes a succession of short sharp snaps, each of great power but very rapid. Those who have been chased by the wolves compare the snapping of their jaws to the clash of a steel trap when slipped. When young, the wolf can easily be tamed, and shows as great attachment to its master as any dog will. It is very tenacious of life. Parry relates an anecdote of a wolf that was caught in a trap, and after being pierced with three bullets, and several thrusts of a sword, sprang at one of the officers, and actually succeeded in escaping, although its hind-legs were firmly tied together.

Almost every traveller who has journeyed out of the regular railroad and diligence line has stories of the wolves, and in almost every case the feeling appears to be more of irritation than anything else. Either they have hunted the wolf until their horses were knocked up, while the animal still continues his provoking and lasting gallop, or a wounded wolf has succeeded in escaping, or the wolves have eaten all the leather straps from their carriages, or some such misfortune, appears to excite the most vindictive feelings against the wolves.

The flesh of the wolf is considered rather good eating by those who have tried it, and at all events, as one hunter said, "it is much nicer than lean deers' meat."

The JACKAL.—This animal is found in North Africa, Persia, and India. It derives its name "aureus" from the yellow tint of its skin. It, like the wolf, unites in bands to hunt, and the prey which the pack has taken so much pains to secure is not unfrequently confiscated by the lion, who keeps the reluctant hunters at a distance until he has satisfied his own royal appetite. The Jackals, however, often retaliate by assisting at the demolition of the larger prey which the lion destroys. It is very useful in the East, as it acts as scavenger, and consumes the offal which, in those not very cleanly towns, is cast into the streets, and would inevitably cause a pestilence, were it not for the assistance of the jackals and other creatures.

CANIS.

Aurĕus (Lat. *golden*), *the Jackal.*

It is excessively fond of grapes, and makes dreadful havoc in the vineyards, so that the fable of the Fox and the Grapes might be quite as appropriately related of this animal. While hunting, it utters most piercing shrieks, which have been compared by those who have heard them to the wailings of evil spirits, an association which the oriental tombs and ruins which it frequents, recalling to mind the mysterious Arabian Nights, are most fitted to produce.

There are several kinds of Jackals, one inhabiting Senegal, and another the Cape of Good Hope. They are rather larger than the fox, but do not possess nearly so bushy a tail as that "brush" wherein sportsmen take so much delight.

The Fox.—This terror of hen-roosts and delight of sportsmen is found in most parts of England, and many other countries. It varies very much in colour and size, according to the country where it lives.

The habits of this animal are mostly nocturnal. It lies by day concealed in its burrow, if it be fortunate enough to possess one, or in the depths of some thicket, if it is not a householder. Towards evening it sallies out in search of food, and woe to the unfortunate hare, rabbit, pheasant, or fowl that comes in its way! Reynard does not attempt to chase the hare, for it is too swift for him, nor the rabbit, as it would immediately dive into its hole; nor does he run at the pheasant, which would fly away, and probably only leave a tail feather in the fox's mouth. He knows his business too well. He creeps very quietly and slowly to some place where hares or rabbits are likely to pass, and then springs on them as they run by him. One of his most ingenious devices to catch rabbits is really quite mathematical. When the rabbit has young, she keeps them safely at the end of the burrow. Now the burrow is too small for the fox to creep into, and it is so deep that even the six or seven young rabbits would not recompense him for the toil. So he traces them by scent, fixes on a spot over the termination of the burrow, and by digging perpendicularly, attains his object with comparatively little trouble.

Vulpes.—(Lat. *a Fox.*)

Vulgāris (Lat. *common*), *the Fox.*

Sometimes he steals into the hen-roost, destroys and carries off most of its inmates, some of which he devours on the spot, others he carries home, and the remainder he buries for a future repast.

When irritated, the fox gives out a strong disagreeable scent, which lies so long on the ground that it may be perceived for nearly an hour after the fox has passed. Partly on this account, and partly on account of its speed, endurance, and cunning, the chase of the fox is one of the most admired English sports.

Many tales are related of its cunning when pursued, such as driving another fox out of its home, and forcing it to substitute itself as the chase; diving into a heap of manure, so that the dogs could not perceive its scent; jumping over a wall, running a little way, coming back again, and lying under the wall until all the dogs had passed, and then leaping a second time over the same place where it had passed before, and making off on its old track. A fox has been known to leap through a kitchen window, and hide itself behind the plates on the dresser, without the observation of the cook, to whose terror and consternation six or seven dogs have leaped through the same window, and dragged the fox from its lurking-place.

The Fox appears to be the same crafty animal in all countries, there being quite as many American tales of vulpine sagacity as we can find in England. There are so many stories of this nature that it is impossible to give even a hasty account of them. I therefore select one of the shortest.

On the banks of the Kentucky river rise huge rocky bluffs several feet in height. A fox that lived near this river was constantly hunted, and as regularly lost over the bluff. Now, nothing short of wings would have enabled the animal to escape with life down a perpendicular cliff. At last, a hunter, being determined to discover the means by which the animal baffled them, concealed himself near the bluff.

Accordingly in good time the fox came to the top of the cliff as usual and looked over. He then let himself down the face of the cliff by a movement between a leap and a slide, and landed on a shelf not quite a foot in width about ten feet down the cliff. The fox then disappeared into a hole above the shelf. On examination, the shelf turned out to be the mouth of a wide fissure in the rock, into which the fox always escaped. But how was he to get out again? He might slide down ten feet, but he could never leap ten feet from a ten-inch shelf up the face of a perpendicular rock. This impossibility struck the hunter's mind, so he instituted a search, and at length discovered an easier entrance into the cave from the level ground.

The fox was too wise to use that entrance when the hounds were behind him, so he was accustomed to cut short the scent by dropping down the rock, and then when all the dogs were at the edge of the cliff, he walked out at his leisure by the other entrance.

The Arctic Fox changes its fur, and becomes white during the winter.

The Mustelina, or Weasels, are easily distinguished by their long slender bodies, short muzzle, sharp teeth, and predatory habits. They inhabit almost every part of the world, and procure their food by creeping on the unsuspecting victim, generally a rabbit, rat, or bird, and then suddenly darting at it and piercing its neck with their sharp teeth. Almost all the weasels devour the brain and suck the blood of their prey, but seldom touch the flesh, unless they are pressed by hunger.

Two kinds of MARTENS inhabit England, named, from their favourite haunts, the Pine and the Beech Marten. Some naturalists assert that these two martens are not distinct animals, but only varieties of the same species. The Pine Marten is not uncommon in Derbyshire, where it is much too fond of chickens and ducklings to be a desirable neighbour. This animal, as well as the Sable, is much sought after on account of its skin, which furnishes a beautiful fur, not much inferior to that of the Sable.

Sub-family *e. Mustelīna.*—(Lat. *Mustēla*, a Weasel.)

MARTES.—(Lat. *a Marten.*)

Abiĕtum (Lat. *of the Pine-tree*), *the Pine Marten.*

The SABLE, long famous for its costly fur, which is thought worthy to adorn the coronation robes of a monarch, inhabits Siberia. The chase, or rather the search, after these animals is attended with dreadful hardships and great danger. Sometimes a sable will not be seen for days; sometimes the bait of the trap is eaten by other animals, such as gluttons, &c.; sometimes the hunter's provisions fail; he spends days and nights in the midst of snow, surrounded by interminable pine forests, and exposed to the piercing blasts of the tempest. Many hunters lose their lives in these terrible solitudes, overwhelmed by snow-storms, or famished with hunger.

MARTES.

Zibellīna (from a Sclavonic word, from which is derived our "Sable"), *the Sable.*

A species of Sable (*Martes leucŏpus*) inhabits North America. The hair of the sable will turn either way, and in this respect differs from the fur of other animals. The skins are very valuable, varying from one to ten pounds in price, according to the quality.

The POLECAT, fitchet, foulmart, or "fommard," as the farmers call it, is very common in most parts of England. It is dreadfully destructive to the poultry, and destroys both old and young. William Howitt relates an interesting anecdote of his dog unearthing a polecat, and afterwards bringing out of its hole an entire brood of ducklings that had most unaccountably disappeared from the premises of a farmer. Winter is the usual time for its appearance in the farmyard, as in the summer it obtains its food with less risk among the warrens.

PUTORĬUS.—(Lat. from *puteo*, to stink.)

Fœtĭdus (Lat. *fetid*), *the Polecat.*

The FERRET is supposed to be a domesticated variety of the polecat, and a mixed breed is generally preferred by rat-catchers, who use the ferret, first muzzling it carefully, to drive the rats out of their holes, when they are either struck down with sticks or killed by terriers, who keep a sharp watch for them. This animal may be rendered very tame, but it is at all times to be handled with caution. I knew a lad who was very proud of a beautiful white ferret, and used to boast of its tameness. One day he was caressing the animal and letting it kiss him as he called it, when he started back and threw down the ferret. His face was streaming with blood, the ferret having bitten both his lips through, and made four cuts like those of a sharp knife. The hair, called Fitch, is much used for making paint brushes.

The STOAT, or ERMINE, is also another common English animal. It is less than the polecat, but its habits are scarcely less predacious. Hares and rabbits fall easy victims to their little enemy, who dispatches them with a single bite, penetrating the brain. During the winter, the stoat becomes partially white, in northern countries wholly so, except the tip of the tail, which remains black. In this state it is called the Ermine, and is killed in great numbers for the sake of its beautiful and valuable fur.

MUSTĒLA.—(Lat. *a Weasel.*)

Erminĕa.—*The Stoat.*

The WEASEL is the least of this tribe. It is excessively useful to farmers, as it wages unrelenting war on rats and mice, and in an incredibly short space of time extirpates them from a barn or stack. It hunts by scent like dogs, and tracks the unfortunate rat with the most deadly certainty. On this account some farmers encourage it on their premises, but the generality destroy it, and nail its body on the barn door, forgetting that although it does sometimes abstract a chicken or an egg, yet it will not touch them as long as it can find rats or mice. It is a most courageous little animal, and will even attack men, who have found it by no means a despicable antagonist, as its instinct invariably leads it to dash at the throat, where a bite from its long sharp teeth would be very dangerous.

MUSTĒLA.

Vulgāris (Lat. *common*), *the Weasel.*

Like many other little beings, it is very passionate and easily irritated, often indeed taking offence where none was intended. On one occasion a colony of weasels attacked an old woman who was returning from market with her basket of provisions, and had sat down upon a heap of stones in which the weasels had taken up their residence.

The RATEL is a native of South Africa, and lives principally on the combs and honey of the wild bee, although it is very probable that much of its subsistence is derived from flesh and roots. It is said to be guided to the bee's nest by a bird called the Honey-guide, which, as the natives assert, being very fond of honey and unable to attack the hive by itself, seeks for the honey-ratel, and admonishes it by a peculiar cry that the desired honeycomb is not very far distant.

MELLIVŎRA.—(Lat. *mel*, honey; *voro*, I devour.)

Ratel (Native name), *the Honey Ratel.*

The WOLVERINE, Glutton, or Carcajou, inhabits North America. Accounts vary respecting the habits of this animal. The older naturalists say that it ascends trees, and drops on the neck of any unfortunate deer which happens to pass beneath, and that having once secured its prey, it never leaves it until the last morsel is consumed.

Be this as it may, the Glutton is known to hunt after its prey, which it follows for many miles at a slow but persevering pace, and seldom fails of bringing it down at last. It is especially hated by the sable hunter, as it will follow him in his rounds, robbing the traps of the baits as it proceeds, and should a sable be caught it generally tears it to pieces, or buries it in the snow. The hunter has some slight revenge in robbing it of its skin, as the fur is in some request, but the mischief it does him is not by any means counterbalanced by the value of its hide.

GULO.—(Lat. *a Glutton.*)

Luscus (Lat. *blinking*), *the Wolverine.*

It is a very determined animal, and when attacked defends itself vigorously, proving more than a match for a dog. The length of the glutton, without the tail, is about two feet six inches.

Several genera are omitted.

The BADGER.—This harmless and much injured animal (which is often subjected to such ill treatment that the term "badgering" a person is used to express irritating him in every possible way) is found throughout Europe and Asia. It is not now very common in England, but is frequently found in Scotland, where it is termed the "Brock," a name familiar to us all, through the means of Dandie Dinmont, who also immortalized the pepper and mustard terriers.

MELES.—(Lat. *a Badger.*)

Vulgāris (Lat. *common*), *the Badger.*

The Badger lives at the bottom of deep burrows, which it excavates, and in which it passes all the day, sleeping on a very comfortable bed of hay and grass. When the evening approaches it seeks its food, consisting of roots, fruit, insects, and sometimes young rabbits. It is also said to attack the wild bee, and boldly to devour the honey and combs, its thick hair and skin rendering it utterly regardless of the stings of the enraged bees, who "might as well attack a barber's block."

The cruel sport of baiting the badger is still continued, although not so openly or frequently as a few years back. The poor creature is placed inside a kennel, and dogs set at it, who are not unfrequently worsted by the badger, as its bite is terrific, and its skin so tough, and hair so thick, that the bites of the dog do not take full effect. The pleasure of this "sport," as in many other diversions of the sporting world, appears to consist in trying whether the dogs or the badger will be most mangled in a given time.

But this sport, cruel as it may seem, is not so cruel as the method employed for training young dogs to attack the badger, or as it is technically called, entering them. The under jaw of the unfortunate animal is sawn off, and most of the remaining teeth drawn. In this state it is put into a barrel lying on its side, and the dogs are encouraged to attack the poor maimed animal, in order that they may be so inspirited by their impunity as to attack badgers armed with their full means of defence.

The power of the badger's bite is caused principally by the manner in which the under jaw is set on. Not only are its teeth sharp, and the leverage of its jaw powerful, but the jaw is so contrived, that when the creature closes its mouth, the jaw locks together as it were, and is held fast without much exertion on the part of the badger.

Its skin is rather valuable, the hair being extensively employed in the manufacture of brushes, and its fur being in some request for holsters. The omnivorous and thrifty Chinese eat its flesh, as indeed they will that of most animals, and consider its hams a very great dainty. The length of the badger is about two feet three inches.

The OTTER seems to play the same part in the water as the polecat and the other weasels on the land. Like the polecat, it is excessively rapacious; like the polecat, it destroys many more creatures than it can devour; and as the polecat only eats the brain and sucks the blood, so the other daintily eats the flakes at the back of the fish's neck, and leaves the remainder for less fastidious animals. In Scotland, where the otter abounds, it is not uncommon to find a large fish, such as a salmon, lying on the bank, perfectly fresh and entire, except a few inches along the back, which the otter has bitten out.

So well do the poor people know this custom of the otter, that in some places they consider the otter's shelf as their larder, and go to look on the bank daily for their salmon, which is none the worse for them because the otter had previously helped itself to a piece of the shoulder.

It is extremely interesting to watch the actions of this almost amphibious creature. It slides noiselessly into the water, turns and twists about below the surface with the same or greater ease than a fish, then, with a graceful sweep of the body, it glides to the surface and ascends the bank with almost the same motion. While below the surface it bears a great resemblance to the seal, the method in which it disposes its hind feet greatly assisting the effect. Its rapid and easy movements in the water are mostly performed by the assistance of its powerful tapering tail.

Otter hunting is a very favourite sport in Scotland, where almost every stream is furnished with its otter. At the sight of the footsteps of the animal the population round is in a commotion, the dogs are assembled, guns and spears provided, and the hunters go out in sufficient numbers and with sufficient arms to kill a tiger; and from all accounts it is quite as difficult an animal to destroy; for by diving, and biting, and hiding among stones, added to its great tenacity of life, it gives the hunters no little trouble to secure it.

Lutra.—(Lat. *an Otter.*)

Vulgāris (Lat. *common*), *the Otter.*

The otter is easily tamed, and its predatory habits have been occasionally turned to account, as it is sometimes trained to catch fish and bring them to shore, precisely as the falcon is trained to catch terrestrial game.

The Hindoos have brought the art of otter training to great perfection, and keep their otters regularly tethered with ropes and straw collars on the banks of the river.

Several genera are omitted.

Family II. . . Ursĭdæ (Lat. *ursus,* a bear. Bear kind).
Sub-family *a.* *Ursīna.*

The Bears and their allies are mostly heavy, and walk with the whole foot placed flat on the ground, unlike the cats, dogs, &c., who walk with merely their paws or toes. All the bears are omnivorous, that is, they can eat either animal or vegetable food, so that a leg of mutton, a pot of honey, a potato, or an apple, are each equally acceptable.

The Brown Bear inhabits the north of Europe, Switzerland, and the Pyrenees. It has been extirpated from England for many centuries, but is recorded to have been found in Scotland so late as 1057. The inhabitants of Northern Europe hunt it with much skill, and take it in traps and pitfalls, availing themselves of its love for honey. It is said that there exists a practice of placing the hive in a tree, and planting long spikes round its foot. A heavy log of wood is then suspended by a cord just before the entrance of the hive, and the trap is complete. The bear scents the honey and comes to look at the tree. The spikes rather astonish him, but he sniffs his way through them, and commences the ascent. When he has reached the hive, he is checked by the log hanging before the entrance; this he finds is movable, and pushes aside, but it is just so long that a mere push will not entirely remove it, so he gives it a tremendous pat, and looks in at the entrance. Just as he has succeeded in putting his nose to the hive, the log returns and hits him very hard

URSUS.

Arctos (Gr. Ἄρκτος, a bear), *the Bear.*

on the head. This makes him exceedingly angry, and he pokes it away harder than ever, only to return with a more severe blow than before. He now has a regular fight with the log, hitting it first to one side and then to the other, the perverse block invariably striking his head every time, until at last a severer blow than usual knocks him fairly off the tree on to the spikes below.

In the time of Queen Elizabeth the bear used to be baited, that is to say, the bear was tied to a pole, and several dogs were set at him, the object being, to see whether the bear could bite the dogs, or the dogs bite the bear with the greater force; but this cruel sport is now happily extinct.

Two genera are omitted.

The Grizzly Bear.—"Bernardin de Saint Pierre dit, 'A la vue de l'homme, les animaux sont frappés d'amour ou de crainte.' Il laissait de côté une troisième impression qu'éprouvent beaucoup d'animaux, à l'aspect de l'homme; c'est la faim, et l'envie de le manger."* This observation applies most fully to the Grizzly Bear, a native of North America. It is the most ferocious and powerful of its family, and is an animal which must either be avoided or fought, for there is no medium. If a Grizzly Bear once sees a man, it will probably chase him, and will do so with great perseverance. An American traveller told me lately, that he had been chased nearly thirty miles by one of these bears, who would probably have kept up the chase as many miles more, had not my informant crossed a wide river, over which the bear did not choose to follow him.

URSUS.

Horribĭlis (Lat. *horrible*), *the Grizzly Bear.*

Some writers have said, that the Grizzly Bear will run away if he comes across the scent of men. This, my informant, who is a practical man, strenuously denies, and states that the man is more likely to run away from the bear, than the bear from the man. The American Indians fear it so much, that a necklace of its claws, which may only be worn by the individual who destroyed the bear, is a decoration entitling the wearer to the highest honours. These formidable claws are five inches long, and cut like so many chisels, so that the Indian of former days,

* "Bernardin de St. Pierre said,—'At the sight of man, all animals are struck either with love or fear.' He forgot to mention a third impression made on many animals when they see a man, namely 'hunger, and a great desire to eat him.'"

armed only with bow, spear, and knife, fully deserved honour, for overcoming so savage and powerful a brute. Since the introduction of firearms, the Grizzly Bear affords a rather easier victory, but even to one armed with all advantage of rifle and pistols, the fight is sure to be a severe one, for when the bear is once wounded, there is no attempt to escape, but life is pitted against life. Before the hunter commences the struggle he must have considerable confidence in his presence of mind, for every one knows how the least tremor of hand or eye, causes a rifle ball to wander far from its intended path, and a ball that does not penetrate a vital part only serves to irritate the bear.

Sometimes, it is said, after a party of hunters have been combating one of these bears, it is impossible to find four square inches of sound skin in the animal's body, a ball through the brain, or heart, appearing to be the only safety on the part of the hunter.

When a traveller is passing through a part of the country where he is likely to fall in with these animals, he provides himself with a quantity of meat strongly impregnated with some perfume. If a bear sees the traveller, and charges him, he throws down a small piece of his prepared meat. The bear stops and sniffs at it, and is dubious about it for some time, but at last finishes by eating it. During the time in which he is undecided, the traveller has gained considerable ground, and by a repetition of the same ruse, either tires the bear out, or meets with a sufficient body of friends to render him independent of the animal.

It is rather singular that this bear has the power of moving each claw separately, as we move our fingers. It is able to overcome and carry off the enormous bison, and to dig a pit in which to bury it.

The Polar, or White Bear, called Nennook by the Esquimaux, lives in the Arctic regions, where it feeds on seals, fish, and even the walrus, but it dares not attack the latter animal openly. It is a formidable antagonist either by land or water, as it dives with great ease, and is able to chase the seal amid the waves. Nelson nearly lost his life by imprudently attacking one of these animals with no weapon but a rusty musket, which could not be induced to fire; and indeed had he not been separated from the infuriated bear by a cleft in the ice, he could hardly have escaped its claws. As the seals frequently crawl out of the water upon rocks or fragments of ice, the Polar bear is forced to swim after them; but lest they should observe him he makes his approaches by a succession of dives, and contrives that the last dive brings him directly under the unsuspecting seal, who is immediately grasped and killed. Richardson states that these bears are often drifted from Greenland to Iceland on fields of ice, and that they find the flocks and herds so very delicious after a long course of seal diet, that the inhabitants are forced to rise in a body and put an end to their depredations.

To give this animal, who is constantly running over fields of ice, a firm footing, the soles of its feet are thickly covered with long hair, on

THALARCTOS.—(Gr. from *θάλασσα*, the Sea, and *ἄρκτος*, a Bear.)

Maritĭmus (Lat. *belonging to the sea*), *the Polar Bear.*

the same principle that induces elderly gentlemen to tie list round their shoes in the winter months, when they have reason to dread the slides that idle boys always have made on the pavement, and we suppose always will make.

The RACOON is an animal about the size of a large fox, and an inhabitant of Canada and other parts of America. It derives its name, *lotor*, from the habit it is said to possess, of washing its food before eating it. Its skin is very valuable, and is much sought after by American hunters, who pride themselves on their skill in shooting this active and wary animal. There is a story related by the Americans, of a hunter who was so excellent a marksman that when he entered a wood, the 'coons came down of their own accord, knowing that escape was impossible; but we must class this tale with the account of the man who could grin the bark off gum-trees, and the swift Indian, who could run so fast round a tree that he sometimes caught sight of his own back.

The food of the Racoon is principally small animals and insects. Oysters are also a very favourite article of its diet. It bites off the hinge of the oyster, and scrapes out the animal in fragments with its paws. Like a squirrel when eating a nut, the racoon usually holds its food between its fore-paws pressed together, and sits upon its hind quarters while it eats. Poultry are very favourite objects of its attack, and it is said to be as destructive in a farm-yard as any fox, for it only devours the heads of the murdered fowl. Like the fox, it prowls by night.

When taken young it is easily tamed, but very frequently becomes blind soon after its capture. This effect is supposed to be produced by the sensitive state of its eyes, which are only intended to be used by night; but as it is frequently awakened by daylight during its captivity, it suffers so much from the unwonted glare, that its eyes gradually lose their sight.

It has been mentioned that the name in general use among the Americans of the present day is "'Coon," a word which strangely contrasts with its ancient Mexican name of Cioatlamacazque. —Two genera are omitted.

Sub-family *c. Procyonīna.*

PROCȲON.—(Gr. Προκύων, a constellation.)

Lotor (Lat. *a washer*), *the Racoon.*

The peculiarly long snout of the COATIS distinguishes them at once from the Racoons, which they resemble in some other respects. Their snout is very moveable, and is of great use to them in routing out the worms and insects which they dig up. The nostrils are placed on a sort of disk at the end of the snout, and give the whole head a most extraordinary aspect. The Coatis live upon birds, eggs, insects, and worms, and sometimes they will eat roots. They are nocturnal in their habits, spending most of the day in sleep, rolled up in a

Sub-family *d. Cercoleptina.*

NASUA.—(Lat. from *nasus*, a nose.)

Fusca (Lat. *fuscus*, dusky), *the Coati-mondi.*

ball.* In descending a tree they walk with their heads downwards, like the cat, which, however, they surpass in activity. These animals inhabit the warmer parts of America, but do not appear to be much sought after by hunters. The Brown Coati-mondi is the species represented in the engraving; there is another species, the Red Coati.

The KINKAJOU is also an inhabitant of Southern America. It is not unlike the Coati in its habits, but is more active, as it possesses a prehensile tail, which it uses in the same way that the Spider Monkeys use theirs. The tongue of the Kinkajou is capable of being inserted into crevices, and drawing out any insects that may be lying concealed beyond the reach of its paws. The Spanish missionaries give it the name of Honey Bear, because it is a great devastator of the nests of the wild bee, using its long tongue to lick the honey out of the cells. When in captivity it is very tame and gentle, and will play with an acquaintance as a cat will. It displays great address in capturing flies and other insects with its tongue, and it is amusing to watch how its eyes gleam directly that a fly settles within its reach. During the earlier part of the day it will not move, but towards dusk it becomes very brisk and animated, climbing about its cage, and swinging from the top bars by its tail and hind paws.

CERCOLEPTES.—(Gr. Κέρκος, a tail; λεπτός, thin.)

Caudivolvŭlus (Lat. *twisted tail*), *the Kinkajou.*

A sub-family is omitted.

Family III. . . Talpĭdæ. (Lat. *talpa*, a mole. Mole kind.)
Sub-family *a.* . *Talpīna.*

The MOLE.—Many ridiculous stories of the Mole and its habits may be found in several authors, among whom Æsop stands very conspicuous. This much maligned animal is said to be deprived of eyes, to undergo unheard-of tortures in forcing its way through the earth, and to spend a life of misery in subterranean damp and darkness. But so far from being a miserable animal, the Mole seems to enjoy its life quite as much as any other creature. It is beautifully fitted for the station which it fills, and would be unhappy if removed from its accustomed damp and darkness into warmth and light.

The eyes of the mole are very small, in order to prevent them from being injured by the earth through which the animal makes its way;

* There are several fine specimens of this animal in the Jardin des Plantes at Paris. They are tolerably lively even during the day-time.

indeed, larger eyes would be useless underground. When, however, the mole requires to use its eyes it can bring them forward from the mass of fur which conceals and protects them when not in use. The acute ears and delicate sense of smell supply the place of eyes. Its fur is very fine, soft, capable of turning in any direction, and will not retain a particle of mould. But the most extraordinary part of the mole is the paw or hand with which it digs. The two fore paws are composed of five fingers, armed with sharp, strong nails, in order to scrape up the earth; and to prevent the accumulated mould from impeding the mole's progress, the hands are turned outwardly, so as to throw the earth out of its way.

TALPA.

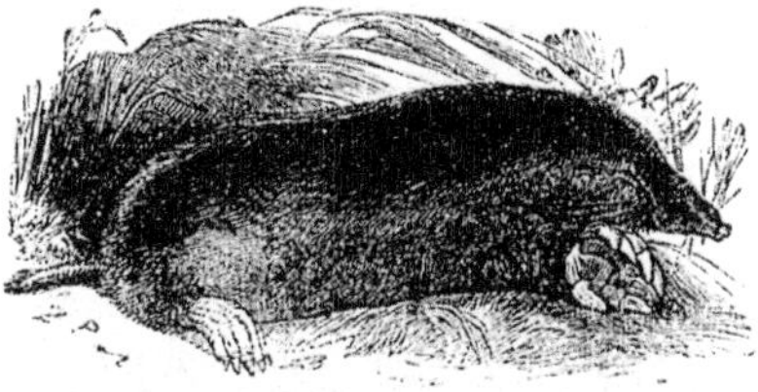

Europæa (Lat. *belonging to Europe*), *the Mole*.

The Mole is a most voracious animal, and is incapable of sustaining even a slight fast. Its principal food is the earth-worm, in chase of which it drives its long galleries underground; but it also will eat insects, bits of meat, and is said sometimes to catch birds, which it takes by surprise, and then rapidly tears to pieces with its powerful claws. This ravenous appetite causes it to suffer from thirst if a supply of water is not at hand. For this reason the mole always makes a tunnel towards a pond or brook, if there is one near. If no water is near, it digs a number of little wells, which receive the rain or dew, and enable it to quench its thirst.

It is a good swimmer, and can pass from bank to bank, or from the shore to an island, and when the fields are inundated by floods it can save itself by swimming.

The construction of the mole's habitation is very singular and interesting. Each mole has its own habitation and hunting ground, and will not permit strangers to trespass upon its preserves, which it guards, not by "man-traps and spring-guns," but by its own claws and teeth.

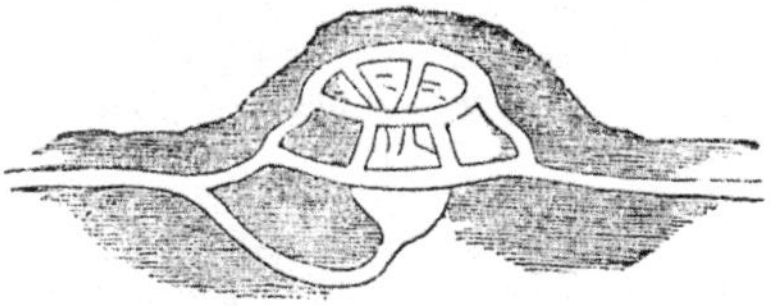

MOLE HILL.

In order to construct a fortress, the mole selects a secure place, as the foot of a tree, or the side of a high bank. It then throws up a heap of earth, which it presses firmly together, as within this mound its fortress has to be made. It commences by running a circular gallery near the summit of the mound, and another larger one near the bottom. These two galleries it connects by five descending passages. In the very centre of the mound, and at the level of the ground, it now digs a circular hole, which it connects with the upper gallery by three ascending passages. Lastly, it makes a number of passages from the lower gallery, and connects the circular chamber with the largest of them, or high road, by a passage that first bends downwards, and then rises into the high road a little outside the large gallery. In the circular chamber the mole sleeps, and can escape into the high road either by the upper gallery or by the road from the bottom of its dormitory.

I have already stated that each mole has its own hunting ground, and permits no intruder. If a strange mole should happen to trespass upon the domains of another, there would be a furious fight, and the conqueror would devour his vanquished foe.

Although each mole has its own hunting ground, yet there are mostly high roads, which connect the different hunting grounds with each other, and which are used by many individuals in common, the only precaution taken being, that if two moles should happen to meet, the weaker immediately retreats into one of the numerous side galleries which open from the high road, and permits its aristocratical neighbour to pass.

All the passions of the mole seem to be furious. Even its passion for work, *i.e.* search after its food, has something fierce in it. The animal works desperately for several hours, and then rests for as many hours. The country people about Oxford say that it works at intervals of three hours each.

The mode of burrowing by this animal, is by rooting up the earth with its snout, and then scooping it away with its fore feet. I have often seen this operation performed.

The depth at which this animal works depends almost entirely on the time of year. In the summer, the worms come to the surface, and the mole accordingly follows them, making quite superficial runs, and sometimes only scooping trenches on the surface. But in the winter, when the worms sink deep into the ground, the mole is forced to follow them there, and as it cannot fast above an hour or two, it is forced to work at the comparatively hard and heavy soil, as it did in the light earth nearer the surface.

Moles vary in colour, the usual tint being a very deep brown, almost black, but they have been seen of an orange colour, and a white variety is not uncommon. I have a cream-coloured skin in my possession. There are several moles known,—the Shrew Mole, the Changeable Mole, the Cape Mole, and the Star-nosed Mole, are the most conspicuous.

Sub-family *d. Erinacīna.*—(Lat. from *Erinacĕus*, a Hedgehog.)

The Shrew Mouse.—This pretty little animal is very like the common mouse, but is easily distinguished from it by the length of the nose, which is used for grubbing up the earth in search of earth-worms and insects.

The reader must not imagine that the Shrew has any connexion with the true mice. It belongs to an entirely different class of animals, its teeth being sharp and pointed, not unlike those of the mole and the hedgehog, whereas those of the mouse are broad and chisel-shape like the teeth of the rabbit.

A peculiar scent is diffused from these animals, which is possibly the reason why the cat will not eat them, although she will readily destroy them.

SOREX.—(Lat. *a Rat.*)

Araněus (Lat. *a Shrew*), *the Shrew Mouse.*

Many species of shrews are known, inhabiting various countries. There are, besides the common species, the Oared and the Water Shrew, all three inhabiting England. The formation of their hair as seen under a powerful microscope, is very beautiful, but quite distinct from the hair of the mouse or rat. In the autumn, numbers of these little animals may be seen lying dead, but what causes this destruction is not known.

This is one of the numerous animals that have suffered by false reports, and have been treated with great cruelty on account of those fables. Rustics formerly believed that the poor little harmless creature paralysed their cattle by running over them, and that the only way to cure the diseased animal was to place a bough of shrew-ash on the injured part. The shrew-ash was made by boring a hole into an ash-tree, and then plugging up in the hole a living shrew-mouse. By the same process of reasoning a shrew cut in half, and placed on a wound supposed to be caused by its bite, was considered a certain remedy.

CROSSŎPUS. (Gr. *κροσσοὶ, fringe; ποῦς, a foot.*)

Fodĭens (Lat. *digging*), *the Water Shrew.*

The WATER SHREW frequents brooks and clear running ditches, in the banks of which it lives. It swims and dives with great ease, and when under water appears as if it had been speckled over its entire surface with silver, from the bubbles of air which adhere to its fur. It eats the grubs of various aquatic insects, digging them out of the muddy banks with its snout. It is not very common, but I have seen numbers of them inhabiting a brook near Little Hinton in Wiltshire, and often watched their elegant movements and gambols through the water.

Its localities may be discovered by searching for its "runs," which are like those of the common water rat, but much smaller.

One or two genera are omitted.

The HEDGEHOG is remarkable as being our only English animal that is guarded with spikes. These spikes are fixed into the skin in a very beautiful and simple manner. When the Hedgehog is annoyed it rolls itself up, and the tightness of the skin causes all its spines to stand firm and erect, bidding defiance to an unprotected hand. While rolled up, even the dog and the fox are baffled by it; but their ingenuity enables them to overcome the difficulty by rolling it along until they push it into a puddle or pool, when the astonished hedgehog immediately unrolls itself to see what is the matter, and before it can close itself again is seized by its crafty enemy.

ERINACĔUS.—(Lat. *a Hedgehog.*)

Europæus (Lat. *belonging to Europe*), *the Hedgehog.*

Many more fortunate animals have outlived the aspersions cast upon their character by ignorant persons, but the prejudice against the hedgehog is still in full vigour in the agricultural districts. Scarcely a farmer or labourer will be persuaded that the hedgehog does not suck the cows. Now this is an impossibility for the hedgehog, but I have seen pigs—not hedgepigs, but real bacon pigs—suck the cows whilst lying down. Among other creatures accused of this theft, are the slow-worm or blind-worm, a kind of legless lizard with an extremely small mouth, and the bird called the goatsucker. Really when a man relates that a bird sucks a cow, it reminds one of the brother philosophers, one of whom milked a bull while the other held the pail.

The food of the hedgehog consists not of cow's milk, but insects, snails, frogs, mice, and snakes. Dr. Buckland placed a snake in the same box with a hedgehog. The hedgehog gave the snake a severe bite, and then rolled itself up, this process being repeated until the spine of he snake was broken in several places; it then began at the tail, and ate the snake gradually, as one would eat a radish. White has seen it bore down and eat the roots of the plantain, leaving the leaves and stem untouched.

The flesh of the hedgehog is said to be good eating, and the gipsies frequently make it a part of their diet, as do the people in some parts of the continent.

There is a peculiar method of preparing the animal for food, strongly reminding one of the earth ovens used by the Polynesians. The hedgehog is simply wrapped up in a mass of clay and put on the fire. In process of time the clay is thoroughly baked, and cracks open, when the hedgehog is supposed to be cooked. On opening the clay, the skin comes off with it, while the insides of the animal have formed themselves into a hard ball, and are taken out entire. By this method of cooking, the juices are retained, and not suffered to dissipate as they would if it were roasted.

During the winter it lives in a torpid state, in a hole well lined with grass and moss, and when discovered looks like a round mass of leaves as it has rolled itself among the fallen foliage, which adheres to its spikes. The engraving of the spine, or quill, of this animal shows the method by which it is retained in the skin. The quill is as it were pinned through the skin, and retained by the head. The curvature is such, that when the animal contracts itself, the quills are drawn upright, and form a strong and elastic covering, useful for more purposes than merely defence from foes. The hedgehog has been known to throw itself boldly from a considerable height, trusting to the elasticity of the spring for breaking its fall. It will be seen that when the spines are upright, the force of the fall would not tend to drive the end of the quill upon the animal, but merely test the elasticity of the curved portion.

SPINE OF HEDGEHOG.

Family V. . . Macropĭdæ.—(Gr. Μακρὸς, long; πoῦς, a foot.)
Sub-family *b. Macropĭna.*

The Kangaroo.—In the mole we saw that the power of the body was placed chiefly in the fore legs; we now come to a family which has the principal power placed in the hinder part of the body. In the Kangaroos the hind legs are very long and immensely powerful; the fore legs are very small, and used more as hands than for walking; the tail also is very thick and strong, and assists the animal in its leaps.

The Great Kangaroo inhabits New Holland and Van Diemen's Land. Its singular formation, peculiarly adapted to the country, calls forth a corresponding degree of ingenuity on the part of the natives, who live much on its flesh. Its method of progression is by immense leaps from its long hind legs assisted by its tail. So rapidly are these leaps made, that the animal appears to alight on its hind legs after every leap, and never to touch the ground with its fore legs at all. But this is only a deception caused by the rapidity with which the animal performs its movements. On alighting from a leap, it comes to the ground on its fore legs, and immediately draws up its hinder limbs.

Indeed, the natural walking position of this animal is on all four legs, although it constantly sits up on the hinder legs, or even stands on a tripod composed of its feet and tail, in order to look out over the tops of the luxuriant grass among which it lives. The leaping movements are required for haste or escape, the length of each leap being about fifteen feet.

Of course this swiftness would soon leave its pursuers behind, but the Australian is able to break one of its limbs or strike it insensible to the

MACRŎPUS.

Major (Lat. *larger*), *the Kangaroo.*

ground with his boomerang, the most wonderful weapon that uncivilized man ever produced. This extraordinary missile is a flat curved piece of wood, which the Australian natives can wield with wonderful skill, making it describe circles in the air, or rush at an object, and then return to its owner's feet; or throw it at the ground and make it leap over a tree and strike an object at the other side.

Many boomerangs have been made in England from models brought from Australia, and it is not very difficult to learn the turn of the wrist necessary to make them describe a circle and return, but no one except an Australian can perform the complicated evolutions which the natives force the weapon to describe.

The English boomerangs are never certain. A purchaser may lay out his money on a dozen before he finds one that will fly, and when he has found a successful one, it is liable to lose its powers by damages received in its fall. In general, the English boomerangs are too much curved, and they are made with too little care, a very small shaving more or less, making or destroying the character of the weapon.

Hunting this animal is a very favourite sport with both colonists and natives. The natives either knock it down with the boomerang, spear it from behind a bush, or unite together and hem in a herd, which soon fall victims to the volleys of clubs, spears, and boomerangs which pour in on all sides. The colonists either shoot it or hunt it with dogs, a pack of which is trained for that purpose just as we train fox-hounds. The "old man," or "boomer," as the colonists call the Great Kangaroo, invariably leads the dogs a severe chase, always attempting to reach water and escape by swimming. It is a formidable foe to the dogs when it stands at bay, as it seizes the dog with its fore-legs, and either holds him under water until he is drowned, or tears him open with a well-directed kick of its powerful hind feet, which are armed with a very sharp claw.

The female Kangaroo carries its young about in a kind of pouch, from which they emerge when they wish for a little exercise, and leap back again on the slightest alarm. All the kangaroos and the opossums have this pouch, from which they are called "marsupiated" animals, from the Latin word marsupium, a purse or pouch.

The length of the Great Kangaroo is about five feet without the tail, the length of which is about three feet.

There are many species of kangaroo, the most extraordinary being the Tree Kangaroo, which can hop about on trees, and has curved claws on its fore-paws, like those of the sloth, to enable it to hold on the branches.

Several genera and two sub-families are omitted.

The Opossum.—This animal inhabits North America, and is hunted with almost as much perseverance as the racoon, not, however, for the sake of its fur but of its flesh. When it perceives the hunter, it lies still between the branches, but if disturbed from its hiding place, it attempts to escape by dropping among the herbage and creeping silently away.

Its food consists of insects, birds, eggs, &c., and it is very destructive among the hen-roosts. The Opossum uses its tail for climbing and swinging from branch to branch as the spider monkeys use theirs; but the Opossum uses its tail in a manner that the monkeys have never yet been observed to do, that is, making it a support for its young, who sit on its back and twist their tails round their mother's in order to prevent them from falling off. Lawson, in a passage quoted in the Museum of Animated Nature, gives the following quaint account of this animal:—"If a cat has nine lives this creature surely has nineteen; for if you break every bone in their skin and mash their skull, leaving them for dead, you may come an hour after, and they will be quite gone away, or, perhaps, you may meet them creeping away. I have for necessity in the wilderness eaten of them. Their flesh is very white and well-tasted; but their ugly tails put me out of conceit with that fare."

In Audubon's delightful work is a passage exhibiting exactly the same character on the part of the Opossum:—

"Suppose the farmer has surprised an opossum in the act of killing one of his best fowls. His angry feelings urge him to kick the poor beast, which, conscious of its inability to resist, rolls off like a ball. The more the farmer rages, the more reluctant is the animal to manifest resistance; at last there it lies, not dead, but exhausted, its jaws open, its eyes dimmed; and there it would lie until the bottle-fly should come to deposit its eggs, did not its tormentor walk off. 'Surely,' says he to himself, 'the beast *must* be dead.' But no, reader, it is only ''possuming,' and no sooner has its enemy withdrawn, than it gradually gets on its legs, and once more makes for the woods."

Sub-family *e. Didelphīna.*—(Gr. Δὶς, double; δελφὺς, a pouch.)

DIDELPHYS.

Virginiāna (Lat. *belonging to Virginia*), *the Opossum*.

The length of the Opossum is about twenty-two inches, and its height about that of an ordinary cat. When disturbed or alarmed, it gives out a very unpleasant odour.

Several genera are omitted.

The Common Seal inhabits the coast of Europe, and is not unfrequently found in many parts of the Scottish coasts, where seal-hunting is a favourite amusement. The young are taken by stretching nets across the narrow straits which they frequent, but the older and stronger animals are shot or knocked down with clubs when they attempt to scramble into the sea, as a blow on the nose instantly disables them.

The fore-feet of the Seal are used as fins, and the two hinder feet almost as the tail of a fish, to assist and direct its course. On land the movements of this animal are very clumsy; it shuffles along by means of its fore-feet, or rather paddles, and drags its hind-feet after it.

This seal, when taken young, is easily tamed. Edmonston gives the following amusing account of a seal named Finna, which he kept for about six months. "We had her carried down daily in a hand-barrow to the sea-side, where an old excavation admitting the salt water was abundantly roomy and deep for her recreation and our observation. After sporting and diving for some time, she would come ashore, and seemed perfectly to understand the use of the barrow. Often she tried to waddle from the house to the water, or from the latter to her apartment; but finding this fatiguing, and seeing preparations by her chairmen, she would of her own accord mount her palanquin, and thus be carried as composedly as any Hindoo princess." This interesting animal, after living in the house for about six months, at last was decoyed away by some wild seals and did not return again. A young seal was tamed by the guard of a small island in the Frith of Forth above Edinburgh. It seemed quite to consider itself one of the party, would accompany their boat across the water, and when the vessel was made fast, it used to take its station inside, and watch until the owners returned. It had the playful manners of a water dog, and would snatch a stick from its master's hand and dash into the sea with it, where it would toss and tumble about, sometimes approaching close to the shore, and swimming off again when its master attempted to grasp the stick, but it invariably brought back whatever it had taken. It would also bring fish out of the water and give them to its owners.

Family V. . . Phocīdæ.—(Gr. Φώκη, a Seal. Seal kind.)

Sub-family *b. Phocīna.*

PHOCA.

Vitulīna (Lat. *belonging to a calf*), *the Seal.*

The length of the Common Seal is about four or five feet, and its weight often two hundred and twenty-four pounds. When surprised basking on the shore, it scrambles off towards the water; but if intercepted, dashes at its antagonist, oversets him if possible, and makes its escape as fast as it can.

The Elephant Seal inhabits the Atlantic, Pacific, and Southern Oceans. It is very much larger than the Common Seal, being from twenty to thirty feet long. It derives its name from the long snout, something like the proboscis of the elephant, or rather the tapir, which it thrusts forward when angry, and snorts loudly. Only the males have this proboscis, and they do not attain it until they are three years old. Although its appearance is very formidable, it does not attempt to attack men; but if it cannot frighten them by opening its mouth and displaying its teeth, it makes off towards the water, but with great deliberation, as when in good condition it is so fat that its body trembles like a mass of jelly, and will furnish seventy gallons of oil. This oil is the principal object of the South Pacific seal fisheries: but the skin of this seal is also very valuable for its strength, and is used in making harness. The seal skin is often used as fur.

It is a migratory animal, and changes its residence several times in the year, the first migration taking place in June.

There are many seals known, among which are the Sea Leopard, a spotted species; the Harp Seal, so called because the markings on its back something resemble a lyre; and the Sea Lion.

MORUNGA.

Proboscidĕa (Gr. that has a proboscis or trunk).

The WALRUS inhabits the northern seas, but has been known to visit our coasts. Three instances of this have happened, one in 1817, one in 1825 at the Orkney Isles, and a third in 1839 at the mouth of the Severn. The most remarkable point in the Walrus is the great length of two of its upper teeth, which extend downwards for nearly two feet, and resemble the tusks of the elephant. They furnish very fine ivory, and are extensively used by dentists in making artificial teeth, as teeth made from them remain white much longer than those made from the tusks of elephants. These tusks are used by the Walrus for climbing the rocks or heaps of ice, and also for digging up the seaweeds on which the animal mostly subsists. It will also eat shrimps and young seals.

Sub-family *c. Trichecīna.*

TRICHĪCUS.—(Gr. τριχικός, hairy.)

Rosmărus (Scandinavian, *Rosmar*,* the Walrus), *the Walrus, or Morse.*

*** In the Scandinavian tongue, the word 'Ros' signifies horse, and 'Mar' sea. The meaning of the word 'Rosmar' is thus "Sea horse." Sometimes the two syllables are transposed, making the word "Mar-ros," which we contract into "Morse."**

In the same manner "Walrus" is an Anglicism of Hval-ros, or Whale-horse. The reader will notice the resemblance of these words to the corresponding words in German.

The Walrus is often hunted for the sake of its oil, its flesh, its skin, and its teeth. It is generally found in troops; and if one is wounded, its companions rush to its rescue, and attack the enemy with their sharp tusks, which they have been known to drive through the bottom of a boat. Their skin is so strong and slippery that it is very difficult to drive the harpoon through it, and even a sharp weapon frequently glides off without injuring the animal. The great enemy of the walrus is the polar bear, who does not always venture on an open battle, as, when a combat takes place, the walrus defends himself most vigorously with his curved tusks, and often inflicts fearful gashes on the bear, forcing it to abandon the contest.

The head of this animal is very small in proportion to the remainder of its body, and often deceives people as to its size, which is difficult to ascertain without examination. The stuffed specimen in the British Museum, although in bad preservation, will give a tolerable idea of the animal. The expression of its countenance is very ferocious, principally on account of the enormous size of the upper lip and the thick bristles with which it is covered. The length of the Walrus is about fifteen or sixteen feet, and it yields from twenty to thirty gallons of excellent oil.

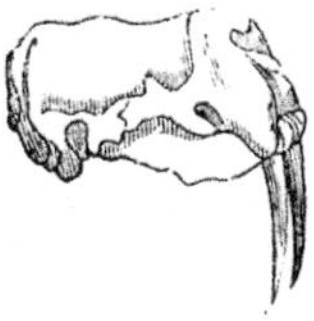
WALRUS'S SKULL.

The Cetacea, or Whale tribe, closely resemble the fishes, and have often been placed among these animals by naturalists. They, however, are distinguished by possessing warm blood, and in consequence, being forced to rise at intervals in order to breathe the air, instead of separating from the water, by means of their gills, sufficient oxygen for supporting life.

Yet the whale remains under water for a time so much longer than could be borne by any other warm-blooded animal, that the most indifferent observer cannot fail to perceive that the whale is furnished with some plan for supporting life during its stay beneath the water.

The manner in which this object is attained is at once beautiful and singular. Every one knows that the object of breathing is to oxygenize the blood, which in its course through the body becomes deprived of its native qualities, and is actually poisonous. If the blood is not renewed, it causes apoplexy and death, as is the case when a person is strangled or drowned. The most natural way to supply this want in the whale would be to give it much more lungs, in order that it might take into its body a reservoir of air, from which the blood might be renewed. But if this were the case, the animal would be seriously inconvenienced by such an amount of air, which would make it too buoyant, and prevent it from diving into the depths of the sea. But there must be a reservoir somewhere, and, therefore, instead of a reservoir of air to arterialize the blood, there is a reservoir of blood already arterialized.

Along the interior of the ribs there is a vast collection of blood-vessels, ramifying from one another, and capable of containing a large quantity of blood, having no immediate connexion with that portion of the blood which is already circulating in the body. As fast as the exhausted and poisonous blood returns from its work, it passes into another reservoir adapted for its necessities, while a portion of the arterialized blood in the arterial reservoir passes into the circulation. It will be seen from this statement, that the whale, and others of the same order, possess more blood in proportion than any animals. By means of this wonderful apparatus, a whale can remain below the water for more than half an hour at a time.

Order III. . . *CETE.*—(Gr. Κῆτος, a Whale, or sea monster.)
Family I. . . Balænidæ.—(Gr. Βάλαινα, a Whale. Whale kind.)

BALÆNA

Mysticetus (Gr. Μύσταξ, a moustache; κῆτος, a sea monster), *the Whale.*

The depths to which the whale can descend are astonishing, wounded whales having been known to take down perpendicularly nearly 800 fathoms of line. The pressure of the water at this depth is very great, amounting, according to Scoresby's calculation, to 211,200 tons. This pressure would certainly cause the water to burst through their nostrils, and enter the lungs, were it not that the nostrils are formed so as to close themselves more firmly as the pressure of water increases.

The great Greenland Whale is found in the Northern Oceans, living amid ice and perpetual cold. Many ships are annually fitted out for the capture of this creature, which, unhappily for itself, furnishes oil and whalebone. The oil is obtained from the thick layer of fatty substance, called blubber, which lies immediately under the skin; and the whalebone—which, by the way, is not bone at all—is obtained from the interior of the mouth, where it fringes the jaws, and acts as a sieve for the Whale to strain his food through. The throat of the Greenland Whale is so small that the sailors, who always use forcible expressions, say that a penny loaf would choke a whale. The greater proportion of its food consists of a little creature about an inch and a half long, called Clio borealis, one of the marine Mollusca, belonging to the class Pteropida, or wing-footed creatures, so called because it propels itself through the water with two wing-like organs. The Whale, when it wishes to feed, rushes through the water with its immense jaws wide open, enclosing a host of little sea animals, and a few hogsheads of water. As the Whale only wants the animals, and not the water, it shuts its mouth, and drives all the water out through the fringes of whalebone, leaving the little creatures in its jaws.

For the capture of this animal, a number of ships leave England, France, and other countries, reaching the Polar Seas about the end of April. When arrived at their destination, a careful look-out is kept from the mast-head for "fish," which are usually first observed by the column of steam and water that the whale sends into the air from its nostrils. At the welcome sound, "There she blows," the whole crew starts into activity; the boats, which are always kept hanging over the side of the ship, furnished ready for action, are instantly manned and lowered into the water, and the boat springs off in chase of the whale. The harpooner, whose station is in the bow, examines his implements carefully, tries the edge of the harpoon, and sees that the rope is properly coiled, as the slightest entanglement would upset the boat, or might even drag it below water.

It will be as well just to notice the different weapons used in the whale-fishery. The first and most important is the harpoon, a kind of spear with a large barbed head, the shape of which is not very unlike the flukes of an anchor. The edges of the barbs are kept very sharp, as otherwise the harpoon would not penetrate beyond the blubber, and the whale would consequently escape. The head of the harpoon is not made of steel, as inexperienced persons would imagine, but of soft iron, so soft that it can be scraped to an edge with a knife. This is fixed to a wooden handle, by which the harpooner holds it. In some vessels, the harpoon is fired at the whale from a small cannon placed in the bow of the boat. There are some very ingenious harpoons in the United Service Museum, one of which, intended to be fired from a gun, has its barbs joined to the head by a hinge, and held apart with a spring, so that when a whale is struck the barbs collapse until the force of the blow is expended, when the spring expands them and holds the whale firmly. The common harpoon, however, is the weapon usually employed.

To the harpoon is fastened a long and very tough line, about 4,000 feet in length. This line is kept ready coiled in a tub at the head of the boat, and great care is taken to prevent it from being entangled. It runs over a kind of pulley, as the friction is so great when the alarmed whale starts off, that the rope when out of its place has repeatedly set the gunwale of the boat on fire. At Deptford, some years back, might be seen a boat, the head of which had been quite cut off by the rope. A bucket of water is therefore always kept at hand to throw on the rope. When a whale is struck, it sometimes runs out with the whole of the line, in which case the line of another boat is fastened to it, and sometimes a whale has carried off three miles of line with it. When the whale begins to slacken the line, it is immediately recoiled in the tub, so as to be always under the command of the pursuers.

The use of the harpoon is merely to hold the whale; it does not enter deeply, and causes the animal but little inconvenience, as a whale has often broken its line and escaped with the harpoon sticking in its back, and been afterwards recaptured, apparently none the worse for its adventure. In order to kill the whale the fishermen have another weapon, called a "lance." This is a long, slender, steel weapon, with a very sharp head, without barbs, as the men have to withdraw the lance as fast as they can after it has pierced a vital part. With these few and simple weapons the fishers contrive to secure the monster of the waters—a beautiful instance of the superiority of reason over brute strength; for, as the expert angler secures a large and strong fish with a single hair, utterly inadequate to bear half the weight of the creature it holds, so the whale-fisher, with a few small weapons, achieves a task which may be compared to a mouse attacking and killing a wolf with a reel of thread and a crotchet needle.

The boats always approach the whale from behind, lest the expected prey should see them and escape. When within a few yards the harpooner throws his weapon at the whale, so as to pierce through the mass of blubber, and hold fast in the flesh. The wounded animal instantly dashes off, taking the line with it. When it has been under water for some time, it is forced to come to the surface to breathe. The fishers knowing the time that it can remain under water, and calculating from long practice the place where it will rise, are at the spot ready to receive it as its huge body reaches the surface, and thrust their long lances deeply into its body, inflicting mortal wounds. Blood mixed with water is now discharged from the whale's nostrils or "blow-holes," a sure sign that it will soon die. Presently streams of blood are thrown up, colouring the sea and frequently drenching the crews of the boats, and after a few violent struggles the whale turns over on its side and dies.

The enormous carcass is now joyously towed to the ship, and preparations are made for "flensing," or cutting off the useful parts. When the carcass has been brought alongside the ship, men wearing shoes armed with spikes, to prevent them from slipping off the oily back of the monster, commence the process by fastening ropes to its head and tail. A strong hook is then fixed into the fat near the neck, called the "kent," as it is used for "kenting," or turning over the whale. To this hook is fastened a rope passing through a pulley at the mainmast head, and fixed to a windlass on deck. The blubber is taken off the upper side by "blubber spades." The blocks of blubber, called "slips," are then hauled up on deck by means of ropes called "speck tackles," speck being the German word for fat or bacon. When the blubber is all stripped from the upper side, the men turn the whale partly round by hauling at the rope fastened to the "kent." They then cut out the whalebone with knives made for that purpose. Lastly, the "kent" itself is stripped off, and the whale left to the sharks and gulls, who have been helping themselves very liberally while the flensing was going on, the shovel-nosed shark sometimes scooping out semicircular pieces as large as a man's head. The birds and fish hold grand festival on the body of the whale until it is so stripped that it sinks, when the sharks have it all to themselves.

When the crew have leisure, the blubber, which has been meanwhile stowed away in a place with a not very polished name, is "made off," that is, carefully stripped of the pieces of skin and muscle adhering to it, cut into moderately sized pieces, and packed in casks until wanted. The oil is extracted by boiling the blubber in large coppers; a most unsavoury occupation, but a very pleasant one to the crew, if they take that duty upon themselves. The refuse blubber is used as fuel, so that there is no waste.

It is impossible to calculate on the amount or mode of resistance which may be met with from a whale when struck, for one whale will yield to a single harpoon loosely fixed, while another will break away and escape with five or six in his back, and two miles or so of rope trailing behind him. Some instances have been related of whales being killed without being struck at all. Scoresby tells us that on one memorable occasion, after a whale had been killed, it sunk as whales sometimes will do. While they were hauling it up, the line sometimes resisted, and sometimes came in easily. At last they drew up a whale with a coil of the rope round it, which they naturally thought to be the animal struck by them. After disentangling it they found to their surprise that the line still descended into the sea, and dragged as if there was a weight at its end; and so there was, for they found their harpooned whale still fixed to the weapon, and discovered that the other unfortunate animal had contrived to entangle itself in the line, and had thus drowned itself. Another anecdote of a similar character is related by Scoresby. "A whale was struck from one of the boats of the ship Nautilus, in Davis' Straits. It was killed, and as is usual after the capture, it was disentangled from the line connected with the first 'fast boat,' (the first boat which had struck it,) by dividing it at the splice of the foreganger, (the part of the rope fastened to the harpoon,) within eight or nine yards of the harpoon. The crew of the boat from which the 'fish' was first struck, in the meantime were employed in heaving in the lines by means of a winch fixed in the boat for the purpose. On a sudden, however, to their great astonishment, the lines were pulled away from them with the same force and violence as by a whale when first struck. They repeated their signal indication of a whale being struck; their shipmates flock towards them, and while every one expressed a similar degree of astonishment with themselves, they all agreed that a 'fish' was fast to the line. In a few minutes they were agreeably confirmed in their opinion by the rising of a large whale close by them, exhausted with fatigue, and having every appearance of a 'fast fish.' It permitted itself to be struck by several harpoons at once, and was speedily killed. On examining it after death, they found the line belonging to the boat in its mouth, where it was still firmly fixed by the compression of its lips. The occasion of this happy and puzzling incident was therefore solved. The end of the line, after being cut, was sinking in the water—the 'fish' in question, engaged in feeding, was advancing with its mouth open, and accidentally caught the line between its extended jaws—a sensation so utterly unusual as that produced by the line, had induced it to shut its mouth and grasp the rope which was the cause of its alarm so firmly between its lips as to produce the effect just stated. This circumstance took place many years ago, but a similar one occurred in the year 1814."

The dangers undergone in this pursuit are very great. Sometimes the boat is dashed to pieces by a blow from the tail of the enraged whale; sometimes the crew are left on the ice for many hours, wet and frozen; sometimes the ice-fields strike together, and crush the ship between them, although the vessel is strengthened in every possible way by cross-beams and treble sheathing; sometimes a fog comes on, and the boat and ship are separated, neither having any means of knowing where the other lies, for sound is much impeded by fog, and even cannon are not heard when fired comparatively close.

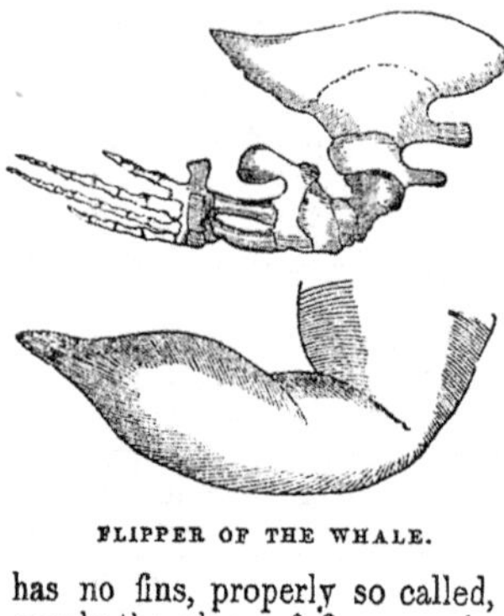

FLIPPER OF THE WHALE.

The Whale shows great attachment to its young, which is called the cub, and on the approach of danger, seizes it with its fin or flipper, and carries it down out of danger. The Whale has no fins, properly so called, as it is not a fish. Its flippers, which supply the place of fins, are in fact fore legs, furnished with a kind of hand covered with a thick skin. They seem to be principally employed in balancing the animal. The hind legs are concealed under the skin, as are those of the boa constrictor. The length of this Whale averages sixty feet. Its tail is placed transversely, and not vertically, as in the fishes.

The CACHALOT.—The chase of the Cachalot is similar to that of the Greenland whale, and need not be described. It is attended with more danger, as the terrific row of teeth with which the lower jaw of the Cachalot is armed, is not unfrequently employed in biting the boat. In the Ashmolean Museum at Oxford is an under jaw-bone of this whale, sixteen and a half feet in length, containing forty-eight huge teeth. Besides this method of defence, it has a very unpleasant habit of swimming off to a distance, and then rushing at the boat with its head, thereby knocking it to pieces. One of these whales actually sank a ship by three or four blows from its head.

PHYSĒTER.—(Gr. Φυσητήρ, a blow-pipe, or bellows.)

Macrocephălus (Gr. Μακρός, long; κεφαλή, a head), *the Cachalot, or Spermaceti Whale.*

Spermaceti is obtained from the head of the Cachalot, and it is this substance that causes the immense size of the head. When the whale is killed, a hole is made in the upper part of the head, and the spermaceti is baled out with buckets. When just procured it is almost fluid, but is rendered solid and transparent by being first drained of its oil, then boiled in water, and lastly set to cool in wide pans, where it soon assumes the white flaky appearance so well known in this country. The skull of the Cachalot occupies a comparatively small portion of the head, the huge mass at the end of the mouth being composed of a gristly kind of substance. The bone of the upper jaw occupies about one-fourth of the distance between the mouth and the top of the snout. It runs backwards nearly straight until just before the eyes, when it joins the remainder of the skull with a bold sweep. That part of the skull is called "Neptune's Chair" by the sailors, and is the part where the spermaceti is found. The layer of blubber is thin, but yields a fine and valuable oil.

Ambergris, so long a riddle to all inquirers, is now found to be produced in the interior of the Cachalot. This substance is of the consistency of wax, inflammable, and gives out a kind of musky odour. It was once in great repute as a medicine, but is now only used as a perfume.

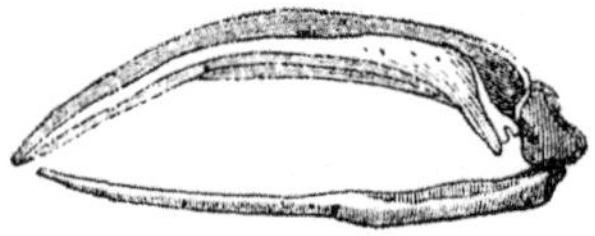

JAW OF GREENLAND WHALE.

The Cachalot, although an inhabitant of the Arctic seas, has sometimes been found and captured off our coasts. The length of this whale is about seventy feet.

Those readers who have formed their ideas of DOLPHINS from the very graceful and elegant creatures represented under that name in the pictures of the "old masters," or the statues of the ancient sculptors, will find that the real animal differs as much from the ideal, as the red and green lions wearing golden collars, represented in heraldry, differ from the lion of Africa. Sad to say, almost the whole history of the Dolphin is imaginary — very poetical, but very untrue. The red and blue colours of the heraldic lion are not less fabulous than the changing tints of the dying dolphin, so dear to poetry. Alas! our unpoetical Dolphin, when we have harpooned and brought him on deck, is only

Family II. Delphinĭdæ.—(Gr. Δελφίς, a Dolphin. Dolphin kind.)

DELPHĪNUS.—(Lat. *a Dolphin.*)

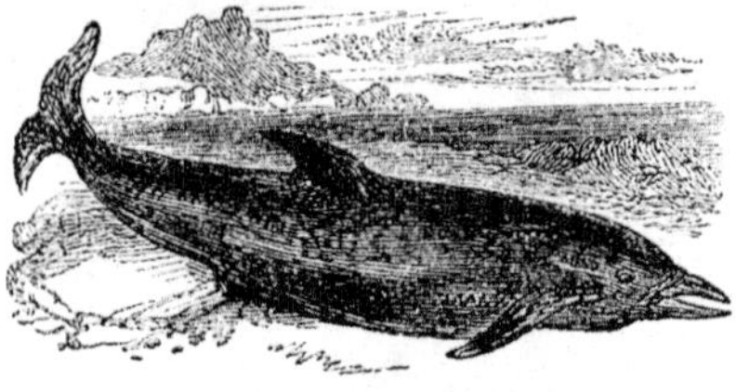

Delphis, *the Dolphin.*

black and white, and all the change that he makes, is that the black becomes brown in time, and the white grey.

The creature that really displays these colours when dying, is a fish called the Coryphene, and not a cetaceous animal of any kind. The sailors generally call them Dolphins, which has led to the mistake.

We will leave poetry and its beautiful errors, and pass on to facts. The Dolphin is, like the whale, a warm-blooded animal, suckles its young, and is forced to come to the surface in order to breathe. Its snout is very long, and is apparently used for capturing such fish, and other animals, as live in the mud.

The length is from six to ten feet. Several species of Dolphin are known, of which the British Museum possesses six.

The Porpoise.—These animals may be observed in plenty playing their absurd antics off every coast of England. There are numbers of them off the Nore, a place which they frequent greatly, as it is the mouth of a river, and they find more food there than in the open sea. They tumble at the surface of the water for the purpose of breathing.

In the olden times, when glass windows were considered an effeminate luxury, and rushes supplied the place of carpets, the flesh of the Porpoise constituted one of the standard delicacies of a public feast, but it has long since been deposed from its rank at the table. Like most of the cetacea, its flesh has a very strong oily flavour, which, however relished by an Esquimaux, is not agreeable to the palate of an European epicure of the present day.

Phocæna.—(Gr. Φωκαινα, a Porpoise.)

Communis (Lat. *common*), *the Porpoise, or Porpesse.*

The voracity of the Porpoise is very great. It feeds on various fishes, but its great feasts are held when the periodical shoals of herrings, pilchards, and other fish arrive on the coasts. In the pursuit of its prey, it frequently ventures some distance up a river, and is then often taken in nets by the fishermen.

The teeth of this animal are very numerous, and interlock when the jaws are closed, so that the fish when once seized cannot escape. Its length is about five feet, its colour a rich black; becoming white on the under side.

The Narwhal.—Although the Narwhal has not suffered from false reports so much as many other animals, yet it has unwittingly contributed to propagate a very old error. The spiral tusk of the Narwhal was accustomed to be sold as the real horn of the unicorn; and as an accredited part of that animal, forming direct proof of its existence, it used to fetch a very high price. Of course, when the whale fishery was established, the real owner of the horn was discovered, and the unicorn left still enveloped in mystery.

The name Monodon is not strictly correct, as the Narwhal possesses *two* of these tusks, one on each side of its head. Only the left tusk projects, the other remaining within the head. Sometimes a specimen has been found with both tusks projecting, and some think that when the left tusk has been broken off by accident, the right one becomes large enough to supply its place.

The use of these tusks is not known; some supposing that they are employed to dig up sea-weeds, and other plants, on which the Narwhal

MONŎDON—(Gr. *Μόνος*, solitary; *ὀδούς*, or *ὀδών*, a tooth.)

Monocĕros (Gr. *Μονος—κέρας*, a horn), the *Narwhal.*

feeds, and some imagining that the living prey is first transfixed and then eaten. Be this as it may, as a weapon the tusk is not to be despised, as the strength and rapidity of the Narwhal are very great. Instances are on record, of the thick oak timbers of a ship being pierced by the ivory tusk of this creature. The Greenlanders employ this ivory in the manufacture of spears, arrows, hooks, &c. They take the Narwhal by a kind of harpoon attached to a line, with a buoy at its extremity. The use of the buoy is to harass and retard the Narwhal when struck, and to give notice when it is about to rise. Immediately on reaching the surface, a lance is thrust into it, which generally proves its death-blow. The adventurous Greenlander finds it a most welcome prey, as he obtains from it oil, food, weapons, and ropes.

Although an inhabitant of the northern seas, it has several times visited our coasts. Its body is from thirty to forty feet in length, and its tusk from five to nine.

The Manatees and Dugong are omitted from want of space.

We now arrive at the Rodentia, or gnawing animals, so called from their habit of gnawing through, or paring away, the substances on which they feed. For this purpose their teeth are admirably formed, and by these teeth it is always easy to ascertain a member of the Rodents. They have none of those sharp teeth called canine, such as are seen in the lions and in those animals which seize and destroy living animals, but in the front of each jaw there are two long flat teeth, slightly curved, and having a kind of chisel edge for rasping away wood, or other articles.

The constant labour which these teeth (called incisors, from the Latin word, *incido*, I cut,) undergo, would rapidly wear them away. To counteract this loss, the teeth are constantly growing and being pushed forward, so that as fast as the upper part is worn away, the tooth is replenished from below. So constant is this increase, that when an unfortunate rabbit, or other rodent, has lost one of its incisors, the opposite one, meeting nothing to stop its progress, continually grows, until sometimes the tooth curls upwards over the lips, and prevents the wretched animal from eating, until it is gradually starved to death. An example of this preternatural growth may be seen in the vignette under the article Rabbit. The sketch was drawn from a specimen in the Anatomical Museum at Oxford.

The Rodentia include the mice, beavers, rabbits, squirrels, and porcupines.

The Brown Rat, sometimes called the Norway Rat, is the species usually found in England. It was some years since imported into this country, and from its superior size, strength, and ferocity, has so completely established itself, and expelled the original Black Rat, that it is very difficult indeed to find a Black Rat in any part of England. Waterton's sympathies are much excited in favour of the original rat, and his anger is great against the invader. He says of the Brown Rat:—

"Its rapacity knows no bounds, whilst its increase is prodigious, beyond all belief. But the most singular part of its history is, that it has nearly worried every individual of the original rat out of Great Britain. So scarce have these last-mentioned animals become, that in all my life I have never seen but one single solitary specimen. It was sent, some few years ago, to Nostell Priory, in a cage, from Bristol, and I received an invitation from Mr. Arthur Strickland, who was on a visit there, to go and see it. Whilst I was looking at the little native prisoner in its cage, I could not help exclaiming, 'Poor injured Briton! hard, indeed, has been the fate of thy family! in another generation, at furthest, it will probably sink down to the dust for ever!'"

Order IV. . . *GLIRES.*—(Lat. *Glis*, a Dormouse.)

Family I. . . . Murĭdæ.—(Gr. Mῦς, a Mouse. Mouse kind.)

Sub-family *a.* *Murīna.*

MUS.

Decumānus (Lat. *tenth* or *large*), *the Rat.*

The same amusing naturalist, being considerably annoyed by the depredations on his provisions, and the unceasing clatter that they kept up behind the panels of his sitting-room, after trying various plans to extirpate them, at last thought of a method, rich in the same humour with which most of his actions are tinged, and as efficacious in its operation as amusing in its idea:—

"Having caught one of them in a box trap, I dipped its hinder parts into warm tar, and then turned it loose behind the hollow plinth. The others, seeing it in this condition, and smelling the tar all along the run through which it had gone, thought it most prudent to take themselves off: and thus, for some months after this experiment, I could sit and read in peace, free from the hated noise of rats. On moving the plinth at a subsequent period, we found that they had actually gnawed away the corner of a peculiarly hard-burnt brick which had obstructed their thoroughfare."

It is at all times difficult to get rid of these dirty, noisy animals, for they soon learn to keep out of the way of traps, and if they are poisoned they revenge their fate by dying behind a wainscot or under a plank of the floor, and make the room uninhabitable. There are, however, two ways recommended to attain the desired object.

Place a saucer containing meal in a room frequented by rats, letting them have free access to it for several days. They will then come to it in great force. When they have thus been accustomed to feed there regularly, mix a quantity of jalap with the meal, and put it in the accustomed place. This will give them such internal tortures that they will not come near the place again.

The second plan is to use the same precautions, but to mix phosphorus with the meal and make it into a ball. The phosphorus is said not to kill the rats, but to afflict them with such a parching thirst that they rush to the nearest water and die there. By this method the danger of their dying in the house is avoided.

I have not proved either of these plans experimentally, but offer them for the benefit of those who are afflicted by the rat pest.

The COMMON MOUSE is so well known, that a description of its form and size is useless. It almost rivals the rat in its attacks upon our provisions, and is quite as difficult to extirpate. It brings up its young in a kind of nest, and when a board of long standing is taken up in a room, it is not uncommon to find under it a mouse's nest, composed of rags, string, paper, shavings, and everything that the ingenious little architect can scrape together. It is a round mass, looking something like a rag ball very loosely made. When opened, seven or eight little mice will probably be found in the interior—little pink, transparent creatures, three of which could go into a lady's thimble, sprawling about in a most unmeaning manner, apparently greatly distressed at the sudden cold caused by the opening of their nest.

MUS.

Muscŭlus (Lat. *a little mouse*), *the Mouse.*

The Mouse is said to be greatly susceptible of music. An anecdote is related of a gentleman who was playing a violin seeing a mouse run along on the floor and jump about as if distracted. He continued the strain, and after some time the mouse, apparently exhausted with its exertions, dropped dead on the floor. An instance occurred to myself very recently, similar in all respects but that of the death of the little animal, which only scampered back to its hole when the music ceased. We afterwards found that it was a partially tame one which had escaped.

Every one has heard of the fable of the Lion and the Mouse, but from the following account from Basil Hall's Fragments, we must conclude that, whatever the lion might have done under the circumstances, the tiger at all events would not have availed himself of the proffered assistance. He relates of a tiger that was kept in a cage at Mysore :—

"But what annoyed him far more than our poking him up with a stick, or tantalizing him with shins of beef or legs of mutton, was introducing a mouse into his cage. No fine lady ever exhibited more terror at the sight of a spider than this magnificent royal tiger betrayed on seeing a mouse. Our mischievous plan was to tie the little animal by a string to the end of a long pole, and thrust it close to the tiger's nose. The moment he saw it, he leaped to the opposite side; and when the mouse was made to run near him, he jammed himself into a corner, and stood trembling and roaring in such an ecstasy of fear that we were always obliged to desist from sheer pity to the poor brute. Sometimes we insisted on his passing over the spot where the unconscious little mouse ran backwards and forwards. For a long time however we could not get him to move, till at length, I believe by the help of a squib, we obliged him to start; but instead of pacing leisurely across his den or making a détour to avoid the object of his alarm, he generally took a kind of flying leap, so high as nearly to bring his back in contact with the roof of his cage."

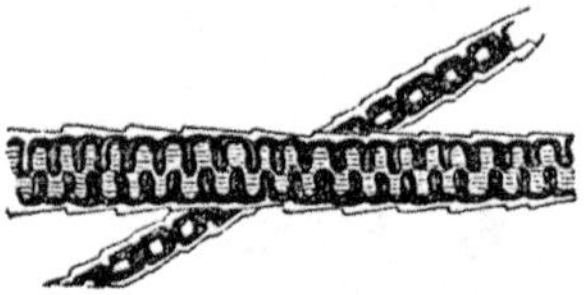

HAIR OF MOUSE.

The two objects here represented are two portions of the same hair, the larger one being the centre and the smaller being taken near the origin. It is worth while to notice that although to external appearance the fur of the mouse exactly resembles that of the bat, yet when they are placed under the microscope they are shown to be very differently formed.

A white variety of mouse is tolerably common, and is usually bred in cages. As it is very tame and beautiful, it is in great repute as a pet.

MICRŎMYS.—(Gr. μικρός, small; μῦς, a mouse.)

Minūtus (Lat. *very small*), *the Harvest Mouse.*

The HARVEST MOUSE, the smallest of the British quadrupeds, discovered and described by White in his "Selborne," is very much smaller than the ordinary mouse, a halfpenny weighing down two of them when placed in a pair of scales. Its nest is raised about a foot and a half from the ground, and supported on two or three straws. It is made of grass, about the size of a cricket-ball, and very compact.

The FIELD MOUSE is as great a pest in the open air as the Common Mouse within a house. It not only devours the corn, but strips the bark off young trees, doing great mischief. The kestrel or windhover hawk lives almost exclusively on the Field Mouse, and ought always to be encouraged.

Several genera are omitted.

Sub-family *b*. *Arvicolīna*.—(Lat. *Arvum*, field; *colo*, I inhabit.)

CRICĒTUS.—(Gr. κρικητὸς, ringed.)

Frumentarĭus (Lat. *belonging to harvest*), *the Hamster*.

The HAMSTER RAT is a native of parts of Germany. It is a terrible pest there, as it not only devours the corn, &c. in the summer, but lays up a large store for the winter. It is a most furious little animal, and will attack a man or a horse and even a waggon wheel if it approaches too near the spot which the Hamster considers its own property. Rats, mice, lizards, birds, and even its weaker brethren, are eaten by this ravenous little animal.

It lives in holes underground, and to escape attack has several passages from its chamber leading in different directions. The skin is of some value, and the hunter who spears it usually opens its granary for the sake of its store, which is far from being inconsiderable, a hundredweight of beans having been found in one granary.

ARVICŎLA.—(Lat. *Arvum*, a field; *colo*, I inhabit.)

Amphibĭus (Gr. 'Αμφι, on both sides; βιόω, I live), *the Water-rat*.

The WATER RAT is a native of England, and very common on the banks of rivers, brooks, &c. It digs holes in the bank, and is reported to eat fish, frogs, &c., but this is very doubtful. These animals exist in great numbers round Oxford, and I have repeatedly watched them feeding. I never saw them eating fish, nor found fish-bones inside their holes, except when a kingfisher had taken possession; but I have frequently seen them gnawing the green bark from reeds, which they completely strip, leaving the mark of each tooth as they proceed. I shot one while feeding, and at first thought that the marks of its teeth were caused by the shot, for until that time I had supposed that the Water Rat fed on fish.

The BEAVER.—North America is the principal country where the Beaver is found, but it is also common on the Euphrates, and along some of the larger European rivers, as the Rhone and the Danube. In former years, when the wolf and bear inhabited England, the Beaver followed its architectural pursuits along the rivers; but it has not been seen in this country since 1188.

The houses of the Beaver are built of mud, stones, and sticks. They are placed in a stream, and their entrance is always below the surface. As a severe frost would freeze up their doors, it is necessary to make the stream deep enough to prevent the frost from reaching the entrances. This object is attained by building a dam across the river, to keep back the water until it is sufficiently deep for the beaver's purposes. The dam is made of branches which the Beaver cuts down with its strong sharp teeth, and mud and stones worked in among the branches. The Beavers throw these branches into the water, and sink them to the bottom by means of stones, and by continually throwing in fresh supplies a strong embankment is soon made.

Sub-family *d*. *Castorīna*.

CASTOR.—(Gr. Κάστωρ, a Beaver.)

Fiber (Lat. *a Beaver*), *the Beaver*.

As many Beavers live together in one society, the formation of a dam does not take very long. By their united efforts they rapidly fell even large trees, by gnawing them round the trunk, and always taking care to make them fall towards the water, so that they can transport the logs easily. The mud and stones used in their embankments are not carried on their tails, as some say, nor do the Beavers use their tails as trowels for laying on the mud, the fact being that the stones and mud are carried between their chin and fore-paws, and the mistake respecting the tail is evidently caused by the slap that Beavers give with that member when they dive. In order that their pond may not be too deep they always leave an opening in the dam to let the water escape when it rises above a certain height.

They cut most of their wood in the summer, taking care to choose trees above their houses, so that the stream floats them down to the place where they wish to use them. They also lay up stores of food for the winter, by cutting a number of green branches and sinking them near the door of their habitations, where they are held firm by stones laid on the summit of the heap.

During the severe winter, their mud-built houses freeze quite hard, and prevent the wolverine, their greatest enemy except man, from breaking through and devouring the inmates. Every year the Beavers lay a fresh coating of mud upon their houses, so that after the lapse of a few years the walls of the house are several feet in thickness. Many of the houses are built close together, but no two families can communicate with each other, except by diving below the walls and rising inside their neighbours' houses.

The fur of the Beaver is exceedingly valuable, especially for the manufacture of hats, and is greatly sought after. The hunting season is in winter, when the beavers are quietly in their houses. The hunters, armed with spears, &c. break the tops of the houses. The alarmed beavers instantly rush out and pass under the ice to certain hiding-places in the bank. The hunter then discovers the position of the hole in the bank by the sound of his spear struck against the ice; he then breaks a hole and spears the animal in its place of fancied security. A substance called Castor was formerly obtained from the Beaver and much used in medicine, but is now discarded.

When in captivity the Beaver soon becomes tame, and will industriously build dams across the corner of a room with brushes, boots, fire-irons, books, or anything it can find. When its edifice is finished it sits in the centre apparently satisfied that it has made a beautiful structure to dam up the river—a proof that the ingenuity of the Beaver is not caused by reason but by instinct.

The fur of the Beaver, like that of many other animals, consists of a fine wool intermixed with long and stiff hairs. The hairs are useless, but the peculiar construction of the fur causes it to penetrate and fix itself into the felt which forms the body of a hat. In making the hat, the only method required to fasten the fur into the felt is to knead fur and felt together. The hair is toothed on its surfaces, and makes its way into the felt, just as an awn of barley will travel all over the body if placed up the sleeve.

The length of the Beaver is about three feet and a half.

Several genera are omitted.

The Common Porcupine is found in Africa, Tartary, Persia, India, and some parts of Europe. It lives in holes which it digs in the ground, and only comes forth at night in order to feed. It eats vegetable substances only, such as roots, bark, and other similar substances. The array of spines or quills with which this animal is covered forms its principal means of defence. If it cannot escape, it suddenly stops, erects all its quills, and runs backwards against its adversary, striking the quills against him by the weight of its body. Occasionally a looser quill than usual remains in the wound or falls on the ground, which evidently gave rise to the foolish error that the Porcupine could dart its weapons at its adversary from a distance. There are two kinds of these quills,—one kind long and curved, the other short, thick, and pointed. These last are the weapons of defence, as the former are too slender to do much service. When the Porcupine walks, its quills make a kind of rustling sound, caused principally by those arranged on the tail, which

Family II. . . . Hystricĭdæ.—(Ὕστριξ, a Porcupine. Porcupine kind.)
Sub-family *a. Hystricīna.*

HYSTRIX.

Cristāta (Lat. *crested*), *the Porcupine.*

are large, hollow, and supported on long slender stalks. When an animal is pierced by one of these quills it is in some danger, for the exterior of the quill is formed on the same principle as the beaver fur, and if not extracted, will gradually work its way into the body, until it buries itself entirely. Dogs are not unfrequently killed in this manner by a neglected quill, which has struck them, but which they are too eager to notice in the excitement of the combat, until it has inflicted a wound upon one of the internal organs.

The American Indians use the quills extracted from the Canada Porcupine, a species living on trees, for ornamenting various parts of their dress, especially their mocassins or skin shoes. In England the quills are much used by anglers for making fine floats for angling. The length of the Porcupine is about two feet, and its spines or quills are from six to fourteen inches long.

The AGOUTI lives in Brazil, Guiana and Paraguay. It is about the size of a rabbit, and like that animal is generally found in company with others. In Brazil and Guiana, the Agouti is much sought after for the sake of its flesh, but it appears that in Paraguay the flesh is not eaten. When pursued, it runs for a short time with much rapidity, but soon endeavours to conceal itself in a hole or under the roots of a tree, when it will suffer itself to be captured without any resistance, merely uttering a plaintive cry. It feeds on vegetables, especially yams and tubers, but in the West India Islands it devours the sugar-canes, and is a great pest to the planters.

Sub-family *c.* *Dasyproctina.*

DASYPROCTA.—(Gr. Δασύς, rough; πρωκτός, hind-quarters.)

Agúti (Native name), *the Agouti.*

The CAPYBARA or CHIGUIRA is the largest of all the Rodentia. At first sight it looks very like a pig, and its skin is covered thinly with hairs like bristles, which add to the resemblance.

It inhabits the borders of lakes and rivers in many parts of Southern America. During the day, it hides among the thick herbage of the banks, only wandering forth to feed at night, but when alarmed, it instantly makes for the water, and escapes by diving. It is hunted for the sake of its flesh, which is said to be remarkably good. The Jaguar appears to be of the same opinion, for he is the most terrible enemy of this creature, destroying immense numbers. The food of the Capybara consists of grass, vegetables and fruits. Its length is about three feet six inches.

Sub-family *d.* *Hydrochœrina.*

HYDROCHŒRUS.—(Gr. Ὕδωρ, water; Χοῖρος, a pig.

Capybára (Native name), *the Capybára.*

The GUINEA-PIG or RESTLESS CAVY belongs to the sub-family Caviina. It was originally brought from South America, and is frequently domesticated in England. Its beauty is its only recommendation, as it shows little intelligence, and is never used for food. Children, however, and particularly schoolboys, are fond of keeping Guinea-pigs, as they are wonderfully prolific, easy to manage, and do not make much noise. They are popularly supposed to keep off rats, and are therefore usually patronised in connexion with rabbit-hutches.

Family III. Leporĭdæ.—(Lat. *Lepus*, a Hare. Hare kind.)

LEPUS.

Timĭdus (Lat. *timid*), *the Hare.*

The Hare is one of our most common quadrupeds. It is constantly hunted both for the sport and for its flesh. When hunted with greyhounds, the amusement is called coursing. Beagles are also used for the same purpose, but they do not catch the Hare by speed, but by patiently following its track, until the wearied animal is no longer capable of escaping. It comes under the denomination of game, and is protected by the Game Laws, as are pheasants and partridges.

When full grown, it is larger than the rabbit and exceedingly like that animal, but its colour is slightly different, and the black spot on the extremity of its ears is a simple method of distinguishing it. The Hare does not burrow like the rabbit, but makes a kind of nest of grass and other materials. In this nest, called a "form," the Hare lies, crouching to the ground, its ears laid along its back, and trusting to its concealment, will often remain quiet until the foot of an intruder almost touches it. Many people can distinguish it when thus hidden by the sparkle of its eye.

Innumerable foes besides man surround this animal. Foxes, ferrets, stoats, and all their tribe, are unmerciful enemies, and sometimes a large hawk will destroy a leveret, as the young Hare is called. Although destitute of all means of defence, it is often enabled to escape by the quickness of its hearing and sight, which give it timely warning of the approach of an enemy, and enable it to escape to a place of safety.

LEPUS.

Variabĭlis (Lat. *varying*), *the Alpine Hare.*

In cold countries the Hare changes its fur during winter, and becomes white, like the Arctic fox and the ermine. The Alpine Hare, inhabiting the northern parts of Scotland, is a good example of this change.

The well-known RABBIT is rather smaller than the hare, but closely resembles it in form. It lives in deep holes, which it digs in the ground. When a number of these holes or burrows occur near each other, the place is called a warren. A loose dry soil, such as the soft red sandstone, is the delight of these animals, who may be seen frisking about in great numbers outside their holes, but diving in on the slightest alarm. Poachers often take them in great numbers by spreading nets over the mouth of the holes, and sending a ferret carefully muzzled down one of the burrows. The terrified rabbits rush out at the sight of their dreaded enemy, and are caught in the nets. If the ferret were not muzzled, it would kill the first rabbit it caught, and remain in the hole, sucking the blood of its victim.

LEPUS.

Cunicŭlus (Lat. *a little Rabbit*).

The female Rabbit forms a soft nest at the bottom of her burrow, composed of fur torn from her body, of hay and dried leaves. Here the young rabbits are kept until they are strong enough to shift for themselves, and make their own burrows.

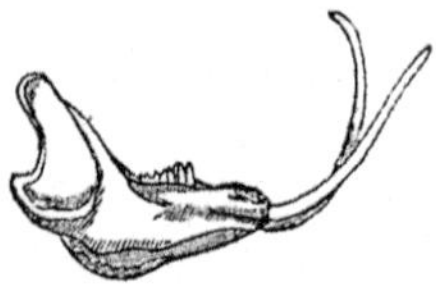

OVER-GROWN TEETH OF RABBIT.*

The tame Rabbit is only a variety, rendered larger by careful feeding and attendance. There are many breeds of domestic rabbits, some, as the fancy or lop-eared rabbits, being often of considerable value, thirty guineas having been refused for a particularly fine one. When tame rabbits are suffered to go free, they speedily return to their wild habits and instincts.

The CHINCHILLA.—This pretty little animal is an inhabitant of the valleys in the mountain districts of South America. In such situations the cold is often very intense; but the long soft fur of the Chinchilla forms an effectual protection against the frosts. The fur is extensively used for clothing, and celebrated for its soft and warm texture. Numbers of these animals are annually destroyed for the sake of their skins, and Coquimbo appears to be the place where they are taken in the greatest numbers.

Family IV. . . Jerboïdæ.—(*Jerboa kind.*)
Sub-family *a.*. . *Chinchillīna.*

CHINCHILLA (Native name).

Lanĭger (Lat. *wool-bearing*), *the Chinchilla.*

The Chinchilla lives in society like the rabbit, and resides in burrows dug in the ground. Its food is entirely vegetable, and principally consists of bulbous roots. In captivity it is quiet and inoffensive, but seems to betray no particular attachment to its keeper; neither does it seem playful. Its tail, covered with long bushy hairs, is usually held turned up over its back, like that of the squirrel, and probably for the same reason. From the various specimens of fur sent to this country it would appear that there are two species of Chinchilla, but it is not quite certain. The length of the Chinchilla is about nine inches, exclusive of its tail, which measures about five.

The JERBOAS are celebrated for their powers of leaping. Their long hind legs enable them to take enormous springs, during which their tails serve to balance them. Indeed, a Jerboa, when deprived of its tail, is afraid to leap. At first sight the Jerboa seems to alight on its hind feet, as well as spring from them, but the fact is, that it alights on its fore feet and draws up the hind legs ready for the next leap with such rapidity that the eye can scarcely follow the movement; as has been before related of the kangaroo.

* From the Anatomical Museum, Oxford.

In the history of the polar bear it was mentioned that its feet were prevented from slipping on the ice by a coating of thick hair. The foot of the Jerboa is defended in the same manner by long bristly hairs, which not only give the creature a firm hold of the ground for its spring, but also defend the foot from the burning soil on which it lives.

Sub-family *c.* *Dipīna.*

Dɪᴘᴜs.—(Gr. Δις, double; πούς, a foot.)

Ægyptĭus (Lat. *belonging to Egypt*), *the Jerboa.*

The timidity of the Jerboa is very great, and on the slightest alarm it instantly rushes to its burrow, but if intercepted, skims away over the plain with such rapidity that it seems to fly, and when at full speed a swift greyhound can scarcely overtake it.

Grain and bulbous roots are its chief food; while eating, it holds the food with its fore paws, and sits upright on its haunches, like the squirrels and marmots. The Jerboa does not bear confinement well; it always appears uneasy and distrustful; it remains hidden during the day, and even when it emerges from its concealment towards the evening is always ready to retreat at the least alarm.

There are many kinds of jerboa; the Egyptian Jerboa is rather small, being about the size of a large rat; its colour is a tawny yellow.

Sub-family *d.* *Myoxīna.*

Mʏᴏxᴜs.—(Gr. Μυοξός, or Μυωξός, a Dormouse.)

Avellanarĭus (Lat. from *Avellāna*, a filbert), *the Dormouse.*

The Dᴏʀᴍᴏᴜsᴇ is very common in all the warmer parts of the Continent, and is often found in this country, especially in the southern and midland counties. It lives in copses and among brushwood, through which it makes its way with such rapidity that it is very difficult to be captured. During the winter it lies torpid, but takes care to have a stock of food laid up, on which it feeds during the few interruptions to its slumbers. A warm day in winter will usually rouse it, but during the cold weather it lies rolled up, with its tail curled round its body. While in this torpid state, a sudden exposure to heat kills it, but a gentle warmth, such as holding it in the hand, rouses it without injury. It lives principally on nuts, acorns, and grain. It brings up its young in a nest composed of leaves and hay, and seems to be fond of society in its household labours, as ten or twelve nests have been seen close to each other.

The SQUIRREL is a very common animal in woods, where numbers may be seen frisking about on the branches, or running up and down the trunks. If alarmed, it springs up the tree with extraordinary activity, and hides behind a branch. By this trick it escapes its enemy the hawk, and by constantly slipping behind the large branches, frequently tires him out. The activity and daring of this little animal are extraordinary. When pursued, it makes the most astonishing leaps from branch to branch, or from tree to tree, and has apparently some method of altering its direction while in the air, possibly by means of its tail acting as a rudder.

Many boys, and men too, know the delights of a squirrel hunt. How the little animal is chased from tree to tree until it is driven into a solitary tree, from which it cannot leap into the branches of other trees; how some venturous climbers ascend the branches and try to shake the squirrel down upon the ground; how the active little creature throws itself from branch to branch, baffling the attempts of its pursuers, until one lucky stroke sends it flying into the air, and it comes softly to the ground, there to be chased amid shouts, tumbles, and laughter, until it is finally overwhelmed by a well-aimed cap, and made prisoner before it can disengage itself from the encumbrance.

It is easily domesticated, and is very amusing in its habits when suffered to go at large in a room or kept in a spacious cage; but when confined in a little cramped box, especially in one of the cruel wheel cages, its energies and playfulness are quite lost. Men often go about with squirrels for sale, and generally cheat those who buy them. In the first place, they constantly try to sell old squirrels for young, but this imposition may be detected by looking at the teeth of the animal, which are nearly white if young, but if old are of a light yellow.

Sub-family *e.* *Sciurīna.*

SCIŪRUS.—(Gr. Σκιά, a shadow; οὐρά, a tail.)

Europæus (Lat. *European*), *the Squirrel.*

In the second place, let the purchaser beware of very tame and quiet squirrels. These are generally animals just caught and perfectly wild, but made sedate by a dose of laudanum, which in many cases causes their death in a short time. One of my friends was deceived in this manner only a few months since, the squirrel dying in the course of the evening of the day on which it was purchased.

The colour of the English Squirrel is a deep reddish brown, and its tail so large and bushy as to shade its whole body when carried curled over its back, from whence it derives its name of Sciurus, or Shadow-tail.

The FLYING SQUIRRELS are well known by their power of making enormous sweeps through the air. They are enabled to make these leaps by a fold of skin at each side, which, when spread by the extended paws, forms a kind of parachute, that supports them in their passage through the air. When they wish to pass from one tree to another, they spring downwards from a lofty branch, stretch out all their legs, and sweep to their mark with an upward curve. The species of Flying Squirrel here represented is a native of the Rocky Mountains in America, where it lives among the dense pine forests that abound there. Its colour is yellowish brown, and its length about a foot.

Webber gives a very happy account of the domestic life of these creatures :—

"Then, when I went out by myself into the deep wood, I sat down on the moss at the root of an old tree to watch for the squirrel. When everything was still again, I would see him after a while poking his head out of the hole, snuff! snuff! Then out his head would pop to rest his chin upon his fore paws, and he would look all around, above and below, very cunningly, to see if all was right. Then out, like a thought, he would glide, and I could see his lovely brush quickly curled and spread so grandly above his head, as he sat upon a limb, still for the moment. Lo! there is another snuffing nose, and then great shining eyes filling the round black knot hole, and out another pops, and another and another. Such whisking of tails, darting along limbs, and bounding from swinging twig to rustling tree tops, until they all meet.

PTERŎMYS.—(Gr. Πτερόν, a wing; μῦς, a mouse.)

Alpīnus (Lat. *Alpine*), *the Flying Squirrel.*

"Now the frolic begins in earnest, and they dart round and round the trunks, rattling the bark down as they chase each other. Their tails are spread now as wide as they can, as if they were scared; and that lady squirrel he makes love to, you may be sure, for now he has chased her out to the very end of a great high limb; and hard pushed, here she comes right off into the air! down almost into my face—the white of her arms underneath, spread wide like her stiffened tail—into the leaves head-foremost, and then up and away, patter! patter! patter! Here he comes too, sailing down after her, plump! and rattles off along the old logs and swinging vines in hot chase.

"So they all would frolic, chasing one another, and one of them would see me, and stop and stamp his tiny feet and bark at me, jerking his tail in comic wrath."

The Alpine Marmot is common in the mountainous districts of Europe. It lives in burrows dug in the ground. These burrows are something in the shape of a Y, one of the forks leading to its habitation, a kind of chamber lined with dry grass and mosses, and the other fork serving as a storehouse for food, as a provision against the winter months, when it retires to its hole, closes the entrance, and becomes torpid until the commencement of spring. When it first retires for the winter, it is very fat, and is then killed and eaten in great numbers. The skin is also of some service.

Many may be seen in England, carried about by the Savoyard boys, who catch them when young, and tame them. When domesticated they are mild and inoffensive, but no instruction entirely overcomes their abhorrence of a dog.

Arctŏmys.—(Gr. Ἄρκτος, a bear; μῦς, a mouse.)

Marmotta (from its native name), *the Marmot.*

When feeding in their native country, the marmots are very suspicious, and always station one marmot as a sentinel, and on his giving the alarm, the remainder instantly seek the protection of their holes, closely followed by their faithful sentinel.

The Ox.—The Ruminantia, or animals that chew the cud, include the oxen, sheep and goats, deer, giraffe, and camels. They have a peculiar construction of stomach, which receives the freshly-gathered food, retains it for some hours, and then passes it back into the mouth to be re-masticated.

The Ox is spread widely over the earth, scarcely any country being without its peculiar breed. In this country, where it is our most useful domesticated animal, there are nearly as many breeds as counties, generally distinguished by the length or shape of their horns. There is the "long-horned breed" from Lancashire, the "short horned" from Durham, the "middle-horned" from Devonshire, and the "polled" or hornless, breed. Each of these breeds has its particular value: some

Order V. . . . *UNGULĀTA.*—(Lat. *possessing hoofs.*)
Family I. . . . Bovĭdæ.—(Lat. *Bos,* an ox. Ox kind.)
Sub-family *a.* *Bovīna.*

BOS.

Taurus (Lat. *a Bull*), *the Ox.*

fatten easily, and are kept especially for the butcher; others give milk, and are valuable for the dairy. The best dairy cow is the Alderney, a small, short-horned animal, furnishing exceedingly rich milk.

In some parts of England, oxen are used to draw waggons, or to drag the plough. They are not so strong as horses, and their movements are much slower.

Formerly, the cruel sport of bull-baiting was much practised in England, and bull-rings, that is, large iron rings firmly fixed in the ground, may be seen in the market-place of many towns. The poor bull was fastened to the ring by a strong rope, and mangled by the repeated attacks of large and fierce dogs. Sometimes the rope did not prove strong enough to restrain his frantic struggles, and the tortured animal chased and scattered the terrified spectators.

In Spain, bull-baiting is a very popular sport. The Spaniards do not confine the animal with a rope, but turn him loose into a large arena, where several men, armed with spears and darts, first goad him into madness, and then slaughter him. The death of the bull is, however, considered as a compliment due to the valour and endurance of the animal; for if a bull is soon overcome, or refuses to attack his opponents, he is driven out of the arena amid the hisses of the spectators, and suffered to prolong an ignominious existence.

Every part of the Ox is of value. We eat his flesh, we wear shoes soled with his skin, our candles are made from his fat, our tables are joined with glue made from his hoofs, his hair is mixed with the mortar of our walls, his horns are made into combs, knife handles, drinking cups, &c., his bones are used as a cheap substitute for ivory, and the fragments ground and scattered over the fields as manure—and soup is made from his tail.

The young ox is called a calf, and is quite as useful in its way as the full-grown ox. The flesh is called veal, and by many preferred to the flesh of the ox or cow, which is called beef: jelly is made from its feet. The stomach is salted and dried, and is called rennet. Cheese is made by soaking a piece of rennet in water, and pouring it into a vessel of milk. The milk soon forms a curd, which is placed in a press, and the watery substance, called whey, squeezed from it. The curd is coloured and salted, and is then cheese.

When a number of cows are kept in the same yard, the oldest cow always takes precedence, and pushes the others with her horns if they interfere with her. She chooses her own rack, and if she sees another rack better furnished, she dispossesses the original proprietor, and with an air of ridiculous complacency appropriates it to herself. None of the junior cows attempt to leave the yard or enter it until she has preceded them; and so jealous is she of her authority, that if any enter before her she refuses to move until they have been turned out. She then looks round in a dignified manner, and marches in, followed by the rest of the troop.

At Chillingham Park there is a breed of wild cattle, apparently the descendants of the original race that overran England in former years. They still retain their wild habits, and when any of them must be killed, thirty or forty men go out armed with rifles. A keeper mounted on a swift horse separates the victim from the herd, and drives it by the concealed marksmen, who speedily lay it prostrate. The colour of the Chillingham breed is always white with dark red ears.

The Zebu or Brahmin Bull is a native of India. It is a very conspicuous animal on account of the hump on its shoulders. There are different breeds of it, some larger than the English cattle, and some hardly larger than an ordinary hog. The Hindoos treat it with great reverence, and will not suffer it to be molested. It is in consequence so tame and familiar that it will often walk down the streets, examine the shops, and perhaps help itself to some sweetmeats; or it will lie down in the narrow street; but no one must disturb it, they must either proceed by another road, or wait until the sacred animal is pleased to rise.

With singular inconsistency, the Hindoo, although he honours the bull with such absurd reverence, yet has no pity on the ox. While the

BOS.

Indĭcus (Lat. *Indian*), *the Zebu.*

consecrated bull wanders with impunity through the streets, walks into shops, (china shops or otherwise,) and resents with a peevish push of its horns the slightest affront, the ox is fastened to the plough, urged on by the goad, and put to every kind of labour. The Zebu-cow, although not quite so well treated as the bull, yet enjoys more forbearance than the ox.

The Asiatic Buffalo is a large and powerful animal with enormous horns. It closely resembles the domestic ox, but is larger and stronger. Its strength is so great that it is a formidable enemy even to the tiger. Captain Basil Hall gives an account of a battle between a buffalo and a tiger. The tiger, however, seemed to have been alarmed at the very unusual scene into which he had been transferred; but the readiness of the buffalo to attack, proves that it did not fear the tiger.

"We were promised a grand day's sport one afternoon, when a buffalo and a tiger were to be pitted against each other. The buffalo entered the ring composedly enough; but after looking about him, turned to one side, and rather pettishly, as if he had felt a little bilious, overturned a vessel of water placed there expressly for his use. The tiger refused for a long time to make his appearance, and it was not till his den was filled with smoke and fire that he sprang out. The buffalo charged his enemy in a moment, and by one furious push capsized him right over. To our great disappointment, the tiger pocketed this insult in the shabbiest manner imaginable, and passing on, leaped furiously at the ropes, with which his feet became entangled, so that the buffalo was enabled to punish his antagonist about the rump most ingloriously. When at length the tiger got loose, he slunk off to a distant part of the area, lay down, and pretended to be dead. The boys, however, soon put him up again, and tried to bring him to the scratch with squibs and crackers, and a couple of dozen dogs being introduced at the same moment, they all set at him, but only one ventured to take any liberty with the enraged

BUBĂLUS.—(Gr. Βούβαλος, a Buffalo.)

Buffĕlus (Latinised corruption of Βούβαλος), *the Buffalo.*

animal This bold dog actually caught the tiger by the tail, but a slight pat of the mighty monster's paw crushed the yelping cur as flat as a board. The buffalo, who really appeared anxious to have a fair stand-up fight, now drove the dogs off, and repeatedly poked the tiger with his nose, and even turned him half over several times with his horns.

"We had then a fight between two buffaloes, which ran their heads against each other with a crash that one could fancy shook the palace to its very foundation; indeed, the only wonder was how both animals did not fall down dead with their skulls fractured. But there appears to be a wonderful degree of thickness or hardness in this part of the animal."

The Buffalo has long been domesticated in India, and from its great strength is exceedingly useful. In its wild state it is always found in marshy grounds, where the air is sufficiently pestilential to destroy most animals. There it will luxuriate through the hottest part of the day, with its entire body immersed in the muddy water, only leaving its muzzle above the surface.

The hide of this animal is particularly thick and strong, and is in great request for making harness.

The CAPE BUFFALO is a native of Southern Africa. It is exceedingly ferocious and cunning, often lurking among the trees until an unsuspecting traveller approaches, and then rushing on him and destroying him. The ferocious creature is not content with killing his victim, but

BUBALUS.

Caffer, *the Cape Buffalo.*

stands over him mangling him with its horns, and stamping on him with its feet. Cumming shot several of these animals, and once or twice had narrow escapes from them, as they are difficult to kill. His description of their aspect is very good, and I cannot do better than give it in his own words:—

"Their horns reminded me of the rugged trunk of an oak-tree. Each horn was upwards of a foot in breadth at the base, and together they effectually protected the skull with a massive and impenetrable shield. The horns, descending and spreading out horizontally, completely overshadow the animal's eyes, imparting to him a look the most ferocious and sinister that can be imagined."

The BISON inhabits the plains or prairies of North America in countless multitudes. Its enormous and heavy mane, its fierce eyes and lowering appearance, give this animal a most terrific aspect. The

American Indians constantly hunt the Bison, which they call by the name of Buffalo. Their weapons are principally bows and arrows, apparently weak and small, but which, when wielded by a skilful hand, will strike the huge bison to the heart. In Catlin's account of his travels among the North American Indians are many most interesting accounts of "buffalo hunts." Mounted on a swift horse, and armed with a spear and bow and arrows, the Indians kill great numbers of these animals. They ride up close to the bison, and with the greatest apparent ease bury an arrow up to its feather in the creature's body. Indeed many instances are known where the slight Indian bow, drawn

BISON.—(Gr. *βίσων, a Buffalo.*)

Americānus (Lat. *American*), *the Bīson.*

without any perceptible effort, has thrown the arrow completely through the body of the huge animal. There are many modes of destroying this animal in vogue among the Indians and white settlers. The skin is so valuable that every exertion is made to procure it. Of the buffalo's hide they make their wigwams or tents, their shields, their robes, their shoes, &c. The Indians can also sell the hides to the traders for a considerable sum, so that an Indian can almost measure his importance and wealth by the number of hides that he takes.

The hunters take advantage of the gregarious instincts of this animal, and hunt them when they are collected together in their vast herds, which blacken the face of the prairie for miles. Sometimes they form in line, and drive the herd to the edge of some tall cliff, over which they fall in hundreds, those behind pushing on those in the van; or sometimes they form a large circle, driving the animals into a helpless and leaderless mass, into which the hunters spring, leaving their horses, and treading with the skill of rope-dancers on the backs of the bewildered bisons, whom they slaughter as they pass, stepping from one to the other, and driving the sharp blade of their spear through the spine of the animal whose back they have just quitted.

When only wounded the Bison is a most dangerous antagonist, and rushes on its enemy with the most determined ferocity. Richardson gives the following instance of its fury when wounded:—

"Mr. Finnan M'Donald, one of the Hudson's Bay Company's clerks, was descending the Saskatchewan in a boat, and one evening, having pitched his tent for the night, he went out in the dusk to look for game. It had become nearly dark, when he fired at a bison bull, which was galloping over a small eminence, and as he was hastening forward to see if his shot had taken effect, the wounded beast made a rush at him. He had the presence of mind to seize the animal by the long hair on its forehead as it struck him on the side with its horn, and being a remarkably tall and powerful man, a struggle ensued, which continued until his wrist was severely sprained, and his arm was rendered powerless; he then fell, and after receiving two or three blows became senseless. Shortly afterwards he was found by his companions lying bathed in blood, being gored in several places, and the bison was couched beside him, apparently waiting to renew the attack had he shown any signs of life."

Despite the wholesale slaughter of this animal which is carried on annually by the Indians, there seems to be no decrease in their numbers. They are more wary than before, and have withdrawn themselves into more distant lands, but their dark masses still crown the plain as of yore, although it is now impossible to judge as men could do in former days of the various migrations which the herds would make. The dreaded fire-arms have had their effect on the bison as on every other animal, and it withdraws as far as possible from the haunts of civilized man.

The improvidence of the Indians is much to be regretted. Myriads of these animals are slaughtered every year, merely for the sake of their skin, their "hump," or their marrow-bones, the remainder of the animal being left to the wolves and the birds.

The principal use of the flesh of the bison is to make "jerked meat" of it. This is made by cutting the meat into long narrow slips, and drying them in the sun. There is a peculiar art in the cutting these slips. The operator takes a large lump of the flesh, and holding his knife firmly in one hand, presses the meat against its edge with the other, continually turning it round and round, until the whole piece is converted into one long strip. The strips thus prepared are pegged out on stays, as washerwomen peg their clothes, or suspended in festoons on the branches of trees, like red snakes, until they are dry enough to be packed up. Three days is considered sufficient for the purpose. The cow is preferred to the bull for conversion into jerked meat, while the skin of the bull is more valuable than that of the cow, from the mass of woolly hair about the shoulders.

Mantles to keep travellers warm during the winter time, when sleighs are in the ascendant, are made from the hide of the bison.

The strength and weight of the bison is enormous, and as it is a fierce as well as a powerful animal, it may seem singular that it should be so easily killed. The explanation of this question will be found in the structure of the animal. It will be seen from the engraving on p. 108, that the bison holds its head low, and cannot see much higher than the legs of a horse without exertion. In consequence, when enraged, it charges in a direct line, and thus the trained horses of the Indian hunters turn aside from its course, and contrive to place themselves close alongside the furious animal, sheering off with admirable dexterity as soon as the deadly arrow has sped from their masters' hands. Indeed a well-trained horse will unhesitatingly dash in among a whole herd of bisons, threading its way among them unscathed.

The flesh of the Bison is tolerable eating, but the "hump" appears from all accounts to be unapproachable in delicacy. It is exceedingly tender, and possesses the property of not cloying even when eaten in

Poephăgus.—(Gr. Πόη, grass; φάγω, I eat.)

Grunnĭens (Lat. *grunting*), *the Yak.**

excess. The fat also is said to be devoid of that sickening richness which is usually met with in our domesticated animals.

The cow is smaller than the bull, and considerably swifter. She is also

* See page 111.

generally in better condition and fatter than her mate, and in consequence the hunters who go to "get meat," always select the cows from the herd.

The YAK inhabits Tartary. Of this animal in a native state little or nothing is known. The name of "grunniens," or grunting, is derived from the peculiar sound that it utters. The tail of the Yak is very long and fine, and is used in India as a fan or whisk to keep off the mosquitos. The tail is fixed into an ivory or metal handle, and is then called a chowrie. Elephants are sometimes taught to carry a chowrie, and wave it about in the air above the heads of those who ride on its back. In Turkey, the tail is called a "horse-tail," and is used as an emblem of dignity.

From the shoulders of the Yak a mass of long hair falls almost to the ground, something like the mane of a Lion. This hair is applied to various purposes by the Tartars. They weave it into cloth, of which they not only make articles of dress, but also tents, and even the ropes which sustain the tents.

The MUSK OX is a native of North America, and is not very unlike the Yak in appearance. It is covered with very long hair, which reaches almost to the ground. Its flesh is tolerably good when fat, but at other times it smells strongly of musk. The horns of this animal are united together at their base, forming a kind of shield or helmet covering the forehead. When the hunters wish to shoot the Musk Ox they conceal themselves, and fire without permitting the oxen to see them. The poor animals seem to fancy that the report of the guns is thunder, and crowd together in a mass, so that they afford a good mark. If, however, they catch sight of one of their assailants, they instantly charge at him, and then are very dangerous enemies. Both this animal and the Yak are small, scarcely equalling in size the small Highland cattle, but the thick hair which covers them makes them look larger than they really are.

OVĬBOS.—(Lat. *Sheep-Ox.*)

Moschātus (Lat. *musky*), *the Musk Ox.*

The GNOO, or WILDEBEEST, inhabits Southern Africa. At first sight it is difficult to say whether the horse, buffalo, or deer predominates in its form. It however belongs to neither of these animals, but is one of the bovine Antelopes. The horns cover the top of the forehead, and then, sweeping downwards over the face, turn boldly upwards with a sharp curve. The neck is furnished with a mane like that of the horse, and the legs are formed like those of the stag. It is a very swift animal, and when provoked, very dangerous. When it attacks an opponent it drops on its knees, and then springs forward with such force that, unless he is extremely wary and active, he cannot avoid its shock. When first alarmed, its movements are very grotesque, and are thus described by Cumming:—

"When the hunter approaches the old bulls, they commence whisking their long white tails in a most eccentric manner; then springing suddenly into the air, they begin prancing and capering, and pursue each other in circles at their utmost speed. Suddenly they all pull up together, to overhaul the intruder, when two of the bulls will often commence fighting in the most violent manner, dropping on their knees at every shock; then quickly wheeling about, they kick up their heels, whirl their tails with a fantastic flourish, and scour across the plain enveloped in a cloud of dust."

CATOBLĔPAS.—(Gr. Κατωβλέπων, looking down.)

Gnu (*Native name*), *the Gnoo.*

When it is taken young, the Gnu can be domesticated, and brought up with other cattle, but it will not bear confinement, and is liable to become savage under restraint.

There are several species of this animal, three being satisfactorily ascertained, namely, the common Gnoo, represented in the accompanying engraving, the Cocoon, (Catoblĕpas Taurīna,) and the Brindled Gnoo (Catoblĕpas Gorgon), as a specimen, all three animals being in the British Museum.

The size of the Gnoo is about that of a well-grown ass, that is, about four feet in height. Its flesh is in great repute both among the natives and colonists.

The NYLGHAU, one of the largest and most magnificent of the Antelopes, inhabits the forests of India. It is extremely vicious, and cannot be approached without danger. Its method of attack is similar to that of the Gnoo, namely, by dropping upon its knees and then springing violently

PORTAX.—(Gr. Πόρταξ, a Calf.)

Picta (Lat. *painted*) *the Nylghau.*

forward. The tiger is its great enemy, and often destroys it in spite of its courage. During the day the Nylghau conceals itself in the forests, and at night leaves its coverts to feed, often doing no inconsiderable harm to adjacent cultivated lands.

The colour of this creature is a slaty blue; it has however several white spots, and from its throat and shoulders hangs a dense bunch of hair. It is about the same size as the gnoo, standing about four feet high at the shoulder.

The KOODOO* is a native of South Africa, living along the wooded borders of rivers. It is chiefly remarkable for its beautifully shaped horns, which are about four feet in length and twisted into a large spiral of about two turns and a half. A bold ridge runs along the horns and follows their curvature. When hard pressed it always takes to the water, and endeavours to escape by its powers of swimming. Although a large animal, nearly four feet in height, it can leap with wonderful activity. The weight of the horns is very considerable, and partly to relieve itself of that weight, and partly to guard them from entanglement in the bushes among which it lives and on which it feeds, it carries its head backwards, so that the horns rest on its shoulders.

* See page 114.

The best and fullest accounts of the ELAND* and the ORYX are to be found in Harris and Cumming's Adventures in South Africa. An extract from Cumming will be both interesting and accurate. Of the Eland, he writes :—

"This magnificent animal is by far the largest of all the antelope tribe, exceeding a large ox in size. It also attains an extraordinary condition, being often burthened with a very large amount of fat. Its flesh is most excellent, and is justly esteemed above all others. It has a peculiar sweetness, and is tender and fit for use the moment the animal is killed. Like the gemsbok, the Eland is independent of water. It is generally diffused throughout all the wooded districts of the interior where I have hunted. Like other varieties of deer and antelope, the old males may often be found consorting together apart from the females, and a troop of these, when in full condition, may be likened to a herd of stall-fed oxen.

STREPSICĔROS.—(Gr. Στρέψις, a twisting; κέρας, a horn.)

Kudu, (native name) *the Koodoo.*

"I have repeatedly seen an eland drop down dead at the end of a severe chase, owing to his plethoric habit. The skin of the eland I had just shot emitted, like most other antelopes, the most delicious perfume of trees and grass."

The height of the eland is fully five feet at the shoulders, and its weight from seven to nine hundredweight. The horns of the male are about a foot and a half in length, while those of the female are smaller, and sometimes without the spiral wreathing.

The ORYX, also a South African animal, is well known among hunters

* See page 115.

Boselăphus.—(Gr. *Ox-stag.*)

Orēas (Gr. 'Ορειάς, belonging to the mountains), *the Eland.*

as the only antelope that revenges itself on the lion. When it sees the lion in the act of springing on it, it lowers its head, receiving the lion on the points of its sharp horns. It invariably perishes by the shock, but the lion also perishes with it. Their skeletons have more than once been seen lying together bleached on the plain. The description given of this animal by Cumming is highly graphic. "The oryx, or gemsbok, to which I was now about to direct my attention more particularly, is about the most beautiful and remarkable of all the antelope tribe. It is the animal which is supposed to have given rise to the fable of the unicorn, from its long straight horns, when seen in profile, so exactly covering one another as to give it the appearance of having but one. It possesses the erect mane, long sweeping black tail, and general appearance of the horse, with the head and hoofs of an antelope. It is robust in its form, squarely and compactly built, and very noble in its bearing. Its height is about that of an ass, and in colour it slightly resembles that animal. The beautiful black bands which eccentrically adorn its head, giving it the appearance of wearing a stall collar, together with the manner in which the rump and thighs are painted, impart to it a character peculiar to itself. The adult male measures 3 feet 10 inches in height at the shoulder."

The sharp horns of the oryx stand it in good stead, when pursued by

ORYX.—(Gr. Ὄρυξ, a word from Herodotus, denoting a gazelle.)

Leucŏryx (Gr. *the white Oryx*), *the Oryx.*

dogs, as it generally kills several of them before it is vanquished, and if the hunter's rifle is not at hand, drives off the dogs and escapes.

The SPRINGBOK is one of the smaller South African antelopes. Its colour is a light cinnamon red on the back, fading into white on the under part of the body, a narrow band of reddish brown separating the two colours.

For a description of the habits of the animal, I must again refer the reader to Cumming. During his early travels in South Africa, the first object that met his eyes on waking one morning, was a herd of Springboks, which he thus describes:

"On the 28th I had the satisfaction of beholding, for the first time, what I had often heard the Boers allude to, viz. a "trek-bokken," or grand migration of springboks. This was, I think, the most extraordinary and striking scene, as connected with beasts of the chase, that I have ever beheld. For about two hours before the day dawned I had been lying awake in my waggon, listening to the grunting of the bucks within two hundred yards of me, imagining that some large herd of springboks was feeding beside my camp; but on my rising when it was clear, and looking about me, I beheld the ground to the northward of my camp actually covered with a dense living mass of springboks, marching slowly and steadily along, extending from an opening in a long range of hills on the west, through which they continued pouring, like the flood of some great river, to a ridge about a mile to the north-east, over which they disappeared. The breadth of the ground they covered might have been somewhere about half a mile. I stood upon the fore-chest of my waggon for nearly two hours, lost in wonder at the novel and wonderful scene which was passing before me, and had some difficulty in convincing myself that it was reality which I beheld, and not the wild and exaggerated picture of a hunter's dream. During this time their vast legions continued streaming through the neck in the hills in one unbroken compact phalanx.

"Vast and surprising as was the herd of springboks which I had that morning witnessed, it was infinitely surpassed by what I beheld on the march from my vley to old Sweir's camp; for on our clearing the low range of hills through which the springboks had been pouring, I beheld the boundless plains, and even the hill sides which stretched away on every side of me, thickly covered, not with herds, but with one vast herd of springboks; as far as the eye could strain the landscape was alive with them, until they softened down into a dim red mass of living creatures."

GAZELLA.—(Arabic, Ghazil, *light or elegant.*)

Euchŏre (Gr. Εὖ, well; χορος, dance), *the Springbok.*

The Springbok is very fearful of man, and if it has to cross a path over which a man has passed before, it does not walk over, but takes a tremendous leap, ten or twelve feet high, and about fifteen long, at the same time curving its back in a most extraordinary manner. It is from this habit of leaping that the Dutch Boers who inhabit the Cape have given it the name of Springbok.

The GAZELLE, so famous in Oriental poetry, inhabits Arabia and Syria. Its eyes are very large, dark and lustrous, so that the Oriental poets love to compare the eyes of a woman to those of a gazelle, just as Homer constantly applied the epithet ox-eyed (βοῶπις) to the more majestic goddesses, such as Juno and Minerva. It is easily tamed when young, and is frequently seen domesticated in the courtyards of houses in Syria. Its swiftness is so great that even a greyhound cannot overtake it, and the hunters are forced to make use of hawks, which are trained to strike at the head of the gazelle, and thus confuse it, and retard its speed, so as to permit the dogs to come up. In several parts of Syria, the gazelle is taken by driving a herd into a large enclosure surrounded by a deep ditch. A few gaps are made, through which the terrified animals leap, and fall into the ditch, when they are easily taken.

GAZELLA.

Ariel (Gr. *proper name*), *the Gazelle.*

The height of the gazelle is about one foot nine inches; its colour a dark yellowish brown fading into white on the under parts.

The CHAMOIS is found only in mountainous regions, especially the Alpine chains of Europe and Western Asia. It lives on the loftiest ridges, displaying wonderful activity, and leaping with certainty and security on places where the eye can hardly discern room for its feet. The Chamois hunters are exposed to the most frightful dangers, to the chance of falling from terrific precipices, to hunger and cold, and every imaginable hardship that days spent among Alpine precipices can suggest. Yet a kind of fascination urges them on, although few Chamois hunters finally escape the dangers that surround them. The skin of the Chamois is used extensively by shoemakers.

RUPICĀPRA.—(Lat. *Rock-goat.*)

Tragus (Gr. Τράγος, a He-goat), *the Chamois.*

Several genera are omitted.

The IBEX inhabits the Alpine regions of Europe and Western Asia. It is instantly recognised by its magnificent horns, which curve with a bold sweep from the head almost to the haunches. The horns are surrounded at regular intervals with rings, and are immensely strong, serving, as some say, to break the fall of the Ibex when it makes a leap from a height.

When chased it is a dangerous animal, as after it has led its pursuer over dangerous heights and fearful chasms, it will frequently turn on him, and unless he can shoot it before it reaches him, will hurl him over the precipice. It is very wary, and, like many other animals, posts a sentry to keep watch, who, when he sees a suspicious object, gives notice by a kind of whistle, as a warning, on hearing which the whole of the herd instantly dash off to the highest point they can find. The height of the Ibex is two feet six inches; the length of its horns often three feet.

CAPRA.—(Lat. *a Goat.*)

Ibex, *the Ibex*, or *Steinbok.*

CAPRA.

Hircus (Lat. *a He-goat*), *the Goat.*

The common GOAT is not in much request in England, but in some other countries, as Syria and Switzerland, large herds of goats are kept for the sake of their milk, and in fact almost entirely take the place of the cow. The most celebrated variety of this animal is the Cashmir goat, which furnishes the beautifully fine wool from which the costly Cashmir shawls are made. The shawls bear a high value even in their own country, but in Europe the price is much increased by the various taxes which are paid in every stage of the manufacture,—the average number of taxes paid on each shawl being about thirty, several of which are limited only by the pleasure of the collector.

The SHEEP.—There are many kinds of Sheep, among which the common sheep, the long-tailed sheep, and the Wallachian sheep are the most conspicuous. Next to the cow, the sheep is our most useful animal. England produces better wool than any country; for although the wool of the Spanish sheep is finer than ours, it is much less in quantity. The Merino, as this sheep is called, is annually conducted from one part of the country to another, and back again. The distance traversed is upwards of four hundred miles, and the time necessary to complete the journey about six or seven weeks. The proprietors of the flocks think that these periodical journeys improve the wool; but it is in all probability a mistaken notion, as the stationary flocks of Leon and Estremadura produce quite as fine a fleece. Of course such a body of sheep—nearly six millions—do great damage to the lands over which they pass, and many fall victims to fatigue or are destroyed by wolves.

OVIS.—(Lat. *a Sheep.*)

Aries (Lat. *a Ram.*)

The long-tailed sheep inhabits Syria and Egypt. Its tail is so large and so loaded with fat, that to prevent it from being injured by dragging on the ground, a board is fastened to the under side of it, and wheels are often attached to the board. The peculiar fat of the tail is considered a great delicacy, and is so soft as to be frequently used as butter. The weight of a large tail is about seventy pounds.

The Wallachian or Cretan sheep is found in Crete, Wallachia, Hungary, and Western Asia. Its horns are exceedingly large, and are twisted in a manner resembling those of the Koodoo. It is very strong, and extremely vicious and unruly. In this and several other sheep the fleece is composed of wool and hair mixed. The hair of the Wallachian sheep is long and silky like that of a spaniel, and of great length, falling almost to the ground.

The GIRAFFE.—This beautiful and extraordinary animal is found only in South Africa. As the gnoo seems to combine the properties of the antelope, horse, and buffalo, so the Giraffe appears to bear the characteristics of the antelope and the camel. In the opinion of modern naturalists, it holds a place by itself between the deer and antelopes;—it forms, at all events, a group to which no other animals belong.

Sub-family *b. Camelopardīna.*

CAMELOPARDĂLIS.—(Gr. Κάμηλος, a camel; πάρδαλις, a pard.)

Giraffa (Arabic, *Zarapha*), *the Giraffe.*

The height of the Giraffe varies from thirteen to eighteen feet. Its beautiful long neck enables it to browse on the leaves of the trees on which it feeds. It is very dainty while feeding, and plucks the leaves one by one with its long and flexible tongue. On its head are two very remarkable projections, closely resembling horns. These projections are not horns, but only thickenings of the bone of the skull, covered with skin, and bearing a tuft of black hair at the extremity of each. The fore legs at first sight appear longer than the hind ones, but this apparent difference is only caused by the great length of the shoulder-blades, as both pair of legs are of the same length at their junction with the body. Its eyes are very large and prominent, so that the animal can see on every side without turning its head. Just over and between the eyes is a third bony prominence, resembling the projecting enlargements of the skull, called horns. The use of these projections is not very well known, as although in play the Giraffe will swing its head round and strike with it, yet when it wishes to repel an assailant it has recourse to violent and rapid kicks from its hind legs. So light and swift are these kicks that the eye can scarcely follow them, and so powerful are they that the lion is not unfrequently driven off by them. Vaillant relates that a Giraffe which he was hunting, kept off his pack of dogs by its rapid kicks. Indeed, if it were to venture its head too near the lion, a blow from his tremendous paw would in all probability lay the animal prostrate.

The skin of this animal is an inch and a half in thickness, so that it is necessary for the hunter to make very sure of his aim before he fires at an animal so well defended.

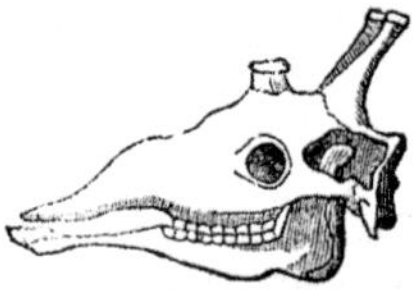

SKULL OF THE GIRAFFE.

The Giraffe has much difficulty in reaching the ground with its mouth, nor does it often attempt to do so, unless it is bribed with something of which it is very fond, such as a lump of sugar. It then straddles widely with its fore legs, and with some trouble succeeds in reaching the object aimed at. This attitude was noticed and copied in the Prænestine pavement.

The appearance of this animal in its native haunts is very magnificent. "These gigantic and exquisitely beautiful animals, which are admirably formed by nature to adorn the forests that clothe the boundless plains of the interior, are widely distributed throughout the interior of Southern Africa, but are nowhere to be met with in great numbers. In countries unmolested by the intrusive foot of man, the Giraffe is found generally in herds varying from twelve to sixteen; but I have not unfrequently met with herds containing thirty individuals, and on one occasion I counted forty together; this, however, was owing to chance, and about sixteen may be reckoned as the average number of a herd. These herds are composed of Giraffes of various sizes, from the young Giraffe of nine or ten feet in height, to the dark chestnut coloured old bull of the herd, whose exalted head towers above his companions, generally attaining to a height of upwards of eighteen feet. The females are of lower stature, and more delicately formed than the males, their height averaging from sixteen to seventeen feet. Some writers have discovered ugliness and a want of grace in the Giraffe, but I consider that he is one of the most strikingly beautiful animals in the creation; and when a herd of them is seen scattered through a grove of the picturesque parasol-topped acacias which adorn their native plains, and on whose uppermost shoots they are enabled to browse by the colossal height with which nature has so admirably endowed them, he must, indeed, be slow of conception who fails to discover both grace and dignity in all their movements. There can be no doubt that every animal is seen to the greatest advantage in the haunts which nature destined him to adorn, and among the various living creatures which beautify creation, I have often traced a remarkable resemblance between the animal and the general appearance of the locality in which it is found.

"In the case of the Giraffe, which is invariably met with among venerable forests, where innumerable blasted and weather-beaten trunks and stems occur, I have repeatedly been in doubt as to the presence of them, until I had recourse to my spy-glass; and on referring the case to my savage attendants I have known even their optics to fail, at one time mistaking these dilapidated trunks for camelopards, and again confounding real camelopards with these aged veterans of the forest." *

The first living Giraffes, in the possession of the Zoological Society, were brought by M. Thibaut in 1835. He succeeded in taking four, all of which he brought with him. One of them is still living. From this stock, several Giraffes have been born, some of which are now in England, and others have been sent to other countries.

One of the four originals killed himself soon after his arrival, by striking his head against a wall as he was rising from the ground. An accident of the same nature happened recently to another animal, one of its horns being broken off, and bent backwards; but owing to the presence of mind of the keeper, who immediately pulled the horn into its place again, no bad results followed, the fractured parts uniting naturally.

The tongue of the Giraffe is one of the most remarkable parts of its structure. It is very flexible and capable of great changes of form, the Giraffe being able to contract it so that its tip could enter an ordinary quill. The animal is very fond of exercising its tongue, and sometimes pulls the hairs from its companions' manes and tails, and swallows them; no very easy feat, as the hair of the tail is often more than four feet long.

The movements of the Giraffe are very peculiar, the limbs of each side appearing to act together. It is very swift, and can outrun a horse, especially if it can get among broken ground and rocks, over which it leaps with a succession of frog-like hops.

In this country it endures the climate well. The Giraffes in the Zoological Gardens which were born and bred in this country seem very healthy and are exceedingly tame, examining the hands of their visitors, and following them round the enclosure. They eat herbs, such as grass, hay, carrots, and onions. When cut grass is given to them, they eat off the upper parts and leave the coarse stems, just as we eat asparagus.

The Camel.—There is much confusion about the names of the Camels.—The Bactrian Camel is distinguished by bearing two humps on its back, the Arabian Camel by bearing only one. The Arabian camel

* Cumming's Adventures, vol. i. pp. 269, 270.

is sometimes, but erroneously, called the Dromedary, as the Dromedary, or El-Heirie, is a lighter variety of that animal, and only used when despatch is required.

The Camel forms the principal wealth of the Arab; without it he could never attempt to penetrate the vast deserts where it lives, as its remarkable power of drinking at one draught sufficient water to serve it for

Sub-family *c. Camelīna.*

CAMĒLUS.—(Gr. Κάμηλος, a Camel.)

Arabĭcus (Lat. *Arabian*), *the Camel.*

several days, enables it to march from station to station without requiring to drink by the way. The peculiar structure of its stomach gives it this most useful power. In its stomach are a great number of deep cells, into which the water passes, and is then prevented from escaping by a muscle which closes the mouth of the cells. When the Camel feels thirsty, it has the power of casting some of the water contained in these cells into its mouth. The habits of this animal are very interesting. A recent traveller, the Rev. J. H. Pollen, most kindly forwarded to me the following interesting and amusing account of the habits of the Camel:—

"My principal experience in camels has been during my travels through the Arabian desert. I followed, after some interval of time, the route of the Hajji—the Mecca pilgrimage.

"The temper of the Camel is in general not very amiable. It is unwilling, jealous, and revengeful to the last degree. Of this latter quality curious tales are told: one, which was fully believed by the Arab that narrated it to me, was as follows. A certain camel driver had bitterly insulted (*i.e.* thrashed in some ignominious way,) the animal under his charge. The camel showed no disposition to resent, but the driver, knowing from the expression of its eye what was passing within, kept on the alert for several days. One night he had retired for safety inside his tent, leaving his striped abbaya or cloak spread over the wooden saddle of the camel outside the tent.

"During the night he heard the camel approach the object, and after satisfying himself by smell or otherwise that it was his master's cloak, and believing that the said master was asleep beneath it, he lay down and rolled backwards and forwards over the cloak, evidently much gratified by the cracking and smashing of the saddle under his weight, and fully persuaded that the bones of his master were broken to pieces. After a time he rose, contemplated with great contentment the disordered mass, still covered by the cloak, and retired.

"Next morning, at the usual hour for loading, the master, who had from the interior of his tent heard this agreeable process going on, presented himself to the camel. The disappointed animal was in such a rage, said my informant, on seeing his master safe before him, that *he broke his heart, and died on the spot.*

"I had once to cross a very high range of rocks, and we had very great difficulty in getting our camels to face the steeper part of the ascent, though any horse would have made very light of it. All the riders had to dismount, and the laden animals made the bare rocky solitudes ring to the continual and most savage growls with which they vented their displeasure. It is well on these occasions to keep out of reach of their long necks, which they stretch out and bring their teeth within dangerous proximity to the arm or side of any one but their master.

"While being laden they testify their dislike to any packet which looks unsatisfactory in point of size or weight as it is carried past them, although when it is once on their backs they continue to bear it with the patient expression of countenance which I fear passes for more than it is worth. All camels are loaded kneeling, and can go from twenty-four to sixty hours without rest, or more than a few mouthfuls of food, which they can crop off a thorny bush as they pass, or a handful of barley given them by their master. Parts of the desert are strewn with small, dry, drab-coloured plants, thorny and otherwise, which the camels continue to crop as they walk, jerking the rider not a little.

"They are very sparing of drinking. I have taken camels for eleven or twelve days without a drop of water. All of them did not drink even when we came to water, nor did any drink a large quantity, or seem disturbed by the want of it, although the sun was very powerful, and we travelled twelve or thirteen hours daily.

"At first they are difficult to ride. The rider mounts while the animal is kneeling, and sits like a lady, with the right leg round the fore pommel of the saddle. In rising, the Camel suddenly straightens its hind legs before moving either of the fore legs, so that if the rider is unprepared, he will be jerked over its ears. It moves the legs of each side alternately, occasioning a long undulating motion, which sways the rider to and fro from the loins. The motion, however, is soon learned, and when fatigued, the rider can change sides, or shift his posture in various ways.

"Sometimes a traveller places his whole family, wife and children, in one pannier fastened to the saddle, puts himself in another pannier

CAMĒLUS.

Bactriānus (Lat. *Bactrian*), *the Bactrian Camel.*

fastened on the opposite side, and then falls in with a caravan and accompanies it.

"DROMEDARIES—the finer and better bred Camels—have sparer frames and more endurance, and are principally led by the Bedouins of the desert. They also object either to going up or down a hill.

"They are fond of kneeling at night just behind the ring of Arabs who squat round the fire, and they stretch their heads over their masters' shoulders to snuff up the heat and smoke, which seems to content them vastly.

"Between Cairo and Suez I saw more than one camel dead or dying. They seem very tenacious of life, as they remain unable to rise from a broken limb or other cause for very many days. I more than once wished to go up and shoot the poor creatures to put them out of their misery, but the Arabs have superstitious notions on this point, and would not suffer it. I did once find a camel that had been stabbed by its master, and once only. The poor beast had been exhausted, and the long, broad dagger struck into his heart. It must have been a very short time before I reached the spot, as the blood was almost fresh.

"The Camels at Grand Cairo are remarkably large and powerful, and my informant told me that they are very *proud*, and will only eat their food from their master's hand—preferring to starve rather than receive it from any other source."

The foot of the Camel is admirably adapted for walking on the loose sand, being composed of large elastic pads, which spread as the foot is placed on the ground. To guard it from injury when it kneels down to be loaded, the parts of its body on which its weight rests are defended by thick callosities. The largest of these callosities is on the chest, the others are placed on the joints of the legs.

The Bactrian Camel inhabits Central Asia, Thibet and China. It is distinguished from the Arabian camel by possessing two humps.

LLAMA.—(*Peruvian name.*)

Pacos (*Peruvian name*), *the Llama.*

The LLAMAS, of which there are several species, inhabit America, and are used for the same purposes as the camel. When wild they are very timid, and fly from a pursuer the moment that they see him; but their curiosity is so great that the hunter often secures them by lying on the ground and throwing his legs and arms about. The Llamas come to see what the extraordinary animal can be, and give the hunter an opportunity of firing several shots, which the astonished animals consider as part of the performance.

The Llamas, like the camels, have a series of cells in the stomach for containing water, and can go for several days without requiring to drink. If too heavily laden, or when they are weary, they lie down, and no threats or punishment will induce them to rise, so that their masters are forced to unload them. When offended they have a very unpleasant habit of spitting at the object of their anger. Formerly it was supposed that their saliva was injurious, and produced blisters if it touched the skin.

The fleece of the Llama is very long and fine, more resembling silk than wool. It is very valuable, and is extensively imported into this country for the purpose of making cloth and other fabrics. The fleece of the Alpaca is considered the best, as it is sometimes twelve inches in length, and very fine.

In Chili and Peru the natives domesticate the Llama, which in a state of captivity frequently becomes white. It is by no means a large animal, as it measures about four feet six in height. In general shape it resembles the camel, but has no hump on its back, and its feet are provided with sharp hoofs for climbing the rocky hills among which it lives. In Peru, where it is most commonly found, there are public shambles established for the sale of its flesh.

Sub-family *d. Moschīna.*

MOSCHUS.—(Gr. Μόσχος, Musk.)

The MUSK-DEER inhabits many parts of India, and is famous for the scent which it produces. This scent, called Musk, is secreted in a kind of pouch, and is so very strong when recent, that the hunter, after killing the animal, is forced to bind his mouth and nostrils with linen before he ventures to open the pouch, as the scent is so intolerably powerful that it causes violent bleeding at the nose. When the merchants traffic for musk, they remain in the open air, holding a handkerchief over their faces, and even with these precautions it often causes headaches. The musk is never imported

Moschifĕrus (Lat. *musk-bearing*), *the Musk-deer.*

SKULL OF THE MUSK DEER.

pure into this country, being always adulterated by the merchants. It is very costly, and forms an important article of commerce in the East. The Musk-deer is about two feet in height at the shoulders. The male possesses two extraordinarily long teeth in the upper jaw, which project from the lips at each side of the mouth.

Sub-family *e.* *Cervīna.*

CERVUS (Lat. *a Stag.*)

Capreŏlus (Lat. *a Wild Buck*), *the Roebuck.*

The ROEBUCK was formerly common throughout the whole of England, but is now only found in Scotland, north of the Forth. It is the smallest and most beautiful of our British deer. It is not at all adapted for confinement, as it is never induced to be familiar with its keeper, and will sometimes attack any object which it dislikes with its horns and hoofs. It does not live in herds like the Fallow-deer, but singly, or in pairs, driving off its young when they are about nine or ten months old. It is very cunning, and, when hunted, sometimes baffles the dogs by making a few enormous leaps, waiting until the dogs have passed, and then returning on its previous track. Its height is about two feet; its horns are divided into three small branches, and are seldom more than a foot in length.

The RED-DEER, or STAG, is the largest of our deer. In the language of hunters, it bears different names according to the size of its horns, which increase year by year. All the male deer have horns, which they shed every year, and renew again. The process of renewal is most interesting. A skin, filled with arteries, covers the projections on which the horns rest. This skin, called the "velvet," is engaged in continually depositing bone on the footstalks, which rapidly increase in size. As the budding horns increase, the velvet increases also, and the course of the arteries is marked on the horn by long furrows, which are never obliterated. When the horn has reached its full growth, it cannot be at once used, as the velvet is very tender, and would bleed profusely if wounded. The velvet cannot be suddenly removed, as the blood that formed the arteries would rush to the brain and destroy the animal. A ring of bone forms round the root of each horn, leaving passages through which the arteries pass. By degrees, these passages become narrow, and finally close

CERVUS.

Elăphus (Gr. Ἔλαφος, a Stag), *the Stag.*

entirely, thus gradually shutting off the blood. The velvet, being deprived of its nourishment, dies and is peeled off by the deer, by rubbing against a tree, leaving the white hard horn beneath.

Hunting the Stag is a very favourite amusement in this country, and packs of hounds, called stag-hounds, are kept expressly for that purpose.

The Wapiti is one of the largest of the deer tribe, often growing to the height of our largest oxen. It inhabits Canada and other parts of North America, and has been confounded with the Moose. Its horns are very large, measuring nearly six feet from tip to tip.

The hunters are acquainted with its peculiarities, and chase it from their knowledge of its character. It is very fond of salt, and comes in great numbers to the saline marshes, for the purpose of licking the salt off the soil upon which it has settled. Such places are called "licks," and to them the hunters resort, lying in wait for the deer, who are sure to visit these places.

It frequents the woods and copses, in which it lies so well concealed, that an inexperienced eye cannot perceive the animal even when it is pointed out to him, so well does its colour agree with the tints of the brush among which it hides. From the branching horns which it bears, one might suppose that it would find great difficulty in forcing its way through the woods; but, in fact, its horns are a defence instead of an incumbrance, and as it lays them flat on its back before plunging among the trees, they defend its back from the branches through which it forces its way.

The skin of the Wapiti is very useful to the hunters, as they have a method of dressing it so that it does not become stiff and harsh after being wetted, but retains its original flexibility. This property makes it very valuable for hunting dresses, which are generally made of leather.

It is very fierce, and boldly attacks an antagonist, especially if wounded. An example of its ferocity, when wounded, is given by Palliser in the following passage:—"We were now about 150 yards from the nearest of the band. I chose a fine old stag, while Boucharville, with an eye to superior meat, singled out a doe. We drew up our rifles slowly, and both shots went off together. The smoke hung heavily for a second or two; when it cleared away, we espied one of the Wapiti lying down: the next instant down rolled the stag also. We agreed to advance at the same moment, lest one or other of the animals should be able to get up and escape. On coming near my stag, he struggled to rise, but unable to regain his feet, rolled back again. I looked towards the other, when what was my surprise at witnessing a regular combat between Boucharville and his wounded elk,* now transformed into a very formidable antagonist! Springing on her haunches, she was striking furiously at him with her fore-feet; one hoof missed him, but the other fell on his rifle, which he held up for his protection, and smashing both his ramrod and his loading-stick, beat him down on his knees. Rising a second time, she was about to repeat the attack, when my ball caught her in the side of the head, behind the eye, and with a splendid bound she fell lifeless on the broad of her back. I had made a quick and necessarily a rather dangerous shot; but I was in luck that day. 'Ah!' exclaimed Boucharville, as he half rose from the ground, but looking at nothing till he had satisfied himself that his rifle was uninjured, 'Mais qui l'aurait cru? Ma foi!' continued he, 'j'ai bien échappé; une biche à une côté et une balle à l'autre!'"

CERVUS.

Canadensis (Lat. *belonging to Canada*), the Wapiti.

* This animal is often called an elk by the hunters.

The Axis.—This beautiful Deer is an inhabitant of India, especially of parts by the Ganges. It has frequently been domesticated in England,

Axis.—(Lat. *the Axis Deer.*)

Maculāta (Lat. *spotted*), *the Axis.*

and thrives well even in open parks. The horns are slender, and are divided into three branches. Its usual colour is a fawn yellow, spotted regularly with white, and a black stripe runs down the back.

The Fallow-deer are usually seen in parks, where they congregate in large herds, and form a most pleasing addition to the landscape when they are seen reposing under the trees, or chasing one another in graceful play. One peculiarly large buck always takes the lead, and suffers none but a few favourite does to approach his regal presence, all the other bucks running humbly away directly he makes his appearance. They are generally tame, and will suffer people to come very close to them; but at certain times of the year they become savage, and will not permit any one to approach their domains. If an intruder is bold enough to venture within the proscribed distance, the buck will instantly charge upon him, and if he does not make his escape, will in all probability inflict considerable damage upon him. They soon

HAIR OF DEER.

become familiar with those who treat them with kindness, and will eat from their hands. At Magdalene College, Oxford, where there are some of these deer, it used to be a common practice to let down a crust of

DAMA.—(Lat. *a Deer.*)

Vulgāris (Lat. *common*), *the Fallow-deer*.

bread by a string from one of the windows that overlooked the park. The deer would speedily approach, and it was singular to see how they would take a large crust in their little mouths, and continue to bite it until they contrived to eat the whole of it without once letting it drop.

The REINDEER is found throughout the Arctic regions of Europe, Asia, and America. The finest animals are those of Lapland and Spitzbergen. The Laplander finds his chief wealth in the possession of the Reindeer, which not only serves him as a beast of burden, but furnishes him also with food and clothing. A Laplander in good circumstances possesses about three or four hundred deer, which enable him to live in comfort. The subsistence of one who only possesses one hundred is very precarious, and he who has only fifty, usually joins his animals with the herd of some richer man, and takes the menial labours upon himself.

A gadfly (*Œstrus Tarandi*) annoys the Reindeer so much, that the Laplander is forced to make periodical migrations to the mountains in order to escape the dreaded gadfly, and the equally dreaded mosquitoes, which are more ferocious in the cold climates than in the Tropics. The

Reindeer feeds principally on a kind of lichen, which it scrapes from beneath the snow. During the winter, its coat thickens, and assumes a lighter hue, many deer being almost white. Its hoofs are divided very

RANGIFER.—(Linnean generic name.)*

Tarandus (Gr. Τάρανδος, *a Reindeer*), *the Reindeer.*

high, so that when the animal places its foot on the ground, the hoof spreads wide, and as it raises the foot, a snapping noise is heard, caused by the parts of the hoof closing together. When harnessed to a sledge, it can draw from 250 to 300 pounds' weight at about ten miles an hour.

The EUROPEAN ELK inhabits the northern parts of Europe. It was considered at one time to be identical with the American Elk, but naturalists now believe it to be a distinct animal. Its usual pace is a high awkward trot, but when frightened, it sometimes gallops. It is very strong, and can destroy a wolf with a single blow of its large and powerful horns. In Sweden it was formerly used to draw sledges, but on account of the facility of escape offered to criminals by its great speed, the use of it was forbidden under high penalties. The skin of the Elk is so tough that a regiment of soldiers was furnished with waistcoats made of its hide, which could scarcely be penetrated by a ball.

* Possibly a Latinized form of the Scandinavian word "Ren-dyr," the word "dyr" signifying "beast," like the Greek θήρ, to which it is evidently allied.

The Scandinavians chase the animal through the deep snow, and succeed in the chase, because they are mounted upon skidor, or wooden skates several feet in length, which glide over the surface of the frozen snow, while the great weight of the Elk forces its legs through the thin crust, and causes it to tire in a short time. Like the reindeer, the Elk makes a great clattering with its hoofs when in rapid motion. It is a good swimmer, and is fond of taking to the water in summer time. It is a rather dangerous antagonist when incensed, as it fights desperately with its horns and hoofs, with which latter weapons it has been known to destroy a wolf with a single stroke. There used to be a curious belief about the foot of the Elk, namely, that it is a preservation against epilepsy and other diseases.

ALCES.—(Gr. Ἄλκη, an Elk.)

Palmātus (Lat. *palmed*), *the Elk.*

The HORSE.—We now arrive at the Pachydermata, or thick-skinned animals, which do not chew the cud. The first on the list is the Horse, an animal too well known in all its varieties to need much description. The ancient war-horse, so magnificently described in the Book of Job, is well represented by that most wonderful head in the British Museum, a fragment from the Temple of Minerva at Athens. The ancients never appeared to ride on the horse to battle, but fought from small open chariots, to which two or more horses were harnessed.

The Arabian Horse is a model of elegance and beauty. The Arab treats his horse as one of his family; it lives in the same tent with him, eats from his hand, and sleeps among his children, who tumble about on it without the least fear. Few Arabs can be induced to part with a favourite horse. The Rev. V. Monro relates that an Arab, "the net value of whose dress and accoutrements might be calculated at something under seventeen pence halfpenny," refused all offers made to purchase a beautiful mare on which he rode, and declared that he loved the animal better than his own life.

The plains of La Plata and Paraguay are tenanted by vast herds of wild horses. These are captured by the lasso, bitted, mounted, and broken, within an hour, by the daring and skilful Gauchos.

The ponderous and powerful dray-horse is of the Flanders breed. These huge animals, as they slowly pace along the streets, conducted by men

Family II. Equĭdæ.—(Lat. *Equus*, a Horse. Horse kind.)

EQUUS.

Caballus (Lat. *a Saddle-horse.*)

who seem to be a Flanders race also, never fail to attract the attention of admiring foreigners.

Wales and the Shetland Isles produce a breed forming a great contrast to the Flanders horse. The Sheltie, as it is called, is very small, its height sometimes being only thirty-four inches; but it is very strong and sure-footed, carrying its rider with perfect safety along the most terrific precipices, and almost invariably choosing to walk on the very edge.

The Race Horse is supposed to have been originally derived from the Arabian breed. The Godolphin Arabian, and the Flying Childers, are two of the most celebrated racers. The skeleton of Eclipse, another celebrated racer, is now in the Ashmolean Museum, Oxford.

The Ass.—The humble and hardy Ass is scarcely less serviceable to man than the more imposing horse. In this country, where it meets with harsh treatment, is scantily fed, and only used for laborious tasks, it is dull and obstinate; but in the East, where it is employed by the rich nobles, and is properly treated, it is an elegant and spirited animal, with good action and smooth coat. White asses are always used in the East for the especial service of bearing persons of distinction, a custom of great antiquity, as appears from Judges v. 10,—"Speak, ye that ride on white asses."

Asĭnus.—(Lat. *an Ass.*)

Vulgāris (Lat. *common*), *the Ass.*

The Dziggueтai.—In Persia and other countries, there are herds of wild asses. They are so fleet that no horses can come up to them, and even with rifles the chase is very uncertain. The Persians esteem its flesh very highly, considering it one of their greatest delicacies. Sir R. Ker Porter gives the following amusing account of an unsuccessful chase after a wild ass:—

"The sun was just rising over the summits of the eastern mountains, when my greyhound started off in pursuit of an animal which my Persians said, from the glimpse they had of it, was an antelope. I instantly put spurs to my horse, and with my attendants gave chase. After an unrelaxed gallop of three miles, we came up with the dog, who was then within a short stretch of the creature he pursued, and to my surprise, and at first vexation, I saw it to be an ass. Upon a moment's reflection, however, judging from its fleetness it must be a wild one, a creature little known in Europe, but which the Persians prize above all other animals as an object of chase, I determined to approach as near to it as the very swift Arab I was on would carry me. But the single instant of checking my horse to consider had given our game such a head of us, that notwithstanding all our speed we could not recover our ground on him. I, however, happened to be considerably before my companions, when, at a certain distance, the animal in its turn made a pause, and allowed me to approach within pistol-shot of him; he then darted off again with the quickness of thought, capering, kicking, and sporting in his flight, as if he was not blown in the least, and the chase was his pastime. When my followers of the country came up, they regretted that I had not shot the creature when he was within my aim, telling me that his flesh is one of the greatest delicacies in Persia. The prodigious swiftness and peculiar manner in which he fled across the plain coincided exactly with the description that Xenophon gives of the same animal in Arabia. But above all, it reminded me of the striking portrait drawn by the author of the Book of Job. I was informed by the Mehmender, who had been in the desert when making a pilgrimage to the shrine of Ali, that the wild ass of Irak Arabi differs in nothing from the one I had just seen. He had observed them often for a short time in the possession of the Arabs, who told him the creature was perfectly untameable. A few days after this discussion, we saw another of these animals, and, pursuing it determinedly, had the good fortune to kill it." This animal, called the Dziggetai, is also found in India, and is quite as difficult to secure as its relations in Persia.

ASINUS.

Dzigguetai (*Native name*), *the Dzigguetai.*

From the skin of the Dzigguetai is made the best shagreen.

There is a mixed breed between the Horse and the Ass, called the Mule, an animal in no very great request in this country, but extensively used in the East for riding, and in Spain it is the established beast of burden. It is very sure-footed, and is on that account employed in the Andes instead of the Llama.

The ZEBRA is found in South Africa. This beautiful animal lives in troops among the mountains, shunning the presence of man. It is a very conspicuous animal, and easily distinguished by the regular stripes of brownish black with which its whole body is covered even down to the hoofs. It is very wild and suspicious, carefully placing sentinels to look out for danger. Notwithstanding these precautions, several zebras have been taken alive, and some, in spite of their vicious habits, have been trained to draw a carriage. In all probability it might be domesticated like the ass, as the black cross on the back and shoulders of the latter

ASĬNUS.

Zebra, *the Zebra.*

animal prove the affinity between them. The voice of the Zebra is very peculiar, and can hardly be described.

ASĬNUS.

Quagga (*Native name*), *the Quagga.*

The QUAGGA is also a native of South Africa. It bears some resemblance to the Zebra, but is at once distinguished from that animal by the paucity and dulness of the stripes, which do not reach to the hind quarters or legs at all, and only faintly mark the back, its head and neck bearing the deepest stripes. It is not formed quite so gracefully as the zebra, — its hind quarters being slightly higher than its shoulders. The natives occasionally tame it for the purposes of draught, but it is not to be depended on, being vicious and very wild.

Family III. . Elephantĭdæ.—(Gr. 'Ελέφας, an Elephant. Elephant kind.)
Sub-family *a. Elephantīna.*

ELĔPHAS.

Indĭcus (Lat. *Indian*), *the Indian Elephant.*

Of this magnificent animal, whose form is familiar to every eye, two species are known, the Indian and the African. The anatomy of this huge quadruped is well worthy of consideration. Its head and tusks are so very heavy that no long neck would bear them; the neck is therefore very short. But this shortness of neck prevents the ELEPHANT from putting its head to the ground, or from stooping to the water's edge. This apparent defect is compensated by the wonderful manner in which its upper lip and nose are elongated, and rendered capable of drawing up water or plucking grass. In the proboscis or trunk there are about forty thousand muscles, enabling the Elephant to shorten, lengthen, coil up, or move in any direction this most extraordinary organ. The trunk is pierced throughout its length by two canals, through which liquids can be drawn by suction. If the Elephant wishes to drink, after drawing the liquid into its trunk, it inserts the end of the proboscis into its mouth, and discharges the contents down its throat; but if it merely wishes to wash itself or play, it blows the contained liquid from the trunk with great violence. Through the trunk the curious trumpet-like voice of the Elephant is produced. At the extremity is a finger-like appendage, with which it can pick up small objects. In order to sustain the muscles of the jaw and neck, the head must be very large: were it solid, it would be very heavy. The skull is therefore formed of a number of cells of bone, forming the necessary expanse without the weight, leaving but a very small cavity for the brain.

This fact will account for the numberless bullet wounds which an elephant will endure in the skull. The ball, instead of penetrating to the brain, merely lodges among the bony cells, and does no great mischief. Not long since, a ball was found firmly imbedded in the tusk of an elephant; it was thoroughly impacted, and there was no apparent opening by which it could have reached the place that it occupied. It was afterwards found that the ball must have struck the elephant at the base of the tusk, so as to have sunk among the soft and as yet unformed ivory. This by degrees was pushed on as the tusk grew in successive years, until it was at last surrounded closely by hard ivory. A spear-head has been also found similarly imbedded.

The Indian Elephant is almost invariably taken from its native haunts and then trained. The Indian hunters proceed into the woods with two trained female elephants. These advance quietly, and by their blandishments so occupy the attention of any unfortunate male that they meet, that the hunters are enabled to tie his legs together and fasten him to a tree. His treacherous companions now leave him to struggle in impotent rage, until he is so subdued by hunger and fatigue that the hunters can drive him home between their two tame elephants. When once captured he is easily trained. Bribes of sugar and arrack, a kind of spirit, are the usual means of inducing an Elephant to attempt some new art, or to labour with particular assiduity. In its wild state it endeavours to gratify its taste for sweets, at the expense of the sugar planters.

"The Elephant has a natural partiality for sugar, which he finds abundant means to gratify in the plantations of sugar-cane. A curious instance is recorded of his liking for sweetmeats, and of a method adopted in his savage state to gratify this propensity. It chanced that a Cooley, laden with jaggery, which is a coarse preparation of sugar, was surprised in a narrow pass in the kingdom of Candy by a wild elephant. The poor fellow, intent upon saving his life, threw down the burthen, which the elephant devoured, and being well pleased with the repast, determined not to allow any person egress or ingress who did not provide him with a similar banquet. The pass formed one of the principal thoroughfares to the capital, and the elephant, taking up a formidable position at the entrance, obliged every passenger to pay tribute. It soon became generally known that a donation of jaggery would ensure a safe conduct through the guarded portal, and no one presumed to attempt the passage without the expected offering."

It has before been mentioned that the Indian elephant is trained for tiger hunting. When the tiger springs, the elephant always raises his proboscis out of reach of the tiger's claws and teeth.

In captivity, it is very docile and gentle, but sometimes, when provoked, will take a very ample revenge. Of this propensity, many anecdotes are told.

"A very characteristic action of D'Jeck, the famous elephant of M. Huguet, was lately near costing the life of a young man, a native of Bruges. The elephant, it is well known, is very fond of sweetmeats, and this young man amused himself at Madame D'Jeck's expense, baulking her by offering her some, which, whenever she reached out her trunk to take, he immediately withdrew. This trick having been noticed by M. Huguet, he observed to the young man how foolish such conduct was towards an animal at once so susceptible and vindictive. But not taking warning from this remark, the Belgian again invited the elephant to approach, and not only again deceived her, but gave the sweetmeats to Mademoiselle Betsy. Madame D'Jeck now lost her patience, and regardless of the presence of her master and a numerous assemblage of spectators, lifted her trunk and knocked the young man down, tearing open his cheek, and rending his clothes to tatters. Happily, M. Huguet interposed his authority, and the elephant left her hold, but the imprudent sufferer was long confined to his bed from the effects of his absurdity."

The tusks and teeth of the Elephant furnish exceedingly fine ivory, which is used for various purposes, such as knife-handles, combs, billiard balls, &c. There is a great art in making a billiard ball. Some parts of the tusk are always heavier than others, so that if the heavy part should fall on one side of the ball, it would not run true. The object of the maker is either to get the heavier portion in the centre, or to make the ball from a piece of ivory of equal weight. In either case, the ball is made a little larger than the proper size; it is then hung up in a dry room for several months, and finally turned down to the requisite dimensions.

It is rather singular that the substance of ivory is almost exactly imitated by a vegetable production, the ivory nut, just about half the size of a billiard ball, which when young is soft, but hardens as it becomes old, until it is as hard as real ivory, and closely resembles it.

All elephants are fond of the water, and sometimes submerge themselves so far, that nothing but the tip of the proboscis remains above the surface. In a tame state, the elephant delights in concealing itself below the water, and deluging the spectators with a stream sent from its trunk.

The following account of Elephant catching in Nepâl was sent to me by a medical gentleman residing at Segouly :—

"The whole batch, tame and wild ones, then rushed into a deep river close by, where it was a splendid sight to see them swimming, fighting, diving, plunging, kicking, and bellowing in a most frantic manner; the mahouts (the riders on the tame ones) sticking to them like monkeys, and dexterously taking the opportunity of the confusion to secure the dreaded noose round their necks.

"One of the wild elephants in the struggle got half-drowned, and then entirely strangled; she just staggered to the shore, and then dropped dead without a struggle. It was really quite piteous to see her poor little young one, about ten days old; she kept walking round the body. pushing it, and trying to coax her dead mother to rise up; then uttering the most heart-rending cries, and lying down by her side as it were to comfort her.

"When the contest was over, and the other elephants, tame ones, were brought up near the corpse, the poor little thing, with the most indignant, though, of course, unavailing valour, charged on all sides at any elephant who came near, determined evidently to defend its mother, even though dead, to the last. The tame ones, of course, were too sagacious to hurt it with their tusks, and looked on with the most curious air of pity and contempt, as they gradually, despite its violent struggles, pushed it away from its mother to a place where it could be properly secured and taken care of. Really its moans and endeavours to remain with its mother were quite affecting. It is too young to be weaned with safety, and will probably die; at least I am very much afraid so. I shall always feel an interest in the poor little animal in future, should it live; it was so devotedly and heroically brave, never attempting to leave its mother in order to procure its own escape, which it might easily have done unseen during the confusion."

On this occasion Jung Bahadoor, the Nepaulese ambassador, distinguished himself greatly by his dexterity and courage, and secured several elephants with his own hands.

The African Elephant.—This species is distinguished from the Indian Elephant by the markings of its teeth and some differences in form. Much interesting information respecting the habits of this animal has been given by Cumming, from whose work the following extracts are taken:—

"The African Elephant is widely diffused through the vast forests, and is met with in herds of various numbers. The male is very much larger than the female; consequently, much more difficult to kill. He is provided with two enormous tusks. These are long, tapering, and beautifully arched; their length averages from six to eight feet, and they weigh from sixty to a hundred pounds each.

"The females, unlike Asiatic elephants in this respect, are likewise provided with tusks. The price which the largest ivory fetches in the English market is from 28*l.* to 32*l.* per hundred-and-twelve pounds. Old bull elephants are found singly or in pairs, or consorting together in small herds, varying from six to twenty individuals. The younger bulls remain for many years in the company of their mothers, and these are met together in large herds of from twenty to a hundred individuals. The food of the Elephant consists of the branches, leaves, and roots of trees, and also of a variety of bulbs, of the situation of which he is advised by his exquisite sense of smell. To obtain these he turns up the ground with his tusks, and whole acres may be seen thus ploughed up. Elephants consume an immense quantity of food, and pass the greater part of the day and night in feeding. Like the whale in the ocean, the Elephant on land is acquainted with, and roams over, wide and extensive tracts. He is extremely particular in always frequenting the freshest and

ELEPHAS.

Africānus (Lat. *African*), *the African Elephant.*

most verdant districts of the forests; and when one district is parched and barren, he will forsake it for years, and wander to great distances in quest of better pasture.

"The Elephant entertains an extraordinary horror of man, and a child can put a hundred of them to flight by passing at a quarter of a mile to windward; and when thus disturbed, they go a long way before they halt. It is surprising how soon these sagacious animals are aware of the presence of a hunter in their domains. When one troop has been attacked, all the other elephants frequenting the district are aware of the fact within two or three days, when they all forsake it, and migrate to distant parts."

"They choose for their resort the most lonely and secluded depths of the forest, generally at a very great distance from the rivers and fountains at which they drink. In dry and warm weather they visit these waters nightly; but in cool and cloudy weather they drink only once every third or fourth day. About sundown the elephant leaves his distant midday haunt, and commences his march towards the fountain, which is probably from twelve to twenty miles distant. This he generally reaches between the hours of nine and midnight; when, having slaked his thirst and cooled his body by spouting large volumes of water over his back with his trunk, he resumes the path to his forest solitudes. Having reached a secluded spot, I have remarked that full-grown bulls lie down on their broadsides, about the hour of midnight, and sleep for a few hours. The spot which they usually select is an ant-hill, and they lie around it with their backs resting against it; these hills, formed by the white ants, are from thirty to forty feet in diameter at their base. The mark of the under tusk is always deeply imprinted in the ground, proving that they lie upon their sides.

"The appearance of the wild elephant is inconceivably majestic and imposing. His gigantic height and colossal bulk, so greatly surpassing all other quadrupeds, combined with his sagacious disposition and peculiar habits, impart to him an interest in the eyes of the hunter which no other animal can call forth. The pace of the Elephant when undisturbed is a bold, free, sweeping step; and from the peculiar spongy formation of his foot, his tread is extremely light and inaudible, and all his movements are attended with a peculiar gentleness and grace.

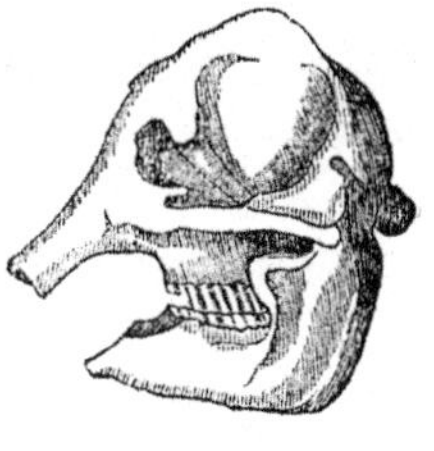

SKULL OF THE ELEPHANT.

"The under skin is of a tough and pliant nature, and is used by the natives for making water bags, in which they convey supplies of water from the nearest vley or fountain (which is often ten miles distant). They remove this inner skin with caution, taking care not to cut it with the assagai; and it is formed into water bags by gathering the corners and edges, and transfixing the whole on a pointed wand."

The TAPIR forms one of the links connecting the elephant with the hog. The snout is lengthened into a kind of proboscis like that of the elephant, but it is comparatively short, and has no finger-like appendage at the extremity. Many of the remaining links are supplied by the various species of the fossil genus Palæotherium.

The Common Tapir is spread throughout the warmer regions of South America. It sleeps during the day, and wanders about at night in search of its food, which consists of water melons, gourds, and other vegetables. It is very fond of the water, and can remain below the surface for a considerable period. It is a very powerful animal, and as it is furnished with a very thick hide, it plunges through the brushwood, breaking its way through any obstacles that may oppose its progress.

Its disposition is gentle, but when annoyed, it sometimes rushes at its antagonist, and defends itself vigorously with its powerful teeth.

Sub-family *b*. *Tapirina.*

TAPIRUS.—(From Native name.)

Malayānus (Lat. *Malay.*)

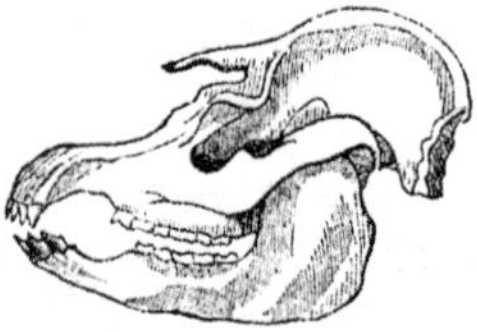

SKULL OF THE TAPIR.

The jaguar frequently springs on it, but is often dislodged by the activity of the Tapir, who rushes through the bushes immediately that its feels the claws of its enemy, and endeavours to brush him off against the thick branches. The height of the American Tapir is from five to six feet. The Malay Tapir is somewhat larger, and is known by the greyish white colour of the loins and hind quarters, which give the animal an appearance as if a white horsecloth had been spread over it.

The BOAR.—The animals composing the Hog tribe are found in almost every part of the globe. Their feet are cloven and externally resemble those of the Ruminants, but an examination of the bones at once points out the difference.

The Wild Hog or Boar inhabits many parts of Europe, especially the forests of Germany, where the chase of the wild boar is a common amusement. It has become extinct in this country for many years. Its tusks are terrible weapons, and capable of being used with fatal effect. They curve outwards from the lower jaw, and are sometimes eight or ten inches in length. In India, where the Boar attains to a great size, the horses on which the hunters are mounted often refuse to bring their

Sub-family c. *Suina.*

Sus.—(Lat. *a Sow.*)

Scrofa (Lat. *an old Sow*), *the Boar.*

riders within spear stroke of the infuriated animal, who has been known to kill a horse and severely injure the rider with one sweep of its enormous tusks.

The Domestic Hog scarcely needs any description. It is by no means the unclean and filthy animal that moralists love to represent it. It certainly is fond of wallowing in the mire, as are the elephant, tapirs, &c., but no animal seems to enjoy clean straw more than the Hog. We shut it up in a dirty narrow crib, give it any kind of refuse to eat, and then abuse it for being a dirty animal and an unclean feeder. While, however, it should be rescued from these unjust imputations, it should bear the weight of an accusation never before made. I have seen pigs suck the cows in a farmyard while they were lying down and chewing the cud, nor did the cows attempt to repel them.

SUS.

Babyroussa (Native word, *Hog-deer*), *the Babyroussa.*

The Babyroussa inhabits the Molucca Islands and Java. It is remarkable for possessing four tusks, two of which proceed from the upper jaw, and do not pass out between the lips, but through an aperture in the skin, half way between the end of the snout and eyes. The sockets of the two upper tusks are curved upwards, and give a singular appearance to the skull of the animal. It looks a ferocious animal, nor do its looks contradict its habits, as it is very savage, and cannot be hunted without danger. Yet when taken young it can be tamed without much difficulty, and conducts itself much after the manner of a well-behaved pig.

Only the male possesses the remarkable double pair of tusks, the female being destitute of the upper pair, and only possessing those belonging to the under jaw in a rudimentary degree. It lives in troops, as do most of the hog kind, and thus does great damage to the cultivated grounds, especially to the maize, a plant to which it is, unfortunately, very partial. It is a good swimmer, and often voluntarily takes to the water in order to cross to another island. The size of the animal when full grown, is about that of a very large hog.

The Common, or Collared Peccary, is an inhabitant of South America. This animal is both dreaded and hated by the residents, for it is so exceedingly ferocious, and so utterly devoid of all sense of fear, that it will always charge at any object that comes in its way; an elephant would not scare it, if an elephant were to be transported to South America. So it puts to flight those whom it attacks, and they fly before it in mixed fear and wrath against the pugnacious little animals which are pursuing them. It is a small animal, rarely exceeding eighteen inches in height, and yet is not less dreaded than the most savage wild boar would be. Its jaws are armed with tusks, like those of the boar, but they are straight instead of curved, are sharp at the edges, and although only about an inch and a half in length, inflict horrible wounds, on account of the muscular strength of the creature's neck. When a body of them charge against an enemy, fancied or real, they will never be driven away, but will fight till the last is slain. On this account, no one will willingly oppose them; and if a herd of Peccaries comes in the way, men, horses, and dogs, all fly in haste, as even the horses would be soon brought down, for their legs would be cut to pieces. The best method of attacking them is that described by Webber in the following passage:—

DICOTYLES.—(Gr. δίς, double; κοτύλη, a hollow cup)*

Tajaçu (Native name), *the Peccary.*

"But with all its other peculiarities to answer for, the drollest is yet

* In allusion to the hollow gland on the back.

to come. I refer to their mode of sleeping. They usually frequent those heavy canebrakes, through which are scattered, at wide intervals, trees of enormous size and age. These, from their isolated condition, are most exposed to the fury of storms, and therefore most liable to be thrown down. We find their giant stems stretched here and there through the canebrakes of Texas, overgrown with the densest thickets of the cane, matted together by strong and thorny vines. In these old trees the peccaries find their favourite lodgings. Into one of these logs a drove of twenty or thirty of them will enter at night, each one backing in, so that the last one entering stands with his nose at the entrance. The planters, who dread them and hate them, as well on account of the ravages on their grain crops which they commit, the frequent destruction or mutilation by them of their stock—their favourite dogs, and sometimes even their horses, as on account of the ridiculous predicaments, such as taking to a tree, or running for their life, to which they have been subjected themselves, seek their destruction with the greatest eagerness.

"When a hollow log has been found which bears the marks of being used by them, the hunters wait with great impatience till the first dark, cloudy day of rain; a dark drizzle is the best, as it is well known that on such days they do not leave their lodgings at all. The planter, concealing himself just before day carefully out of view, but directly in front of the opening of the log, awaits in patient silence the coming of sufficient light. Soon as the day opens, peering cautiously through the cane, he can perceive the protruded snout, and sharp, watchful eyes of the sentinel-peccary on duty, while his fellows behind him sleep. Noiselessly the unerring rifle is raised, the ring of its explosion is heard, and, with a convulsive spring, the sentinel leaps forward out of the hole, and rolls in its death-struggle on the ground. Scarcely an instant is passed, a low grunt is heard, and another pair of eyes is seen shining steadily in the place the others had just held. Not a sound is heard, the planter loads again with such dexterity that not even a branch of the embowering cane is stirred. Again with steady nerve the piece is fired, out springs the second victim as the first had done; then another takes its place, and so on to the third, fourth, fifth, and twentieth, even to the last of the herd, unless the planter should happen by some carelessness to make a stir in the cane around him, when out it springs with a short grunt, without waiting to be shot this time, and followed by the whole herd, when they make a dash at the unlucky sportsman, who is now glad enough to take to his heels, and blesses his stars if he should be able to climb a tree or a fence in time to save his legs. If during the firing, the sentinel should happen to sink in the hole without making the usual spring, the one behind him roots out the body to take its place. They do not understand what the danger is, or whence it comes. Neither do they fear it, but face its mysterious power to the last. They never charge towards unseen enemies, until guided either by the sight of some disturbance caused by a motion in the thicket, or by those sounds with which they are familiar, indicating their position. Incredible as this account may appear, it is actually the method in which the settlements along Caney Creek and in the Brazos Bottoms have been of late years in a great measure relieved of this dangerous annoyance."

The Peccary alone of all animals appears to have resisted the terrors of the gun, and a herd of them will attack men with fire-arms, and only seem to be more enraged by the report and flash of the guns. The Indians eat the animal, but its flesh is not considered to be particularly excellent, especially as the gland which the animal bears in its haunches has an evil effect on the meat, and causes it to become unfit for use in a very short time. Its colour is a greyish black, caused by the colour of the bristles, which are ringed at intervals with grey, straw-colour, and black.

Sub-family *d. Rhinocerina.*

RHINOCĔROS.—(Gr. Ῥίν, or ῥίς, a nose; *κέρας,* a horn.)

Unicornis (Lat. *Unus,* one; *cornu,* a horn), *the Rhinoceros.*

The RHINOCEROS.—There are, apparently, six species of this formidable animal. Their chief peculiarity, the so-called horn, is a mass of fibres matted together, and closely resembling the fibres of whalebone. Their feet are divided into three toes, incased in hoofs. The best description of the various species of the African Rhinoceros is given in Cumming.

"Of the Rhinoceros there are four varieties in South Africa, distinguished by the Bechuanas by the names of the 'borèle,' or black rhinoceros, the 'keitloa,' or two-horned black rhinoceros, the 'muchocho,' or common white rhinoceros, and the 'kobaoba,' or long-horned white rhinoceros. Both varieties of the black rhinoceros are extremely fierce and dangerous, and rush headlong and unprovoked at any object which attracts their attention. They never attain much fat, and their flesh is tough, and not much esteemed even by the Bechuanas. Their food

RHINOCĔROS.

Bicornis (Lat. *bis*, twice; *cornu*, a horn), *the Two-horned Rhinoceros, or Rhinaster.*

consists almost entirely of the thorny branches of the wait-a-bit* thorns. Their horns are much shorter than those of the other varieties, seldom exceeding eighteen inches in length. They are finely polished with constant rubbing against the trees. The skull is remarkably formed, its most striking feature being the tremendous thick ossification in which it ends above the nostrils. It is on this mass that the horn is supported. The horns are not connected with the skull, being attached merely by the skin, and they may thus be separated from the head by means of a sharp knife. They are hard, and perfectly solid throughout, and are a fine material for various articles, such as drinking cups, mallets for rifles, handles for turners' tools, &c. &c. The horn is capable of a very high

* The reader must know that these are the thorns of a kind of acacia, which being several inches in length, very strong and very sharp, generally do cause any one who has fallen among them, to "wait a bit" until he finds himself free from them.

polish. The eyes of the rhinoceros are small and sparkling, and do not readily observe the hunter, provided he keep to leeward of them. The skin is extremely thick, and only to be penetrated by bullets hardened with solder. During the day, the rhinoceros will be found lying asleep, or standing indolently, in some retired part of the forest, or under the base of the mountains, sheltered from the power of the sun by some friendly grove of umbrella-topped mimosas. In the evening, they commence their nightly ramble, and wander over a great extent of country. They usually visit the fountains between the hours of nine and twelve o'clock at night, and it is on these occasions that they may be most successfully hunted, and with the least danger. The black rhinoceros is subject to paroxysms of unprovoked fury, often ploughing up the ground for several yards with its horn, and assaulting large bushes in the most violent manner. On these bushes they work for hours with their horns, at the same time snorting and blowing loudly; nor do they leave them in general until they have broken them into pieces. All the four varieties delight to roll and wallow in mud, with which their rugged hides are encrusted. Both varieties of the black rhinoceros are much smaller and more active than the white, and are so swift that a horse with a rider on its back can rarely overtake them. The two varieties of the white rhinoceros are so similar in habits, that the description of one will serve for both, the principal difference consisting in the length and set of the anterior horn; that of the common white rhinoceros averaging from two to three feet in length, and pointing backwards; while the horn of the long-horned white rhinoceros often exceeds four feet in length, and inclines forward from the nose. . . . .

"Both these varieties of rhinoceros attain an enormous size, being the animals next in magnitude to the elephant. They feed solely on grass, carry much fat, and their flesh is excellent, being preferable to beef. They are of a much milder and more inoffensive disposition than the black rhinoceros, rarely charging their pursuer. Their speed is very inferior to that of the other varieties, and a person well-mounted can overtake and shoot them."

Several travellers have mentioned that there are certain birds which constantly attend the rhinoceros, and give him warning of approaching danger. Their accounts were either received with silent contempt, or treated with open ridicule, as preposterous extensions of the traveller's privilege of romancing. Here, however, we find the same author who has just been quoted, corroborating in every respect these disbelieved tales. His description of these birds is so interesting that I give it entire.

"Before I could reach the proper distance to fire, several 'rhinoceros birds,' by which he was attended, warned him of his impending danger, by sticking their bills into his ear, and uttering their harsh, grating cry. Thus aroused, he suddenly sprang to his feet, and crashed away through the jungle at a rapid trot, and I saw no more of him. . . . .

"These rhinoceros birds are constant attendants upon the hippopotamus and the four varieties of rhinoceros, their object being to feed upon the ticks and other parasitic insects that swarm upon these animals. They are of a greyish colour, and are nearly as large as a common thrush: their voice is very similar to that of the misseltoe thrush. Many a time have these ever-watchful birds disappointed me in my stalk, and tempted me to invoke an anathema upon their devoted heads. They are the best friends the rhinoceros has, and rarely fail to awaken him, even in his soundest nap. 'Chukuroo' perfectly understands their warning, and, springing to his feet, he generally first looks about him in every direction, after which he invariably makes off."

The organs of scent of the rhinoceros are very acute, and as the creature seems to have a peculiar faculty for detecting the presence of human beings, it is necessary for the hunters to use the greatest circumspection when they approach it, whether to avoid or to kill, as in the one case it may probably be taken with a sudden fit of fury, and charge at them, or in the other case, it may take the alarm and escape.

The Indian Rhinoceros is chiefly remarkable for the very deep foldings of the skin, which is so hard at the folds, that an ordinary leaden ball will often fail to penetrate sufficiently to cause a deep wound, and the hunters are consequently forced to harden their bullets with tin or solder.

It may be here remarked, that the so-called "horn" is not a true horn, being nothing but a process from the skin, and composed of a vast assemblage of hairs. If a thin transverse section of the horn of a rhinoceros is placed under a microscope, it will be seen that it is composed of a series of minute tubes, exactly like those of hair. Indeed it is possible to make an artificial rhinoceros horn, by gluing together a bundle of hairs, and if a transverse section of this composition is placed under the microscope, it almost exactly resembles the section of the rhinoceros horn. Round the base of the horn, and wherever it has been injured, the broken ends of the hairs stand out separately, and can be cut off.

The contempt with which the hunters, at all events the British hunters, appear to treat the rhinoceros is very amusing. If they want the skin or the horn, why, then they shoot it; but if they do not stand in need of such articles, and a rhinoceros comes in their way, they are obliged to pelt it away, or drive it off as they would a cow.

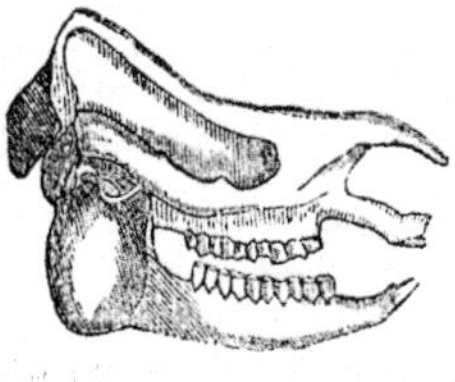
SKULL OF THE RHINOCEROS.

Goblets, made of the horn, were formerly in high estimation as perservatives against poison. The Indian kings were accustomed to have their wine served up in these goblets, as they imagined that if any poison were introduced into the cup, the liquid would boil over, and betray its presence.

The upper lip is used by the rhinoceros as an instrument of prehension, with which it can grasp the herbage on which it feeds, or pick up small fruit from the ground. The very tame rhinoceros in the Zoological Gardens will take a piece of bun or biscuit from a visitor's hand by means of its flexible upper lip.

The HIPPOPOTAMUS.—There is, in all probability, but one species of the Hippopotamus. It inhabits Africa exclusively, and is found in plenty

Sub-family *e. Hippopotamīna.*
HIPPOPOTĂMUS.—(Gr. Ἵππος, a Horse; ποταμός, a River.)

Amphibĭus (Gr. Ἀμφί, on both sides; βιόω, I live), *the Hippopotamus.*

on the banks of many rivers in that country, where it may be seen gambolling and snorting at all times of the day.

These animals are quiet and inoffensive while undisturbed, but if attacked, they unite to repel the invader, and have been known to tear several planks from the side of a boat, and sink it. They can remain about five or six minutes under water, and when they emerge they make a loud and very peculiar snorting noise, which can be heard at a great distance.

The hide is very thick and strong, and is chiefly used for whips. The well-known "Cow-hides" are made of this material. Between the skin and flesh is a layer of fat, which is salted and eaten by the Dutch colonists of Southern Africa. When salted it is called Zee-koe speck, or Sea-cow's bacon. The flesh is also in some request.

The Hippopotamus feeds entirely on vegetable substances, such as grass and brushwood. The fine animal now in the possession of the Zoological Society eats all kinds of vegetables, not disdaining roots. This individual is peculiarly interesting from being the first Hippopotamus brought to Europe for many hundred years, and is in all probability the first that has ever reached this country.

A young Hippopotamus has been introduced into the Jardin des Plantes at Paris, but it is not so fine an animal as ours. Our specimen is fast losing its docility and gentle bearing, and is apparently likely to become a savage animal, which from its great strength will be troublesome to manage. There is rather a curious story respecting its capture, which is too long to be related here at length, but which will exhibit the character of eastern despots. Its actual capture was caused by a boathook run into its flank, the mark of which wound was very conspicuous for a year or two after its arrival in England, but which is gradually becoming fainter. It was very young when taken, and was fed on milk, of which article it consumed so much, that all the cows on its route had to be forced into its service, and the tax which it seemed disposed to levy on the milk of the cows, was hardly less oppressive than that laid by the ruler on the person and purses of their owners.

In Harris's Sports of South Africa, a very good and accurate account is given of the habits of the Hippopotamus.

"This animal abounds in the Limpopo, dividing the empire with its amphibious neighbour the crocodile. Throughout the night the unwieldy monsters might be heard snorting and blowing during their aquatic gambols, and we not unfrequently detected them in the act of sallying from their reed-grown coverts, to graze by the serene light of the moon; never, however, venturing to any distance from the river, the stronghold to which they betake themselves on the smallest alarm. Occasionally, during the day, they were to be seen basking on the shore, amid ooze and mud; but shots were most constantly to be had at their uncouth heads, when protruded from the water to draw breath; and, if killed, the body rose to the surface. Vulnerable only behind the ear, however, or the eye, which is placed in a prominence, so as to resemble the garret window of a Dutch house, they require the perfection of rifle practice, and after a few shots become exceedingly shy, exhibiting the snout only, and as instantly withdrawing it. The flesh is delicious, resembling pork in flavour, and abounding in fat, which in the colony is deservedly esteemed the greatest of delicacies. The hide is upwards of an inch and a half in thickness, and being scarcely flexible, may be dragged from the ribs in strips like the planks from a ship's side."

From the construction of the head, the animal is enabled to raise its eyes and nostrils above the water at the same time, so that it can survey the prospect and breathe without raising more than an inch or two of its person from the water. In order to attain this object, the eyes are very small, and placed very high in the head. while the muzzle is very large, and the nostrils open on its upper surface.

Cumming relates that the track of the Hippopotamus may be readily distinguished from that of any other animal by a line of unbroken herbage which is left between the marks of the feet on each side, as the width of the space between the right and left legs causes the animal to place its feet so considerably apart, as to make a distinct double track.

The teeth of the Hippopotamus are the mainstay of the dentist, who cuts from the tusk of a Hippopotamus those series of elegant teeth which replace those that age or accident has struck out of the human mouth. The ivory is exceedingly hard, and does not readily lose its beautiful whiteness, being properties which render it especially valuable for such purposes.

This is supposed by many to be the animal called Behēmoth in Scripture.

Family IV. Bradypĭdæ—(Gr. Βραδύς, slow; πούς, a foot.)

BRADЎPUS.

Tridactўlus (Gr. Τριδάκτυλος, three-fingered), *the Sloth.*

The Sloth.—The Edentata include the ant-eaters and the pangolins which possess no teeth at all, and the sloths, armadillos, &c., whose teeth are small and of peculiar structure.

The Sloths form the first division of the Edentata—the leaf-eaters.

The Sloth or Aï, is another example of the errors into which even great naturalists are led from hasty observation. The great Cuvier himself condemns the Sloth as a degraded and miserable animal, moving with pain, and misshapen in form. Yet no animal is more fitted for its position than the Sloth. "The Sloth," says Waterton, "in its wild state spends its whole life in the trees, and never leaves them but through force or accident, and what is more extraordinary, not upon the branches, like the squirrel and monkey, but under them. He moves suspended from the branch, he rests suspended from the branch, and he sleeps suspended from the branch." In fact, as Sydney Smith observes, he passes a life of suspense, like a young clergyman distantly related to a Bishop.

To render it fit for this singular mode of life, its long and powerful arms are furnished with strong curved claws, which hook round the branches, and keep the animal suspended without any effort. When on the ground, these claws are very inconvenient, and it can barely shuffle along; but when it is among its native branches, it moves with exceeding rapidity, particularly in a gale of wind, when it passes from branch to branch and from tree to tree with an activity which its movements on the ground by no means portend.

It is gifted with great tenacity of life, surviving under injuries which would have proved instantly mortal to any other animal. It even surpasses the opossum in endurance. Waterton gives an interesting account of a sloth which he kept in his house for some time. The animal usually lived on the back of a chair to which it slung itself by its curved claws. After keeping it for some time, he was desirous of killing it, as its skin was required for the purpose of stuffing, and the death warrant was issued against the sloth. But how to kill it was the difficulty: and its owner, being a naturalist and therefore a merciful man, in spite of popular prejudices on the subject, was much perplexed in his mind. At last he determined on trying the effect of the wourali poison, used by the Indians to give their weapons of war and the chase a more deadly effect. Even a sloth could not resist the wourali. A very small wound was made through the animal's skin, and inoculated with the poison. Soon the sloth began to droop, its head sunk upon one side, and after a few minutes one of its feet lost its hold of the chair on which it was hanging. The other foot soon gave way under the influence of the poison, and the dying animal fell to the ground. It lay there perfectly quiet, and after a few minutes had elapsed, gently closed its eyes and was dead. Its whole demeanour was that of an animal overcome with sleep, and it never appeared to suffer the slightest pain.

Such indeed seems to be the effect of this singular composition upon any living creature. If an animal is wounded, although slightly by a weapon charged with this poison, it runs a few paces, staggers, and lies down as if to sleep, and in a few minutes is dead. The effect is the same upon man. Two Indians were hunting after birds, and one of them had just launched a poisoned arrow at a bird nearly above him. The arrow missed its mark, glanced against a bough, and in its fall struck into the arm of the man who had thrown it. He looked at his arm, took off his quiver of arrows, remarked that he should never use them again, laid himself down, and was dead almost immediately.

Family V. . Dasypĭdæ.—(Gr. Δασύς, hairy; πούς, a foot. Hairy-footed.)
Sub-family *a. Manīna.*

MANIS (Native name).

Tetradactȳla (Gr. Τετραδάκτυλος, four-fingered), *the Phatăgin, or Long-tailed Manis.*

The MANIDÆ or PANGOLINS are immediately known by the peculiar, strong, horny plates with which their bodies are defended, giving them the appearance of animals enveloped in a suit of scale armour. When attacked, they roll themselves up, wrap their tails round them, and raise the whole array of sharp-edged scales with which their body is covered, and bid defiance to almost any enemy except man. They live on ants, and termites, or white ants, as they are called, which they take by thrusting their long slender tongue among the ants, which adhere to it by a gummy saliva. When the tongue is covered it is rapidly retracted, and the ants swallowed. To obtain the ants, the Pan-

MANIS.

Pentadactȳla (Gr. Πενταδάκτυλος, five-fingered), *the Short-tailed Manis.*

golins are furnished with powerful claws to tear down the dwellings of their prey.

The Long-tailed Manis is widely scattered through Africa, but is not very common. The length of its body is about two feet, and that of its tail rather more than three.

The Short-tailed Manis, or Bajjerkeit, is very common in India. Its entire length is about four feet.

The ARMADILLOS live exclusively in the warmer parts of America. They eat carrion, insects, and sometimes fallen fruit. The great mainstay of these animals lies in the number of bisons annually slaughtered for the sake of their hides. The carcases of these animals are left to rot on the plain, and would speedily do so did not the combined effect of birds and beasts, soon destroy every trace of the animal and only leave a heap of bones. In this work the Armadillo takes his full share.

Sub-family *b. Dasypīna.*

DASȲPUS.—(Gr. Δασύς, hairy; πούς, a foot.)

Sexcinctus (Lat. *six-banded*), *the Armadillo.*

The armour that covers them, instead of resembling scale armour like that of the Manis, forcibly reminds the observer of the modified plate armour worn in the time of Charles I. They burrow with great rapidity, and can only be forced from their refuge by smoke or water. When they are hunted and are very close pressed, they either endeavour to escape their foes by rapidly burrowing into the earth, or try to oppose a partial resistance by rolling themselves up and trusting to the protection of their armour. The natives and colonists consider them great delicacies when roasted in their shells.

The Armadillos are all small except the Gigantic Armadillo, which is well described in the following extract. "I found that an Armadillo of gigantic size had caused the commotion. It was lying a round, misshapen mass, its head partly buried under its armour, the feet drawn together, and its body pierced by numerous arrows. It offered not the slightest resistance to its tormentors, whom I desired to end its sufferings by a heavy stroke of a club. Two men were required to carry it, and Mr. Schomburgh estimated its weight at from 110 to 120 pounds; its height was about three feet, its length five and a half. Its tail was about fourteen or sixteen inches long, and its root nearly as thick as a man's thigh, tapering very abruptly. The middle one of the five toes of the fore foot was seven and a half inches in length. In size it greatly surpasses the largest Giant Armadillo known (*Dasypus giganteus, Desm.*), though Mr. Schomburgh does not mean to assert it is a different species from the giganteus; yet its enormous size will attract the attention of naturalists and geologists to the fossil genera."

Sub-family *c. Myrmecophagīna.*

MYRMECOPHĂGA.—(Gr. Μύρμηξ, an Ant; φαγεῖν, to eat.)

Jubāta (Lat. *crested*), *the Ant-eater.*

The ANT-EATER.—This curious animal inhabits Guiana, Brazil, and Paraguay. As its name imports, it lives principally upon ants and termites, which it procures in precisely the same manner as was related of the Manis. Its short legs and long claws would lead an observer to suppose that its pace was slow and constrained, but when chased, it runs off with a peculiar trot, and with such rapidity, that it keeps a horse to its speed to overtake it.

The tongue of this animal looks exactly like a great red worm, and when the creature is engaged in devouring its food, the rapid coiling and twisting of the tongue add in no small degree to the resemblance.

The claws are very long and curved, and as they are used in tearing down the habitation of the termites or white ants, as they are called, are exceeding strong. They are placed on the foot in such a manner that when the animal is walking, its weight rests on the outside of the fore feet and the outer edge of the claws, which make a great clattering if the ant-eater is walking upon a hard surface.

When it sleeps, it lies on one side, rolls itself up, so that its snout rests on its breast, places all its feet together, and covers itself with its bushy tail. The fur of the animal at all times resembles hay, and when it is thus curled up in sleep, it is so exactly like a bundle of hay, that any one might pass it carelessly, imagining it to be nothing but a loose heap of that substance.

Schomburgh relates that a tame Ant-eater, in his possession, by no means restricted itself to ants, but devoured meat, when minced, with much avidity. The same naturalist also discovered a Julus, or Millipede, in the stomach of an ant-eater which he dissected. The ordinary length of this animal is about three feet seven inches, and its height about three feet.

TAMANDŬA (Native name).

Tetradactyla (Gr. four-fingered), *the Tamandua.*

The MIDDLE ANT-EATER, or Tamandua, is not so large as the preceding animal, from which it is readily distinguished by the tail, which is long and tapering, and almost devoid of hair except at the base. The tail indeed is used as an organ of prehension, to assist it in climbing trees, a feat which it sometimes performs, although not so often as the Little Ant-Eater.

This animal produces a strong scent of musk, which is generally excited when it is enraged. The scent is not pleasant, like that of the musk deer, but very disagreeable, and can be perceived at a considerable distance.

The LITTLE ANT-EATER also inhabits Guiana and Brazil. The principal characteristics of this animal are the shortness of its muzzle, and the prehensile power of its tail, which it twists round the branches on which it principally resides. It often attacks the nests of wasps, pulling them to pieces with its claws, and devouring the grubs. The length of its body is ten inches.

CYCLOTHŪRUS (Gr. *κύκλος, a circle, θύρα, a door, or boards put together like a door*).*

Didactўla (Gr. *Διδάκτυλος*, two-fingered), *the Little Ant-eater.*

The skeleton of this little animal presents a very singular appearance, as its ribs are very flat and very wide in proportion to the size of the creature, and envelop each other in such a manner that the interior of the thorax is protected by a bony mail, no interstices being perceptible between the ribs.

The DUCK-BILLED PLATYPUS.†—Australia, where everything seems to be reversed, where the north wind is warm and the south wind cold, the thick end of a pear is next the stem, and the stone of a cherry grows outside, is the residence of this most extraordinary animal. When it was first introduced into Europe, it was fully believed to be the manufacture of some impostor, who with much ingenuity had fixed the beak of a duck into the head of some unknown animal. It will however be seen by the woodcut representing the skull of the animal, that this duck-like beak really belongs to the animal, and is caused by a prolongation of some of the bones of the head.

Sub-family *d. Ornithorhynchīna.*

ORNITHORHYNCUS.—(Gr. *Ὄρνις*, a bird; *ῥύγχος*, a snout.)

Paradoxus (Lat. *puzzling*), *the Duck-billed Platypus.*

It lives by the banks of rivers, in which it burrows like the water

* In allusion to the form of the ribs.

† The word "Platypus" signifies broad-footed, and is derived from the Greek words, πλατύς, broad, and ποῦς, a foot.

rat. Curiously enough, it finds no difficulty in this labour, although its house is always very deep, for the feet are so constructed that the animal can fold back the web at pleasure, and thus the foot is enabled to perform its task. It feeds upon water-insects and shell-fish, always rejecting the crushed shells after swallowing the inhabitant.

Mr. Bennet attempted to rear some young ORNITHORHYNCI at Sydney, but they died in a short time. They were very fond of climbing between a press and the wall, placing their backs against the press and their feet against the wall. They used to dress their fur with their beak and feet, just as a duck prunes its feathers.

The male has a sharp spur on its hind feet.

The learned have given the animal several names. Some follow Shaw, and call it Platypus Anatinus; some give it the name of Ornithorhyncus rufus or fuscus, or crispus or brevirostris, with other titles. The native name for the creature is "Mullingong," a title which, although not euphonious, is perhaps little less so than the scientific names, while it certainly has the advantage over them in point of brevity.

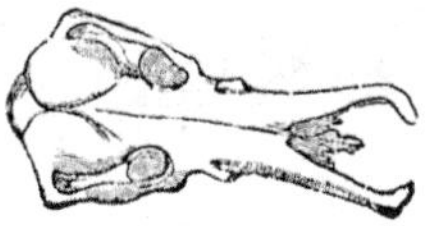

SKULL OF THE PLATYPUS.

Division II. AVES.—(Lat. *Birds.*)
Order I. . *ACCĬPITRES.*—(Lat. *Hawks.*)
Sub-order I. Accipĭtres diurni.—(Lat. *Hawks of the day.*)
Family I. . Gypaëtĭdæ.—(Gr. Γύψ, a Vulture; Ἀετός, an Eagle. Vulture-eagle kind.)

GYPAĔTUS.

Barbātus (Lat. *bearded*), *the Bearded Vulture or Lämmergeyer.*

Birds are immediately distinguished from the Mammalia by their general form, their feathery covering, and by producing their young enclosed in eggs.

The different orders of birds are principally known by the character of the claws and beak, examples of which will be seen in the progress of the work. Before we pay attention to any individual species, we will first examine some of the structures common to all birds.

One of the first great marks of distinction in birds is the wing. This organ is a modification of the arm or forelimb of mammalia, clothed with feathers instead of hair.

The bones of adult birds are not filled with marrow like the bones of mammalia, but are hollow and filled with air, and are therefore rendered very light, a bone of a goose being barely half the weight of a rabbit's bone of the same size, after the marrow has been extracted. In this formation, strength as well as lightness is consulted, as a tubular rod is well known to be very much stronger than the same quantity of matter formed into a solid bar. The bones forming the wing are worthy of notice for the beautiful manner in which they are jointed together, and arranged so as to give great strength together with lightness. Unscientific individuals are apt to make certain mistakes in their ideas of birds, and especially as regards the formation of their legs. Most persons seem to fancy that the foot of the bird is that part which grasps the branch, or by means of which it walks on the ground—that the joint above that member is the knee—and that the thigh is the feathered portion of the limb that proceeds from the bird's body. Now, all these ideas are wrong; with this method of arrangement, the knee of the bird would bend backwards, a thing which no perfectly formed knee ever did or ever will do.

The leg of a bird is formed on much the same principle as the hind leg of a quadruped, the part that grasps the branches being composed of the *toes*, the so-called knee-joint being the heel-bone of the foot, so that the whole foot reaches half-way from the perch to the bird. The knee-joint is placed high up against the body, and is buried in the feathers. In the following figure, the limbs of a man and of a bird are compared, the corresponding divisions of each being marked by similar letters.

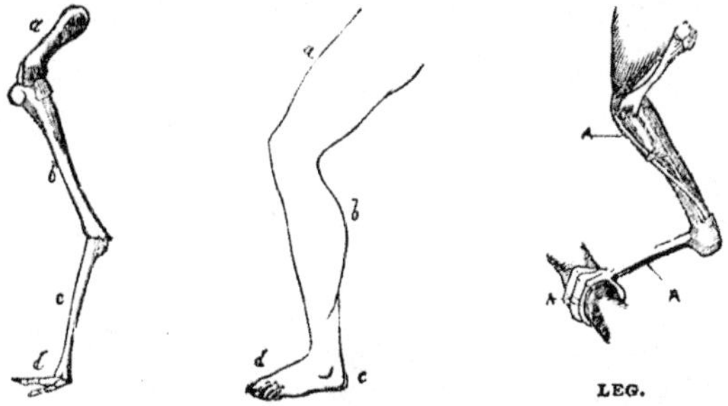

LEG.

As many important characteristics are drawn from the plumage, it will be necessary to give a figure, (p. 166,) exhibiting the feathers of the different parts, together with their names:—A, primaries, or great quill feathers of the wing; B, secondaries; C, tertials; D, lesser coverts; E, greater coverts; F, winglet, or bastard wing; G, scapularies; H, upper tail coverts; I, under tail coverts; K, rectrices or tail-feathers.

In the above engraving is also a figure, showing the method by which birds hold on the perch while sleeping. It will be seen that the great tendon A, which is connected with all the toes or claws, passes over the joints in such a manner that when the leg is bent, the tendon is shortened and the claws drawn together, so that the weight of the bird while perched, pressing on the tendon, holds it firmly on the branch. This action of the tendon is easily observed by watching a common fowl walk.

At each step that it makes, on lifting its foot, the claws are seen to be drawn together. It is partly by this power that the birds of prey are enabled to fix their talons so forcibly into the bodies of their prey. When, for example, an eagle wishes to drive his claws into its prey, he

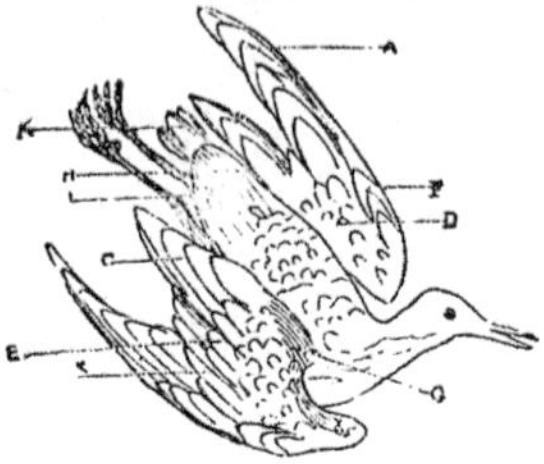

PLUMAGE.

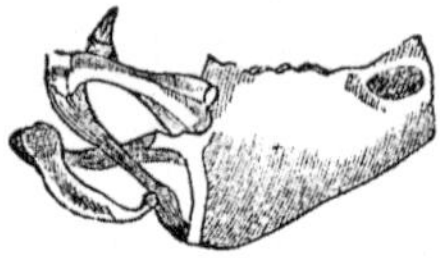

STERNUM.

perches on it, and then sinks down with the whole weight of his body, by which movement the tendon is shortened, and the claws forcibly pressed together.

As the wing presents a very broad surface to the air, it is necessary that very powerful muscles must be used to move it with sufficient rapidity. The pectoral muscles are therefore enormously developed, extending almost the whole length of the body, as every one who has carved a fowl must have seen; and in order to form an attachment for these immense muscles, the ridge of the breastbone is equally enlarged. It is the want of these enlarged muscles that prevents man from flying, even when he has attached wings to his arms.

CLAW AND BEAK OF RAPACIOUS BIRD.

The principal characteristics of birds are taken from their foot and beak. I therefore give the reader a figure representing the general character of the head and foot of the rapacious birds.

The Lämmergeyer, (Germ. *Lamb's-vulture,*) or Bearded Vulture, inhabits most mountain ranges, and is very common in the Alps of Switzerland and Germany, where, from its depredations on the kids and lambs, it has earned its name of Lämmergeyer.

Although called the Bearded "Vulture," it is not strictly a vulture, as its head and neck are feathered, and it rejects putrid flesh, unless hard pressed by hunger.

It destroys hares, and young or sickly sheep and goats, nor, when rendered fierce by hunger, does it fear to attack the adult chamois, or even man. It is said to destroy the larger animals by watching until they are near the brink of a precipice, and then suddenly driving them over the rocks by an unexpected swoop. In this manner the strong and swift chamois falls a victim to the craft of its winged foe, and instances are not wanting where the chamois hunter himself has been struck from a narrow ridge into the valley beneath by a blow from this ferocious bird.

It is exceedingly bold, and shows but little fear of man. While Bruce was preparing his dinner on the summit of a mountain, one of these birds, after scalding its feet in several unavailing attempts to extract some meat out of the boiling water, actually seized a piece from a platter, and went off with it.

The name of "Bearded" Vulture is given to it on account of the long tuft of hairs with which each nostril is clothed. The length of its body is about four feet, and the expanse of its wings from nine to ten. The second and third primary feathers are the longest.

It lays two eggs,—white, marked with brown blotches.

Family II. Sarcorhamphĭdæ.—(Gr. Σάρξ, flesh; ῥάμφος, beak.)

SARCORHAMPHOS.

Gryphon (Gr. Γρύψ, a Griffon), *the Condor.*

The Condor.—The Sarcorhamphidæ are distinguished by a fleshy tuft growing on their beaks, somewhat resembling the wattles of a turkey.

The genius Sarcorhamphos includes the Condor, the King Vulture, and the Californian Vulture. These birds are distinguished by the wattles on their beaks, their naked necks, and the size of the nostrils. The third primary feather is the longest.

The Condor inhabits the Andes of South America, always choosing its residence on the summit of a solitary rock. It appears that this bird does not build any nest, but lays its two white eggs on the bare rock after the manner of many sea birds. It is a very large bird, but by no means the gigantic creature some former naturalists relate, with wings twenty feet in length, and powerful enough to carry off a horse. The real expanse of wing is about nine or ten feet, and the length of the bird about four feet. It is, however, exceedingly strong and tenacious of life. Two Condors will attack and kill the llama, or even the puma; for by their repeated buffeting and pecking they weary it so completely that it yields to their perseverance.

SARCORHAMPHOS.

Papa, *the King Vulture.*

The KING VULTURE is also a native of South America, seldom if ever being seen north of Florida. Travellers relate that this species keeps the other vultures under subjection, and does not suffer them to approach a dead animal until he has completely satisfied his own appetite, which is certainly none of the smallest.

We now arrive at the true VULTURES. These birds are the representatives of the carrion-devouring animals, such as the hyenas, wild dogs, &c. They however do not, as the hyenas and wild dogs, attack living animals. The neck of the Vulture is almost naked, very slightly sprinkled with down, and from the formation of the lower part of the neck, the bird is enabled to draw its head almost under the feathers of its shoulders, so that a hasty observer would conclude that the creature had no neck at all.

The marvellous quickness with which the vultures discover a dead animal has caused many discussions among naturalists as to the sense employed; some, as Audubon, declaring entirely for sight, and others, as Waterton, asserting that the scent of putrid animal matter leads the vultures to their prey.

He especially ridicules one experiment tried by Audubon, by stuffing a deer's hide, and placing it in the open air. The vultures soon came to the stuffed skin, and one of them tore open the skin, just as if the animal had been really lying dead, and continued to pull out large quantities of straw until it became tired and went away. Audubon argues from this experiment, that the vultures were led to the bait by the sight, and not by the scent, or they would not have been so taken in as to

Family III. . Vulturĭdæ.—(Lat. *Vultur.* Vulture kind.)
Sub-family *a. Vulturīnæ.*

GYPS.

Fulvus (Lat. *tawny*), *the Griffon Vulture.*

work for so long a time at a stuffed skin. However, few people see the same incident in the same light, and Waterton's inferences are widely different from those drawn by Audubon. He declares that the experiment of stuffing a deer's hide, and placing it exposed in the open air, was by no means conclusive, as the hide, however dry, must have given out some odour, and the vulture certainly acted very properly in pulling out the straw, and endeavouring to get at the inside.

The probability is that both senses are used, one aiding the other; for in another experiment, where a dead hog was hidden under canes and briars, numbers of vultures were seen sailing in all directions over the spot, evidently directed by the scent, but unable to discover by their eyes the exact position of the animal. The olfactory nerves of the Vulture are beautifully developed, so that Waterton had reason for his pathetic remark,—"I never thought I should have lived to see this bird deprived of its nose."

The GRIFFON VULTURE* is found in almost all parts of the old world. It is one of the largest of its group, measuring upwards of four feet in length. Like most of the vultures, it does not appear to move its wings while flying, but soars on expanded pinions in large circles, apparently gaining the necessary impetus by the movements of its head and body, just as an accomplished skater uses but little force in his various evolutions—an imperceptible inclination of the head, or sway of the body, sufficing to keep up the impetus gained at starting, and to bring him round in any direction he chooses.

Vultures are generally protected by the natives of the countries where they reside on account of their great utility in clearing away the putrid animal matter, which would otherwise be exceedingly injurious as well as disagreeable. The Turkey Buzzard or John Crow (*Cathartes Aura*), or Jamaican Vulture, is protected by a fine of five pounds, inflicted on any one who destroys the bird within a certain distance of the principal towns. Waterton's account of this bird is very interesting, and well worthy of notice. There are many species of vultures inhabiting different countries, but their habits as well as their forms are so similar that a detailed description of each is needless.

EAGLES.—The Vultures seem to hold the same place among birds as the Hyenas among the Mammalia; and in like manner the Falconidæ fill the same position in the ornithological kingdom, as do the Felidæ among the quadrupeds. The beak of this family is strong and curved, and the feet furnished with sharp talons, just as the Felidæ are armed with long sharp teeth and powerful claws. The Falconidæ differ from the Vulturidæ in having feathered necks, and in killing their prey themselves, and devouring it while fresh.

At the head of the Falconidæ the EAGLES are placed. In them the wings are large, powerful, and slightly rounded, the fourth primary feather being longest. The feet of the genus Aquila are feathered to the toes.

The Golden Eagle is found in most parts of Europe, and is not uncommon in Great Britain, especially in the mountainous parts of Scotland and the Hebrides. The flight of this magnificent bird is peculiarly beautiful and imposing, but its gait when on land is rather awkward; for its long talons encumber it in the same manner that the Sloth and Great Ant-Eater are impeded by their long curved claws when they attempt to walk with any rapidity.

* See p. 169

Its food is usually sea birds and the smaller quadrupeds, such as hares rabbits, &c., but it does not hesitate to carry off young lambs, or sometimes to destroy a sickly sheep. Some instances have been related o children that have been carried away by this eagle, but they are ver doubtful. Eagles certainly have pounced upon children, and carried the

Family IV. . . . Falconĭdæ.

Sub-family *a*. . . *Aquilīnæ*.

Aquĭla.—(Lat. *an Eagle*.)

Chrysaëtos (Gr. Χρυσάετος, Golden Eagle), *the Golden Eagle*.

a little way, but there are no authenticated accounts of children having been actually taken to the eagle's nest, although there are many pretty stories founded on such a supposition.

It generally hunts in pairs, one eagle watching from some height while the other courses along the ground, and drives the game from the bushes. The male and female remain together all the year, and very probably for life. It lays two eggs of a yellowish white colour with pale brownish spots, on a nest composed of a great mass of sticks, rushes, and grass, and the young are fledged about the end of July. While the young are in the nest it is very dangerous to approach the spot, as the Eagles are then extremely fierce and daring.

The Rev. Mr. Inglis gives the following account of an adventure with an eagle:—"The farmer of Glenmark, whose name was Miln, had been out one day with his gun, and coming upon an eagle's nest, he made a noise to start her and have a shot. She was not at home, however, and so Miln, taking off his shoes, began to ascend, gun in hand. When about half-way up, and in a very critical situation, the eagle made her appearance, bringing a plentiful supply to the young which she had in her nest. Quick as thought she darted upon the intruder, with a terrific scream. He was clinging to the rock by one hand, with scarcely any footing. Making a desperate effort, however, he reached a ledge, while the eagle was now so close that he could not shoot at her. A lucky thought struck him: he took off his bonnet and threw it at the eagle, which immediately flew after it to the foot of the rock. As she was returning to the attack, finding an opportunity of taking a steady aim, he shot her."

The eye of this bird, and of most of the birds of prey, is provided with an arrangement for enabling it to see an object near or at a great distance. The old tale of the eagle delighting to gaze at the sun is equally poetical and false, the true fact being that the eye is shaded from the sun by the projecting eyebrow. As to the nictitating membrane, which some assert to be given to the Eagle in order to enable it to gaze at the sun, all birds have it, and the owl, who is blinded by ordinary daylight, possesses it in perfection.

The feet of the genus Pandion are naked, armed with very long curved talons, the outermost of which can be drawn together, so as to hold their slippery prey. The wings are ample, and the second and third primary feathers the longest.

PANDĪON (Gr. Proper name).

Haliaëtus (Gr. Ἅλς, the sea, ἀετός, an Eagle), *the Osprey.*

The OSPREY, or FISHING HAWK, is spread over the whole of Europe, part of Asia, and some portions of North America. As its name imports, its food consists entirely of fish, which it obtains by dashing into the water, and seizing them with its curved talons. The Osprey, although it takes the fish, is not the only bird that has a predilection for that diet, as the bald-headed eagle frequently waits until the osprey has seized the prey, and then deprives him of it. I should like to present my readers with the entire spirited passage from Wilson's Ornithology, describing the chase and capture, but as want of space compels me to compress as much as possible, there will only be room for a very small extract. He relates that the eagle, after watching the osprey as it dashes into the water after its finny prey, starts off in pursuit as it emerges, bearing a fish in its talons. "Each exerts his utmost to mount above the other, displaying, in these rencontres, the most elegant and sublime aerial evolutions. The unencumbered eagle rapidly advances, and is just on the point of reaching his opponent, when, with a sudden scream, probably of despair and honest execration, the latter drops his fish: the eagle, poising himself for a moment, as if to take a more certain aim, descends like a whirlwind, snatches it in his grasp ere it reaches the water, and bears his ill-gotten booty silently away to the woods."

When the Osprey plunges after its finny prey, it never attempts to seize them while they are leaping out of the water, but plunges downwards with such force, that it disappears below the surface, throwing up the foam around it.

Its nest is made of an enormous heap of sticks, grass, &c., laid among the branches of a tree, and large enough to make a fair cart-load. In this nest it lays its eggs, which are a yellowish white, sprinkled with brown blotches.

The length of this bird is about two feet, and the expanse of its wings about five feet and a half. Its feet are of a pale greyish blue colour, which unfortunately fades in a stuffed specimen. Indeed, it is generally useless to attempt to obtain the slightest conception of the colours of membrane or skin from stuffed animals. The ears, lips, and eyelids of quadrupeds, always present a most miserable appearance, while the legs and heads of birds look equally wretched from the same cause. The skin not only loses its colour, and assumes a dingy brown hue, but it also shrivels up, and is puckered up into wrinkles, that entirely obliterate it former appearance while the animal was living; even a very few hours will make this change.

HALIAĔTUS.

Leucocephălus (Gr. Λευκοκέφαλος, white-headed), *the White-headed Eagle.*

The WHITE-HEADED EAGLE, or BALD EAGLE, as it is called by Wilson, inhabits most parts of America, and especially frequents the cataract of Niagara. It is very accommodating in its appetite, and preys indiscriminately on lambs, pigs, swans, and the fish which, as related above, it takes away from the unfortunate osprey. Sometimes it can take fish honourably for itself in shallow water, by wading as far as it can, and snatching up the fish with its beak. Audubon gives a splendid description of the chase of a swan by an eagle, but want of space again prevents its insertion.

Like the Golden Eagle, this bird lives constantly with its mate, and hunts in company. It lays from two to four eggs, of a dull white colour, in a huge nest placed on a tall tree.

The claws of this bird are grooved beneath, and the hind claw is the longest. The feet are half-feathered, and the fourth primary feather of the wing is the longest. When full grown, the general colour of the bird is a deep brownish black, but its head, neck, tail, and upper tail-coverts are white.

THE BUZZARD.—The family of the Buzzards are distinguished by their short beaks, large rounded wings, and squared tails. They all live on small animals, reptiles, and various insects.

Sub-family *c. Buteoninæ.*—(Lat. *Butĕo,* a Buzzard.)

BUTĔO.

Vulgāris (Lat. *common*), *the Buzzard.*

The Common Buzzard occurs throughout most of Europe and part of Asia, being frequently found in England. When searching for food, it rests upon some high branch, keeping a keen watch on the ground, and waiting patiently until some small animal, such as a rat or young rabbit, makes its appearance, when it instantly sweeps down from its elevation, seizes its prey without settling on the ground, and returns, if not disturbed, to the same spot, very much in the same manner that the fly-catcher may be observed to act.

It generally builds in high trees, but has been known to make its nest among rocks. Its eggs are usually three in number, of a whitish colour, spotted with pale brown, and almost devoid of the peculiar red tinge that generally characterises the eggs of the diurnal birds of prey. The length of this bird is from twenty to twenty-two inches: the fourth primary feather is the longest.

THE HONEY BUZZARD.—The genus Pernis is distinguished by the feathered lorum or band round the eyes. The HONEY BUZZARD is found in the warmer parts of Europe and in Asia, seldom visiting our shores.

Sub-family *d. Milvinæ* (Lat. *Milvus, a Kite*).

PERNIS—(Gr. *πέρνης, a Hawk.*)

Apivŏrus (Lat. *Bee-eating*), *the Honey Buzzard*.

Its food does not consist of honey, as its name might seem to indicate, but of bees, wasps, and their larvæ. In the stomach of one that was shot in Scotland, a great number of bees and grubs were found, but no honey or wax. It does not, however, refuse small quadrupeds, or sometimes small birds, if pressed by hunger. It is a bird of passage, leaving Europe at the commencement of winter. Its nest is built in high trees, and its eggs are two or three in number, grey, spotted with red at one end, and surrounded with a red band. Its length is about two feet, and the expanse of its wings fifty-two inches. The third primary feather is the longest.

MILVUS.—(Lat. *a Kite.*)

Regālis (Lat. *royal*), *the Kite.*

The KITE, GLEDE, or GLED, is not uncommon in England, and is spread over Europe, Asia, and Northern Africa. It is especially hated by the farmer for its depredations on his poultry, and its appearance is the signal for a general outcry among the terrified poultry, who perceive it long before the keenest-eyed man can distinguish it from a casual spot in the distant sky. The sportsman also detests it for the havoc which it makes among the game, — possibly the kite hates the sportsman for the same reason.

It builds in tall trees, and lays three eggs, white, spotted with reddish brown at the larger end. Its length is rather more than two feet; the fourth primary feather is the longest, the first and seventh nearly equal.

The SWALLOW-TAILED FALCON is an inhabitant of North America, but has been more than once taken in England. It feeds on the wing, like the swallows, pursuing the large moths and other insects with an ease and rapidity for which its formation eminently fits it. These insects are, however, not the only food of this bird. Audubon mentions that "their principal food is large grasshoppers, grass caterpillars, small snakes, lizards and frogs. They sweep close over the fields, sometimes seeming to secure a snake, and holding it fast by the neck, carry it off, and devour it in the air." This act is shown in the figure below. Its nest is built on the summit of an aged pine or oak, and its eggs are from four to six in number, of a greenish white colour, irregularly spotted with brown at the large end. The length of this bird is two feet. It should properly be called the Swallow-tailed Kite, as it belongs to the family of the kites.

ELANOÏDES.—(Gr. ἐλανὸς, *a Kite, like a kite.*)

Furcātus (Lat. *forked*), *the Swallow-tailed Falcon.*

FALCONS.—In the genus Falco, the second primary feather is the longest, the first and third being of equal length. The GYR-FALCON may be considered the type of the British Falconidæ. It is, however, extremely rare in England, those intended for hawking being principally brought from Iceland. On the rocky coasts of Norway and Iceland its eggs are laid. These birds are very courageous in defending their young. A pair of them attacked Dr. Richardson while he was climbing near their nest, flying in circles round him, and occasionally dashing at his face with loud screams. The entire length of the Gyr-falcon is twenty-three inches.

Sub-family e. *Falconinæ.*

FALCO.—(Lat. *a Falcon.*)

Gyr-falco (Lat.) *the Gyr-falcon.*

The PEREGRINE FALCON, an inhabitant of most parts of Europe, Asia, and South America, was, in the palmy days of hawking, one of the favourite falcons chosen for that sport. Its strength and swiftness are very great, enabling it to strike down its prey with great ease; indeed, it has been known to disable five partridges in succession. From its successful pursuit of ducks the Americans call it the Duck Hawk.

There is a peculiarity in the method of attack which this bird employs when pursuing small game. Instead of merely dashing at its prey, and grasping it with its claws, the Peregrine Falcon strikes its victim with its breast, and actually stuns it with the violence of the blow before seizing it with its claws. The boldness of the Peregrine Falcon is so great that it was generally employed to take the formidable Heron. After the Heron had been roused from his contemplations by some marsh or river, the Falcon, who had previously been held hooded on its master's hand, was loosed from its bonds and cast off. A contest then generally took place between the Heron and the Falcon, each striving to ascend above the other. In this contest the Falcon was always victorious, and after it had attained a sufficient altitude, it swept, or "stooped," as the phrase was, upon the Heron. When the Falcon had closed with its prey, they both came to the ground together, and the sportsman's business was to reach the place of conflict as soon as possible, and assist the Falcon in vanquishing its prey. Sometimes, however, the wary Heron contrived to receive its enemy on the point of its sharp beak, and transfixed it by its own impetus.

FALCO.

Peregrinus (Lat. *wandering*), *the Peregrine Falcon.*

It changes the colour of its plumage several times before it arrives at full maturity, and in the days of falconry was known by different names, such as "haggard" when wild, "eyass," "red falcon" when young, "tiercel" or "tassel-gentle" when a full-grown male; a term forcibly recalling the words of Juliet, "Oh for a falconer's voice, to lure this *tassel-gentle* back again!"

It builds on ledges of rocks, laying four eggs of a reddish brown colour. Its length is from fifteen to eighteen inches.

HYPOTRIORCHIS.

(Gr. Ὑποτριόρχης.)

Subbuteo (Lat. *the Hobby.*)

The HOBBY is a summer visitor in this country, appearing in April and leaving in October. It was formerly trained to fly at larks, quails, and other small birds. When wild it seems to feed principally on small birds and large beetles, the common dor-beetle being a very favourite article of food. It builds its nest at the summit of a high tree, usually appropriating the deserted habitation of a crow. The eggs are four in number, of a dirty white colour speckled with reddish brown. The length of the bird is from twelve to fourteen inches.

The MERLIN, the least of our Falcons, was considered in olden times as the lady's bird, every rank being obliged to content itself with the bird allotted to its peculiar station, royalty alone having the privilege to bear an eagle into the field.

The spirited little Merlin seizes with great dexterity small birds such as buntings, thrushes, and blackbirds, itself really hardly larger than its prey, its entire length being barely eleven inches. Even the partridge falls before a trained bird. Its eggs are four in number, of a reddish mottled brown, laid in a rude nest among the heather.

HYPOTRIORCHIS.

Æsălon (Gr. Αἰσάλων), *the Merlin.*

The KESTREL, or WINDHOVER as it is often called, frequently falls a victim to the mistaken zeal of the farmer, who takes every opportunity of destroying it, as he confounds it with the sparrow-hawk. The natural food of the Kestrel is field-mice, so that the farmer should protect instead of remorselessly murdering his benefactor. These birds are not uncommon. Many live close to Oxford, and especially in Bagley Wood, where they may be seen almost daily. They also live in great numbers among the precipices in Dovedale. Their nest is usually built in the deserted mansion of a crow or magpie. The eggs are four in number, of a dark reddish brown. The length is from thirteen to fifteen inches.

TINNUNCŬLUS.—(Lat. *a Kestrel.*)

Alaudarĭus (Lat. *of a Lark*), *the Kestrel.*

The GOSHAWK is found plentifully in most of the wooded districts of Europe, but is comparatively rare in the British Isles. It seldom breeds south of Scotland, but its nest is not unfrequently found in that country, built upon lofty trees, principally firs, and containing three eggs of a bluish white colour with reddish brown marks. When in pursuit of

Sub-family *f. Accipitrinæ*

ASTUR.—(Lat. *proper name.*)

Palumbarius (Lat. *of the Dove*), *the Goshawk.*

prey, it strikes its victim to the ground by the force with which it dashes through the air. Should the terrified quarry hide itself, the Goshawk takes up its station on some elevated spot, and there patiently waits until the game takes wing. Its principal food consists of hares, squirrels, pheasants, and other large birds, which its great strength enables it to destroy. Its length is about two feet; the fourth primary feather is the longest.

The SPARROW-HAWK is common throughout Europe. It displays great pertinacity in pursuit of its prey, which it will chase for a long while, skimming along a few feet above the ground. One of these hawks was known to dash through a window in pursuit of a small bird. When taken young it is easily tamed, and will then associate with the most incongruous companions. A gentleman had a young Sparrow-hawk which used to live in his dovecote among his pigeons, would accompany them in their flights, and was uneasy if separated from its strange friends.

Although a brave little bird in its wild state, it sometimes becomes sadly degenerate when domesticated. I had a tame Sparrow-hawk, which was purchased for the ostensible purpose of frightening birds from the vegetables in the garden. But unfortunately, his presence was rather an attraction than otherwise, for besides the pleasing excitement caused by mobbing him, he was invaluable to the tomtits, who always ate his meat if he were not protected by some human being. He would fly in terror before the wagtails, and the tomtits regularly charged at him when he was fed, and then made a dash at his saucer of provisions.

ACCIPITER.—(Lat.)

Nisus (Lat. *proper name*), *the Sparrow-hawk.*

But all tame Sparrow-hawks are not so demoralised. I know one of these birds, which, although young, is the acknowledged master of an establishment consisting of three dogs and a cat. He flies at the dogs if they approach too near his throne, and sometimes dashes at them without any apparent reason.

The length of this bird is from twelve to fifteen inches. The fourth and fifth primary feathers are the longest. It builds upon lofty trees, laying five eggs, of a whitish colour blotched with variable reddish brown markings, usually collected towards the large end, and often forming a deep reddish irregular band.

The SECRETARY BIRD * derives its name from the tufts of feathers at the back of its head, which bear a fanciful resemblance to pens stuck behind the ear. This extraordinary bird, whose true position in ornithology has been such a stumbling-block to naturalists, inhabits South Africa, Senegambia, and the Philippine Islands. Probably a different species inhabits each of these countries. It feeds on snakes and other reptiles, of which it consumes an amazing number, and is on that account protected. When battling with a snake, it covers itself with one wing as with a shield, and with the other strikes at the reptile until it falls senseless, when a powerful blow from the beak splits the snake's head asunder, and the vanquished enemy is speedily swallowed. In the crop of a Secretary bird that was dissected by Le Vaillant were found eleven large lizards, three serpents, each a yard in length, eleven small tortoises, and a great quantity of locusts and other insects. Besides these, the bird had just overcome another serpent, which would in all probability have been transferred to the same receptacle had it not been killed. The Secretary is easily tamed, and is then exceedingly useful. It builds on high trees, laying three large eggs, almost white. Its length is about three feet.

* See page 181.

Sub-family *g. Circīnæ.*

Serpentarĭus.—(Lat. *of a Serpent.*)

Reptilivŏrus (Lat. *Reptile-eating*), *the Secretary Bird.*

The Harriers are remarkable for the peculiar feathered disk round their eyes, something resembling that of the owl. The Hen Harrier is a native of England, and lives principally about forests and heaths. Its length is about seventeen inches, the first primary feather is very short, the third or fourth is the longest.

Circus.—(Lat. a Falcon.)

Cyanĕus (Lat. *azure*), *the Hen Harrier.*

Owls.—A large round head, with enormous eyes looking forward, is a distinguishing mark of the Owl family. Many species possess two feathery tufts placed on the head, greatly resembling horns. The Owls are nocturnal birds, pursuing their prey by night, and sleeping during the day. In order to enable them to see their prey, their eyes are enormously large, and capable of taking in every ray of light. Their power of vision is also increased by the method in which the eye is fixed in a kind of bony socket, just like the watchmaker's glass. The nictitating membrane is very conspicuous in these birds. The power of hearing is also very delicate, and greatly assists them. In order to protect them from the cold, they are furnished with a dense covering of downy feathers, which also prevent the movements of the wing from being heard by the wary mouse; and so noiseless is their flight that they seem to be borne along by the wind like a tuft of thistle-down.

Sub-order II. Accipĭtres nocturni.

Family I. Strigĭdæ.—(Lat. *Strix*, a Screech-owl. Screech-owl kind.)

Sub-family *a. Surnĭnæ.*

SURNIA.

Ulŭla (Lat. *howling*), *the Hawk-owl.*

The Hawk, or Canada Owl, inhabits the arctic portions of Asia and America. Its head is not so round, nor is its face so broad, as those of the other owls, from which it is also distinguished by its habit of hunting by day. In face it bears some resemblance to the harriers. It builds in trees, and lays two eggs—white, as are those of all owls. The eggs of owls are easily distinguished from other white eggs by a peculiar roughness of surface, which cannot be mistaken. The length of the Hawk-Owl is from fifteen to eighteen inches.

The Snowy Owl* is properly an inhabitant of the north of Europe, but has more than once been discovered in Great Britain. It is also found in North America. Wilson relates that it is a good fisher, snatching its prey from the water by a sudden grasp of the foot. It also preys on lemmings, hares, ptarmigans, &c., chasing and striking at them with its feet. It makes its nest on the ground, and lays three or four white eggs, of which more than two are seldom hatched. Its length is from twenty-two to twenty-seven inches, the expanse of wing four feet; the third primary feather is the longest.

The Burrowing Owl accompanies the prairie dog, and wherever that

* See page 184.

NYCTĔA.—(Gr. Νύκτιος, nightly.)

Nivĕa, (Lat. *snowy*), *the Snowy-owl.*

animal chooses to live, there is the Burrowing Owl. This singular little bird finds that to take possession of the ready-made burrows of the prairie dog is much more agreeable than to dig a hole for itself; so it takes unfurnished lodgings in a deserted dwelling, undisturbed by anything except a casual lizard or rattlesnake.

ATHĒNE.—(Gr. proper name.)

Cunicularia (Lat. *of the Rabbit*), *the Burrowing Owl.*

So numerous are these little owls, that they may be seen in small flocks seated on the tops of the mounds in which the entrance of the burrows is formed. It is said that the owls, marmots (or prairie-dogs), lizards, and snakes, all live harmoniously in one happy family. Such, however, is not really the case, as the Burrowing Owls prefer holes unoccupied by any other tenant, and have been seen with something most suspiciously like a young snake struggling in their mouths. The bottom of its hole is generally comfortably filled with dried hay and roots.

The legs of this bird are longer than those of other owls. It is by no means large, measuring but ten inches in length.

The SCOPS EARED-OWL has been once or twice found in Yorkshire, but usually resides in the southern parts of the Continent. It is remarkable for the regularity with which it utters its monotonous cry, as if a person were constantly repeating the letter Q at regular intervals of two seconds. It does not seem to prey upon mice and other animals, like most of its relations, but feeds on large insects, such as beetles and grasshoppers. The size of this owl is very small, as it only measures seven inches in length; the third primary feather is the longest. It lays from two to four white eggs in a simple nest made in a hollow tree or in a cleft in the rock.

The GREAT EARED-OWL or EAGLE OWL, is the largest of the family. This powerful bird, not satisfied with the "rats and mice and such small deer" which content the English owls, boldly attacks young fawns, hares, and rabbits, together with small birds. It inhabits the

Sub-family *b.* *Buboninæ.*

EPHĬALTES.—(Gr. proper name.)

Scops (Gr. Σκώψ, an Owl), *the Scops Eared-owl.*

BUBO.—(Lat. *an Owl.*)

Maxĭmus (Lat. *greatest*), *the Great Eared-owl.*

north of Europe, but has been several times observed in Great Britain. It lays its eggs in the clefts of rocks or in ruined buildings. The length of this bird is upwards of two feet.

The BARN OWL affords another instance of mistaken persecution This beautiful and most useful bird, whose carcase we so often see triumphantly nailed to the barn, actually feeds upon and destroys in incalculable numbers the rats and mice which bear it company in its undeserved punishment. Waterton remarks, "When farmers complain that the Barn Owl destroys the eggs of their pigeons, they lay the saddle on the wrong horse. They ought to put it on the rat. Formerly I could get very few young pigeons till the rats were excluded from the dovecote. Since that took place, it has produced a great abundance every year, although the barn owls frequent it, and are encouraged all around it. The barn owl merely resorts to it for repose and concealment. If it were really an enemy to the dovecote, we should see the pigeons in commotion as soon as it begins its evening flight, but the pigeons heed it not, whereas if the sparrow-hawk or hobby should make its appearance the whole community would be up at once. . . . I am amply repaid for the pains I have taken to protect and encourage the barn owl; it pays me an hundred fold by the enormous quantity of mice which it destroys throughout the year."

Sub-family *d.* *Striginæ.*

STRIX.—(Lat. *a Screech-owl.*)

Flammĕa (Lat. *flaming*), *the Barn-owl.*

It also devours great numbers of beetles and other insects. It is possible that it may, also, destroy young birds, but not probable, as feathers and birds' bones are never found among the rejectamenta. It will, however, when domesticated, devour a dead sparrow or linnet when presented to it.

Few people know what a little bird this owl really is. The thick loose plumage is so deceptive, that no one unacquainted with the structure of the bird would imagine that it is hardly so large as a pigeon. The head, too, when deprived of its feathery covering, completely loses its previous aspect, being long and narrow, like that of a hawk. In fact, few creatures look more contemptible than an owl stripped of its feathers.

The domestic habits of the bird are very curious. When irritated or alarmed, it has a habit of snapping its beak loudly, and making a hissing sound, something like that of a cat when very much provoked. Indeed, there is something very cat-like in the whole aspect of the owl. Its round soft looking face, in which are set two great eyes that shine in the dusk of the evening with an almost phosphoric gleam, and are capable of taking in every feeble ray of light; its noiseless movements in pursuit of its prey, all strongly remind the observer of the feline character.

The plumage of this bird is very thick, in order to defend it from the cold night air, and very soft, in order to admit of its flight with such silence that even the quick and wary ear of the timid field-mouse may not perceive the approach of its enemy. This noiselessness is caused by the formation of the feathers, which, instead of being stiff and smooth, like those of the diurnal birds of prey, are loose and furnished at their extremities with a delicate fringe, which completely prevents that rushing sound which accompanies the flight of the eagle or the hawk.

White, in his Natural History of Selborne, gives some interesting accounts of the habits of this owl. He watched them leave their homes about an hour before sunset in search of the mice, which begin to run about at that time, and saw them quartering their ground like a pointer dog, with the greatest regularity. He had the curiosity to count the number of mice caught by a single pair, and found that upon the average, one was brought to the nest every five minutes. With his usual accuracy of observation, he noted the manner in which the birds were enabled to reach their nests, although encumbered with a mouse. Their nest was made under the roof of the church, so that the owls had to crawl up by their feet in order to rise under the eaves. This they managed by perching on the church roof, shifting the mouse from their claws to their bill, and thus leaving their feet at liberty.

Its method of devouring a mouse is quite different from the mode in which it eats a bird. If a mouse is given to an owl, the bird seizes it across the back and gives it one or two smart bites, much as a terrier handles a rat. The mouse is then jerked upwards, and caught again head downwards. A second jerk sends the mouse half down the owl's throat, while its tail remains sticking out of the side of its bill, where it is rolled about as if the owl were smoking. After some time has been spent in this amusement, another jerk causes the mouse to disappear altogether, and the owl looks very happy and contented. But if a small bird is presented to it, the owl tears it up and devours it piecemeal. Such, at least, was the conduct of a barn-owl belonging to myself.

This bird is easily tamed, if taken young, and is a valuable assistant in a garden, only a place of refuge must be provided for its retreat during the day-time, or the little birds will mob it unmercifully.

The Barn Owl lays three or four eggs upon a mass of those pellets which all the owls disgorge. There is a rough chalky look about the eggs of the owl, which renders them different from the eggs of all other birds, from which they can be distinguished by the touch alone. There is a peculiarity in the domestic economy of this owl, for it often has at the same time in the nest, young owls almost fledged, and eggs on which the hen bird is sitting.

The length of the bird is rather more than twelve inches; the second primary feather is the longest. Its colour is a bright yellowish brown, marked with dots and lines of various tints, the lines being generally dark, and the dots light. The under parts of the bird are very variable, and seldom exactly the same in any two individuals, the white feathers being sometimes greyish white, sprinkled with brown dots; sometimes pure white, without any marks at all; and sometimes white very slightly dotted indeed.

When it is threatened or attacked, it throws itself on its back and fights vigorously with its claws and bill.

Order II. . . . *PASSĔRES*. (Lat. *Sparrow kind.*)
Tribe I. . . . . FISSIROSTRES.—(Lat. *Split-bills.*)
Sub-tribe I. . . FISSIROSTRES NOCTURNÆ.
Family I. . . . Caprimulgĭdæ.—(Lat. *Goat-sucker kind.*)
Sub-family *a*. . *Caprimulgīnæ.*

CAPRIMULGUS.

Europæus (Lat. *European*), *the Goat-sucker.*

THE Accipitres, it will be remembered, possess strong hooked beaks and sharp curved claws. The foot and head of the Passeres are entirely different;—the beak being without the formidable curved tip, and the claws being of a quiet and peaceful character.

The first tribe of this order, the Fissirostres, are so called from the peculiar formation of their mouths, which appear as if they had been slit up from their ordinary termination to beyond the eyes, much resembling the mouth of a frog. In the insect-eating Fissirostres this formation is admirably adapted for capturing their active prey, and in the Kingfishers it is equally adapted for securing the slippery inhabitants of the waters.

The Caprimulgidæ are nocturnal in their habits, chasing their insect prey by night or at the dusk, when the chaffers and large moths are on the wing. In order to prevent the escape of the insect when taken, the mouth is fringed with long stiff bristles, called "vibrissæ." The name of Goat-sucker is derived from a silly notion that they suck goats, a piece of credulity only equalled by the hedgehog's supposed crime of sucking cows, and the accusation against the cat of sucking the breath of children. The genus Caprimulgus is furnished with a kind of comb on the middle claw of its foot, but for what purpose is not clearly ascertained. The power of wing in these birds is very great, and hardly surpassed by that of the swallow, both birds obtaining their food in a similar manner.

The Nightjar, or Goat-sucker, sometimes called the Fern Owl, is spread over Europe, and is tolerably common in England. It may be seen at the approach of evening, silently wheeling round the trees, capturing the nocturnal moths and beetles; then occasionally settling and uttering its jarring cry. When flying, the bird sometimes makes its wings meet over its back, and brings them together with a smart snap. It arrives in this country at the beginning of May, and leaves in December. It makes no nest, but lays two mottled eggs on the bare ground. Its length is ten inches. The Whip-poor-Will and the Chuck-Will's-Widow both belong to this family.

These two birds derive their singular names from their cry, which is said closely to imitate the words that have been assigned to them as their names. Of course the English language must feel itself highly honoured that an American bird should prefer the language of the "Britisher" to that of the Delaware or the Sioux. Both the birds fly by night, or rather in the dusk of the evening, and like the owl are much distressed by being forced to face a brilliant light. The Chuck-Will's-Widow is partially migratory, and dwells in the more southern parts of America during the winter. Audubon relates that this bird applies its enormous mouth to rather an unexpected use, viz. that of removing its eggs if it finds that they have been disturbed. Of this curious circumstance he was an eye-witness. He saw the bird that first discovered that an intruder had touched the eggs wait for its mate, and then saw each of them take an egg in its mouth and convey it off.

MARTINS.—The Hirundinidæ are remarkable for their great power of

* I have seen it near Amiens, accompanying a train, and apparently hawking after the moths that were attracted by the lights.

wing, their wide mouths, and short legs. In the genus Cypselus, the toes are all directed forward, and the tarsus is thickly feathered. The whole of their plumage is constructed with a view to rapid and active motion. The feathers of their bodies are firm and close, so as not to impede their passage through the air; their wing feathers are long, stiff and pointed, and their tails are long and forked; all which properties we know to belong to great speed.

Sub-tribe II. . . FISSIROSTRES DIURNÆ.

Family II. . . . Hirundinĭdæ.—(Lat. *Hirundo*, a Swallow. Swallow kind.)

Sub-family *a*. . *Cypselĭnæ*.

CYPSĔLUS.—(Gr. κύψελος, a Martin.)

Melba. (*The Alpine Swift.*)

The ALPINE SWIFT has several times been found in England, and is therefore admitted into the calendar of British birds. It may at once be distinguished from the common Swift by the white throat and abdomen, its breast being crossed by a brown band.

Its nest is generally built in the crevices of lofty towers, and is made of straw and moss, glued together and hardened by some matter furnished by the bird, which when dry forms a compact mass.

The length of this bird is rather more than eight inches.

The COMMON SWIFT, popularly called "Jack Screamer," is the largest and swiftest of the British Hirundinidæ. It seems to spend the whole day on the wing, wheeling with wonderful velocity, and occasionally soaring until it is hardly perceptible, but screaming so shrilly that the sound is plainly heard. The number of insects which it destroys is almost incredible; they are retained in a kind of pouch under the tongue, and when taken out, can hardly be pressed into a tea-spoon. These

CYPSĔLUS.

Apus (Gr. ἄπους, without feet), *the Swift.*

are intended for the young, and the supply is constantly renewed. It lays from two to four long white eggs, on a nest composed of grass, straws, feathers, silk, &c. The colour of this bird is a dusky black. The length is eight inches, the expanse of wing eighteen inches, and its weight barely *one ounce.*

The foot of the Swift is of a singular form, unlike that of any other bird. All the toes are directed forward, there being no hinder toe at all. Some naturalists say that the object of this formation is that the bird may be enabled to climb up the eaves under which its nest is made. If, however, that is the case, we may ask, Why is not the same shape of foot found in the sparrows and other birds, which build in precisely similar localities?

The CHIMNEY MARTIN or SWALLOW is the most common of its family, and too well known to need much description. When skimming over ponds or rivers in search of insects, the snap with which it closes its bill may easily be heard. In the course of its flight over the surface, it also dashes up the water with its wings, which action gave rise to the opinion that Swallows passed the winter under water, and rose in the spring. It is so eager after its prey, that it may be easily caught with a rod and line baited with a fly, after the manner of anglers.

Sub-family *b. Hirundininæ.*

HIRUNDO.—(Lat.)

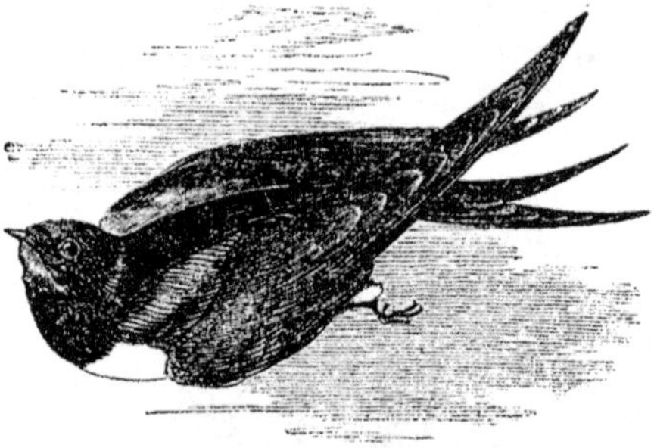

Rustĭca (Lat. *rustic*), *the Chimney Martin.*

When I was at school, we used to knock down plenty of swallows with stones in the following manner. We went to a bridge, or some such place, where many swallows were flying about. Where they appeared in the greatest numbers we threw a small white stone, and immediately hurled a larger one after it. The swallows all dashed at the little stone, taking it for some entomological luxury, and one or two were generally struck down by the big stone following in its wake.

It breeds twice in the year, building a nest of mud against a wall or other convenient situation, and laying five very pale pink eggs, spotted with reddish brown, the pink of which vanishes when the egg is emptied of its contents, as it is caused by the light passing through the yolk, and has to be renewed by artificial means if the egg is placed in a collection.

Such is also the case with most small light-coloured eggs. The bird appears regularly to return, year by year, to its old nest. The whole of its upper surface is a deep purplish black, its forehead and throat chestnut.

Humboldt, in his "Travels," relates that he saw a swallow perch on the rigging of the vessel when it was one hundred and twenty miles from the land.

The SAND MARTIN is the smallest of our British Swallows, but makes its appearance before any of its brethren. It principally builds in cliffs of sandstone, boring holes three feet or more in depth, and often winding in their course, most probably to avoid a casual stone or spot too hard for its bill, which, although small and apparently unfitted for the task, makes its way through the sandstone with extraordinary rapidity. Where a convenient sand-cliff exists, hundreds of these pretty little birds may be seen working away at their habitations, or dashing about in the air, looking at a distance like white butterflies, and occasionally returning to the rock, which is often completely honeycombed by their labours.

Near Ashbourn in Derbyshire there are plenty of these rocks, where the Sand Martins build in myriads, tolerably safe except from the school-boy, who *will* clamber up and down the crumbling surface, and thrust his arm into the holes, perfectly regardless of the danger, and content with grasping a tuft of grass or a root of blackberry as an anchorage. I have seen the Sand Martins there engaged in mobbing a sparrow-hawk, who, after being buffeted about for some time, retaliated by seizing a too daring Martin and carrying it off, when the whole scene was changed—the triumphant jeerings turned into cries of fear, and the place was deserted except by the crafty hawk and his screaming prey. The eggs are five, pinkish white, with an almost imperceptible dotting of red.

COTĬLE.—(Gr. Κωτίλος, twittering.)

Ripărĭa (Lat. *of a bank*), *the Sand Martin.*

The MARTIN or WINDOW SWALLOW reaches this country a little after the swallow, and almost invariably takes possession of its old nest, which it repairs about May. It lays five eggs closely resembling those of the sand martin. About September immense numbers may be seen perched upon houses and trees preparatory to their departure. The dome of the Radcliffe Library at Oxford is a favourite assembling place for these birds, where they may be seen lingering for several days after most of their fellows have vanished. At these times every available point is covered with them. The dome of St. Paul's is also a favoured spot.

CHELĪDON.—(Gr. Χελιδών, a Swallow.)

Urbĭca (Lat. *of the city*), *the Martin.*

The ESCULENT SWALLOW, whose nests are considered such a delicacy among the Chinese, builds its singular habitation in the sides of almost inaccessible cliffs, so that the business of procuring them is a most dangerous task. The nature of the jelly-like transparent material of which the nests are made is not yet known. The nests are found in Java.

Family III. . Coraciĭdæ.

Sub-family *a. Coraciīnæ.*

CORACĬAS.—(Gr. Κορακίας, like a Raven.)

Garrŭla (Lat. *talkative*), *the Roller.*

The ROLLER is plentifully found in most parts of Europe, but has seldom been seen in England. Its mouth is slightly furnished with vibrissæ or long bristly hairs, like those of the nightjar. It is a very shy bird, frequenting the depths of the forests. It builds its nest in hollow trees—some say in banks—and lays from four to seven white eggs, very like those of the kingfisher. Its legs are short, and the upper mandible is bent over the lower at the extremity. The colouring is brilliant, shades of blue and green prevailing. Its length is about thirteen inches.

The TROGON.—The magnificent family of the Trogons stands preeminent in beauty and brilliancy of plumage, the usual tint being a metallic golden green, boldly contrasted with scarlet, black, and brown. The toes are placed two behind and two before, like those of the woodpeckers.

The Resplendent Trogon is the most gorgeous of all this gorgeous family. Its long and gracefully curved tail, nearly three feet long; the whole of the upper surface, and the throat, are a glowing green; the breast and under parts are bright crimson; the middle feathers of the tail black, and the outer feathers white. This splendid bird is an inhabitant of Mexico, and was used by the Mexican nobles as an ornament to their head-dress.

From the feathers of these and other Trogons the mosaic pictures of the Mexicans were made. One of these, most delicately and beautifully

Family IV. Trogonĭdæ.

TROGON.—(Gr. τρώγω, *I gnaw.*)

Resplendens (Lat. *shining*), *the Resplendent Trogon.*

executed, containing many figures, is now in the Ashmolean Museum at Oxford, and is there said to be made of humming-birds' feathers. The subject is "Christ fainting under the cross." The whole picture is about the size of the palm of the hand, and the figures are barely half an inch in height.

This is a very difficult bird to stuff, on account of the delicate texture of the skin, which is so fragile, that it tears almost as easily as wet blotting paper.

The KINGFISHER.—The peculiarities of their form immediately distinguish the Kingfishers from other birds. The disproportionate length of the bill is their chief characteristic.

Family V. Alcedinĭdæ.—(Lat. *Alcēdo*, a Kingfisher.)
Sub-family *a. Alcedininæ.*
ALCĒDO.

Hispĭda (Lat. *rough*), *the Kingfisher.*

The Common Kingfisher is found in most parts of England. Scarcely anything more beautiful can be conceived than the metallic glitter of its plumage as it shoots along the banks of the river, or darts into the water after its struggling prey. Its usual method of fishing is by placing itself on a stump or stone overhanging the water, from which spot it watches for the unsuspecting fish beneath. After a fish is caught, the bird kills it by beating it several times against its resting-place, and then swallowing it, head foremost. Sometimes it does not exercise sufficient caution in its devouring propensities. A heedless Kingfisher was exhibited at the Ashmolean Society, which had been found dead with a peculiarly large minnow firmly fixed in its throat.

It lays its eggs in holes bored in the banks of rivers or ponds, and appears to build no nest. A pair of kingfishers, for two successive years, inhabited a bank of a very small stream, little more than a drain, at Little Hinton, Wiltshire, where no fish lived, nor were there any to be found within a considerable distance.

The eggs are from four to seven in number, of a pearly whiteness, and remarkably globular in shape. In many parts of the country it is fully believed that if a kingfisher is dried and suspended by the beak, the breast will always turn in the direction of the wind. This belief has caused the death of no few kingfishers, whose suspended bodies may be seen in many a cottage, their brilliant blue and red plumage rotating in a most impartial manner. The length of this bird is seven inches.

The BEE-EATER is common on the Continent, but seldom visits this country. In appearance it is not very unlike the kingfisher, both in shape and its brilliant colours. It has long been celebrated for the havoc it causes among the inhabitants of the hive, although it does not restrict itself to those insects, but pursues wasps, butterflies, &c., on the wing, with great activity. Like the kingfisher, it lays its eggs in holes bored in banks. The eggs are white, and from four to seven in number. Its length is eleven inches.

Family VI. . . Meropĭdæ.
Sub-family *a*. . *Meropīnæ.*
MEROPS.—(Gr. Μέροψ.)

Apiaster (Lat. *Bee-eater*).

The HOOPOE,* one of the most elegant birds that visit this country, is unfortunately a very rare guest, and seldom, if ever, breeds here. Its beautiful crest can be raised or depressed at pleasure, but is seldom displayed unless the bird is excited from some cause. Its food consists of insects, which it first batters and moulds into an oblong mass, and then swallows, with a peculiar jerk of the head. In Yarrell's British Birds, there is a very interesting account of a tame Hoopoe in the possession of Mr. Bartlett.

In France Hoopoes are very common, and may be seen examining old and rotten stumps for the insects that invariably congregate in such places. There they may be seen in flocks, but they never seem to come over to England in greater numbers than one pair at a time. M. Bechstein gives a curious account of the attitude assumed by the Hoopoe on perceiving a large bird in the air. "As soon as they perceived a raven or even a pigeon, they were on their bellies in the twinkling of an eye, their wings stretched out by the side of the head, so that the large quill feathers touched the head, leaning on the back with the bill pointing upwards. In this curious posture they might be taken for an old rag!"

* See page 198.

These birds of which he is speaking are two young Hoopoes whom he had taken from the nest and was rearing. They lived for some time, but both died of civilization. The female had a habit of dragging her food about the floor, so that it became covered with rubbish. This formed a hard mass nearly the size of an ordinary nut in the bird's stomach, something like the balls of hair found in the stomach of a cow, and soon killed the poor Hoopoe. The male bird lived through the winter, but becoming attached to the warmth of the stove, its beak became so unnaturally dry, that the two mandibles separated from each other and curved outwards, having an interval of nearly an inch between their tips. The bird of course soon died of absolute starvation.

The Hoopoe lays from four to seven grey eggs in the hollow of a tree. Its length is one foot.

Tribe II. TENUIROSTRES.—(Lat. *Slender-billed.*)

Family I. Upupĭdæ.—(Gr. ὄυποψ, *contracted from* ὁ Ἔποψ, *the Hoopoe, Hoopoe kind.*)

Sub-family *a. Upupīnæ.*

UPŬPA.

Epops (Gr. Ἔποψ), *the Hoopoe.*

The Humming-bird.—These little living gems are exclusively found in the New World, especially about the tropical parts, becoming gradually scarcer as we recede from the tropics in either direction. Only two species are known to exist in the northern parts, but in the central portions and in the islands about Florida they absolutely swarm. They glance about in the sunshine, looking like streaks of brilliant light, and so rapid is the vibration of their fine and elastic wings, that when hovering over a flower, a humming or buzzing sound is produced, from which peculiarity the name of Humming-bird has been given them in almost every language. Waterton's description of the appearance of the Humming-bird in the sun is very characteristic.

"Though least in size, the glittering mantle of the Humming-bird entitles it to the first place in the list of the birds of the New World. It may truly be called the Bird of Paradise; and had it existed in the Old World, it would have claimed the title instead of the bird which has now the honour to bear it. See it darting through the air almost as quick as thought! now it is within a yard of your face—in an instant gone—now it flutters from flower to flower to sip the silver dew—it is now a ruby—now a topaz—now an emerald—now all burnished gold."

Family II. Trochilidæ.

TROCHĬLUS.—(Gr. Τροχίλος.)

Colŭbris (Lat. *like a snake*), *the Ruby-throated Humming-bird.*

It is a singular fact, that a common insect called the Humming-bird Moth is formed on precisely the same principle, and flies in just the same manner. This moth is furnished, like the Humming-bird, with rigid sharp wings; instead of the long slender bill and longer tongue of the Humming-bird, the moth is furnished with an exceedingly long and flexible proboscis, which it uses in the same manner, *i.e.* in thrusting into the interior of flowers while the creature is hovering above them. The moth also possesses a kind of moveable tail, wherewith to direct its course. The description of a Humming-bird hovering over a flower will exactly serve for the moth, save that the moth lacks the brilliant plumage of the bird. Gardens are a great attraction to this moth, and if the observer is very quiet, while looking at a flower, he suddenly sees an insect apparently suspended over it exploring the flower with its proboscis. It moves from blossom to blossom, always balancing itself over them by its wings. Let the observer move but his hand, and it is gone—has vanished as mysteriously as it came.

In the same way, the Humming-bird hovers over flowers, not only to extract the honey and dew, but to search for the little insects that are always to be found in such places. Speaking of the Ruby-throated Humming-bird, Waterton observes:—

"It seems to be an erroneous opinion that the Humming-bird lives entirely on honey-dew. Almost every flower of the tropical climates contains insects of one kind or other; now, the Humming-bird is most busy about the flowers an hour or two after sunrise, and after a shower of rain, and it is just at this time that the insects come out to the edge of the flower in order that the sun's rays may dry the nocturnal dew and rain which they have received. On opening the stomach of the Humming-bird, dead insects are almost always found there."

ORNISMYA.—(Gr. Ὄρνις, a bird; μῦς, a mouse.)

Gouldĭi (Lat. *of Gould*), *Gould's Humming-Bird.*
Sappho (Gr. *proper name*), *the Bar-tailed Humming-bird.*
Cora (*proper name*), *the Cora Humming-bird.*
Chrysolŏpha (Gr. Χρυσός, gold; λόφος, a crest), *the Double-crested Humming-Bird.*

The tongue of the Humming-bird is formed much like that of the woodpecker, being curled round the head, under the skin, and thus capable of being darted to a considerable distance.

There is a fable of a wren and an eagle. The two birds entered into a contest respecting the height to which they could severally attain. A day was fixed, and the birds started. Away went the eagle, soaring in lessening spires, until his form was lost in the clouds. But where was the wren? The eagle had lost sight of his pigmy opponent long ago, but in his pride to show what he could do, he still soared on and on until the lighter air would scarcely bear his weight. As he hovered with wearied and rapidly beating wings, unable to gain another yard, up sprang the wren from among the eagle's feathers, where it had sat very comfortably all the while, and fluttered above his head with a song of triumph.

But truth, as has been often said, is stranger than fiction, as appears from the fact that the eagle can be vanquished by a more insignificant foe than even the wren, by the Humming-bird, which is not content with a mere racing victory, but drives the eagle before it. The Ruby-throated Humming-bird has been seen to dart between the wings of a flying eagle, to perch upon its head, deliberately to strip off the feathers, and send them floating in a stream after the flight of the persecuted eagle, which seemed almost driven to madness by its tiny foe.

Like many other little creatures, the assurance and impudence of the Humming-bird is remarkable. It is easily tamed for that very reason, and has been known to domesticate itself in an hour from the time of its capture, and even when released, it has returned again to partake of the dainties which it had tasted during its captivity.

There are an immense number of species of these exquisite birds, varying from the size of a swift to that of a humble bee. Any description of them is impossible—they must be seen. Fortunately, the magnificent collection brought to England by that most indefatigable and enterprising naturalist, Mr. Gould, places it in the power of every one to view these living gems in all the attitudes of life, and surrounded with the appropriate vegetation. They need nothing but motion. It appears that cold is destructive to the Humming-birds. Wilson says:—

"This little bird is extremely susceptible of cold, and if long deprived of the animating influence of the sunbeams, droops, and soon dies. A very beautiful male was brought me this season, which I put into a wire cage, and placed in a retired shaded part of the room. After fluttering about for some time, the weather being uncommonly cool, it clung by the wires, and hung in a seemingly torpid state for a whole forenoon. No motion whatever of the lungs could be perceived on the closest in spection; though at other times this is remarkably observable; the eyes were shut, and when touched by the finger it gave no signs of life or motion. I carried it out to the open air, and placed it directly in the rays of the sun in a sheltered situation. In a few seconds respiration became very apparent; the bird breathed faster and faster, opened its eyes, and began to look about with as much seeming vivacity as ever. After it had completely recovered I restored it to liberty; and it flew off to the withered top of a pear-tree, where it sat for some time, dressing its disordered plumage, and then shot off like a meteor."

Fear will also produce the same effect, as they have repeatedly died when caught in a common gauze net, which does not injure even the delicate scales of the butterfly's wing. They are very quarrelsome little creatures, and frequently fight with expanded crests and ruffled feathers until they fall exhausted to the ground.

The nests are very neat and beautiful, and, as may be imagined from the diminutive size of the little architect, exceedingly small. They are composed of down, cotton, &c., and are sometimes covered on the outside with mosses and lichens. Waterton relates a curious formation of the nest of one particular species, whose habitations are built at the extremity of thin branches.

"Instinct teaches one species, which builds its nest on the slender branches which hang over the rivers, to make a rim round the mouth of the nest, turned inwards, so as to prevent the eggs from rolling out... The trees on the river's bank are particularly exposed to violent gusts of wind, and when I have been sitting in the canoe and looking on, I have seen the slender branch of the tree which held the Humming-bird's nest so violently shaken, that the bottom of the inside of the nest has appeared, and had there been nothing at the rim to stop the eggs, they must inevitably have been jerked out into the water."

Family III.. Certhĭdæ.—(Gr. Κέρθιος. Creeper kind.)

Sub-family *a.* *Certhīnæ.*

CERTHĬA.

Familiāris (Lat. *familiar*), *the Creeper.*

The CREEPERS are remarkable for their long slender bills and claws, adapted for climbing trees, and capturing insects. The common Creeper may often be seen in this country, running spirally up the trunks of trees, and probing the bark with its bill, and so firmly do the claws hold, that when shot it does not always fall, but remains clinging to the tree. The nest of this elegant little bird is made in a decayed tree. The eggs are from seven to nine in number, grey with dusky spots.

The NUTHATCH.—The term Nuthatch well explains the habits of this interesting little bird. As may be imagined from its name, nuts form a considerable portion of its food, but it also feeds largely on insects, pecked from the bark of trees. While searching after insects, it displays an activity even surpassing the creeper, as it runs up and down the trunk, mostly descending with its head downwards, a feat beyond the capacity of either creeper or woodpecker. In order to break the shell of the nuts, it contrives to fix the fruit in some crevice, and then grasping with its powerful feet, it swings its beak against the nut with the whole force of its body, and soon splits the shell in pieces. In spots frequented by this bird, heaps of nut-shells may be seen, as it usually resorts to a place where it has found a convenient resting-place for the nut, just as heaps of snail-shells may be often found by stones which the thrush has found fitted for breaking them.

It has hitherto been found impossible to keep the Nuthatch in captivity. Its restless spirit and obstinate perseverance in pecking at its prison speedily kill it; and although several have been placed in confinement, none have been recorded to survive beyond the third day. The Rev. Mr. Bree relates that the bill of a Nuthatch which he had taken in a common brick trap was worn away to barely two-thirds of its usual length by the unremitting attempts of the bird to escape from its prison.

Sub-family *b.* . *Sittinæ.*

Sitta.—(Gr. Σίττη.)

Europæa (Lat. *European*), *the Nuthatch.*

The nest of this bird is usually made in a hole in a decayed tree, and as is usual with nests made in holes, it is a very rough fabrication, composed of a few dried leaves. The bird is also remarkable from its habit of plastering up the hole with mud when it is too large to suit it. The eggs are from five to seven in number, of a whitish colour, spotted with reddish brown.

The Wren shares with the robin some immunity from juvenile sportsmen. Although it may be fearlessly hopping about in the hedge, jerking its funny little tail, and playing its antics just at the muzzle of the gun, few boys will fire at it—a privilege for which it is difficult to give a reason, except, perhaps, the very incomprehensible assertion that "The robin and the wren are God Almighty's cock and hen;" although why these two birds, both proverbially quarrelsome and pugnacious, should be selected, to the exclusion of others, is difficult to say. Perhaps the robin enjoys his immunity from the "Babes in the Wood," and the wren makes a convenient rhyme. Be this as it may, it is to be wished that a similar rhyme existed, including the owl and the kestrel.

A singular anecdote is related of this bird.

"In the end of June, 1835, a person was shooting in the neighbourhood of Bandrakehead, in the parish of Colton, Westmoreland: he killed a brace of blue titmice (*Parus cæruleus*), which some time before had been observed to be constructing a nest, in the end of a house belonging to a Mr. Innes of the same place. In the course of the day, it was ascertained that the titmice had completed the time of incubation, and that their death had consequently left their offspring in a state of utter destitution. This, however, was not long permitted to continue, for the chirping of the young birds attracted the attention, and excited the compassion of a wren; which, since that period, adopted the nestlings, and was daily engaged in rearing and feeding them, with the affectionate kindness and unremitting assiduity of a parent bird."

Sub-family *c.* *Menurīnæ.*—(Gr. Μηνύω, I display; *οὐρά*, a tail.)

Troglodȳtes.—(Gr. Τρωγλοδύτης, a creeper into caves.)

Parvŭlus (Lat. *very small*), *the Wren.*

The nest of the Wren is built in any convenient cranny; an ivy-covered tree, the thatch of a barn, or a warm scarecrow, are all used by this fearless little bird. The nest is usually of an oven-like shape, always covered on the outside with some material resembling the colour of the objects round it, such as green moss if built among ivy, or brown lichen if built on a rock or in the fork of a withered branch. The eggs are six or eight in number—white, speckled with reddish brown.

Tribe III. . DENTIROSTRES.—(Lat. *tooth-billed.*)

Family I. . Luscinĭdæ.—(Lat. *Luscinia*, a Nightingale. Nightingale kind.)

Sub-family *a.* *Luscĭninæ.*

Calamodȳta.—(Gr. Καλαμοδύτης, a diver into reeds.)

Locustella, *the Grasshopper Warbler.*

The Grasshopper Warbler.—While walking along the hedges during the spring, an incessant cry, closely resembling that of the grasshopper, and easily to be taken for it, is heard proceeding from the hedge. This cry proceeds from a little bird, called on that account the

Grasshopper Warbler. The little creature keeps so close that it is very difficult to catch even a casual glance at it, as it flits along the bottom of the hedge.

The nest is carefully concealed, and very difficult to find. It is composed of dried grass, and is usually hidden by the tufts of herbage among which it is built. The eggs are from five to seven in number; white, speckled with red. The length of the bird is five inches and a half; the third primary feather is the longest.

The NIGHTINGALE.—

"Tiuu tiuu tiuu tiuu—Spe tiu zqua—
Tiō tiō tiō tio tio tio tio tix—Qutio qutio qutio qutio—
Zquo zquo zquo zquo—Tzü tzü tzū tzü tzū tzü tzü tzü tzü tzi—
Quorror tiu zqua pipiquisi—Zozozozozozozozozozozozo zirrhading! &c. &c."

So does a well-known naturalist endeavour to express the wild and spiritual melody of this most exquisite of British song-birds, the NIGHTINGALE. And in truth it is perhaps as good a description as can be given without the aid of music. Even its own marvellous notes sound comparatively weak unless backed by the accompaniments of night and tranquillity; for the inimitable song of this Mendelssohn among birds loses great part of its beauty when uttered by day, deadened and confused with other sounds. There are some people who cannot appreciate the song of this bird. There is a story that a man who was engaged as gardener in a gentleman's family, was permitted to live within the grounds. In a short time he asked to be allowed to change his house, and on being asked his reason for giving up so good a situation, answered that he could get no sleep at night, because those nasty Nightingales kept up such a continual guggling.

LUSCINĬA.

Philomēla (Gr. Φιλομήλα, proper name), *the Nightingale.*

In some counties of England it is never found, but in many its nightly strains are frequently heard. The fields and College gardens of Oxford are full of Nightingales, whose songs add greatly to the effect of the scene. Well may Isaak Walton say in his delightfully quaint language:

"But the Nightingale, another of my airy creatures, breathes such sweet, loud music out of her instrumental throat, that it might make mankind to think that miracles are not ceased. He that at midnight, when the very labourer sleeps securely, should hear, as I have very often, the clear airs, the sweet descents, the natural rising and falling, the doubling and redoubling of her voice, might well be lifted above earth, and say, '*Lord, what music hast thou provided for the saints in Heaven, when thou affordest bad men such music on earth!*'"

It must be borne in mind, that not only in this bird, but in other singing birds, the male is the vocalist, so that Milton's address to the "sweet songstress" is unfortunately not quite so correct as poetical; a misfortune of frequent occurrence.

SYLVĬA.—(*Proper name.*)

Undāta (Lat. *wavy*), *the Dartford Warbler.*

The WARBLERS are spread over almost the entire globe, and many gladden this country with their pleasant songs. The Dartford Warbler derives its name from the place where it was first noticed as a British bird. It is extremely small, hardly larger than a wren, but the length of its tail increases its apparent size. Furze-bushes form its usual residence. There it may be seen hovering over the tops of the bushes, uttering its curious quavering song, at the same time erecting the feathers of its head, and puffing up its throat. It lays its eggs in a nest carefully concealed in the centre of a furze-bush. The eggs are of a greenish white, speckled with brown spots. Its length is five inches.

SYLVĬA.

Cinerĕa (Lat. *ashy*), *the Whitethroat.*

The pretty little WHITETHROAT is one of the migratory birds, remaining with us during the summer. Few copses are without the singular and pleasing song of the Whitethroat. It derives its name from the white colour of its throat and abdomen, which renders it a conspicuous bird, and even if that proof were wanting, its curious habit of flying upwards from its perch, and again descending on the same spot, immediately points it out.

It builds a small nest, mostly among brambles or on a stump densely covered with weeds. The eggs are five in number—a greyish white thickly spotted with brown. The length of the bird is not quite six inches.

The Blackcap, almost a rival to the nightingale, is at once recognised by the black colour of the crown of the head. Only the males, however, are thus decorated, the crown of the head of the female being dark brown. Its sweet notes are poured forth from the concealment of some thicket or tuft of trees, where it trusts to the density of the foliage to elude discovery. Like the mocking-bird of America, it can imitate the songs of other birds with such perfect inflection that it is almost impossible to detect the imposture.

SYLVIA.

Atricapilla (Lat. *blackhaired*), *the Blackcap Warbler.*

Among bushes and brambles it builds its nest, which is made of dried grass, moss, and hairs. The eggs are five in number—reddish brown, marked with dark spots. The length of the bird is nearly six inches; the third primary feather is the longest.

The Garden Warbler.—This bird is one of our sweetest songsters, and is supposed by some to be little inferior to the Nightingale itself. So we may well pardon its occasional depredations on our garden fruit for the sake of its melody.

It is a migratory bird, arriving in England in April, and leaving towards the end of August or the beginning of September. Almost every part of England is visited by this bird, and especially those counties where are thick woods and plenty of water.

The colour of this Petttichaps is an olive green, shot, as the ladies say, with a greyish shading; while some parts of the body, such as the sides of the neck, the throat, and under parts, are either ash grey or greyish white. The length of the bird is about six inches.

Its nest is built in hedges, and situated near the ground. In it are laid four or five eggs, of a whitish grey colour, spotted with brown, the spots being collected towards the larger end.

SYLVĬA.

Hortensis (Lat. *belonging to the garden*), *the Garden Warbler, or Greater Pettichaps.*

This is the Beccafico of the Italians, so celebrated as a dainty for the table.

The CHIFF-CHAFF, so called from its peculiar cry, is almost the first of the Warblers that visits us in the spring, and one of the last to leave us in the autumn, sometimes remaining here until the middle of October. This little bird is found in most of the southern counties of England, and in Wales, but has not been noticed north of Northumberland. On its first arrival it feeds on the leaf-rolling caterpillars that infest the leaves and early buds of trees, thereby doing great service to the gardener, who especially detests all "blight," as he calls these insidious little devourers.

SYLVĬA.

Rufa (Lat. *ruddy*), *the Chiff-chaff.*

The nest of the Chiff-chaff, like that of the wren, is oval or rounded, and entered by a hole at the side. It is placed near the ground in a bush, or sometimes resting on the hedge-bank. Its eggs are six in number, speckled with purplish red on a white ground. The length of the bird is not quite five inches; the third and fourth primary feathers are the longest.

SYLVĬA.

Trochĭlus (Gr. τροχίλος), *the Willow Warbler.*

The WILLOW WARBLER, or WILLOW WREN, is one of our early visitors, reaching this country at the same time as the blackcap. It much resembles the chiff-chaff and the wood warbler, but may be distinguished from the former by the comparatively light tint of its legs, and from the latter by the yellow tinge of the under parts.

It is an useful bird to the gardener, as it is entirely insectivorous, never touching the fruit or seeds, and should, therefore, be protected by him.

Its nest is oval or roundish in its form, and is entered by a little hole in the side.

The GOLDEN-CRESTED REGULUS, as it ought properly to be called, is one of the smallest of British birds. Fir plantations are its favourite resort, and there it may be seen hopping about the branches, or running round them, head downwards, in search of the insects hidden beneath the bark. Its name is derived from the orange-coloured tuft of feathers on the crown of its head, for which reason it is often called the Kinglet. Its note is weak, but very pleasing, and much resembles that of the common wren. The female is very bold while sitting, and will permit close observation without quitting the nest. The nest itself is an object of great beauty. It is usually placed on the under side of a fir branch sheltered

REGŬLUS.—(Lat.)

Cristātus (Lat. *crested*), *the Golden-crested Wren.*

by the overhanging foliage, and sometimes further protected by a large bunch of cones forming a kind of roof over it. The eggs are from six to ten in number, very small, and of a reddish white colour. The length of the bird is three inches and a half. The fourth or fifth primary feather is the longest.

Sub-family *b. Erythacinæ* (Gr. Ἐρίθακος, a Redbreast).

SAXICŌLA.—(Lat. *Stone inhabiting.*)

Œnānthe (Gr. οἰνανθη), the Wheatear.

The WHEATEAR is one of our early visitors, appearing at the beginning of March. It is a very conspicuous bird, and can be readily distinguished by the black mark that surrounds the eye, and stretches from the base of the bill, to beyond the ear coverts. It is a very pretty songster, its notes being soft and sweet, although wanting in power.

It is killed in great numbers for the table, as its flesh is so delicate as to entitle it to the name of the English Ortolan. In the proper season, the bird is covered with fat to such an extent, that the plumage is often spoiled by the fat running from the holes caused by the shot.

The nest of the Wheatear is made of the usual materials, and is placed in some sheltered spot where it is well concealed from prying eyes. The eggs are five or six in number, of a delicate faint bluish tinge, and very smooth on the exterior.

The REDSTART derives its name from the bright reddish chestnut colour of the upper tail coverts and tail feathers, which appear very conspicuous as the bird flits from one tree to another, or dashes off when startled. It inhabits the skirts of forests, copses, gardens, and especially frequents old ivied walls, where numbers of the nests may be found. In 1847, I found a Redstart's nest built in a hole of a wall, forming one side of a narrow passage in Merton College, Oxford. The eggs were nearly hatched, and the birds did not seem to be disturbed by the constant passing of servants with their paraphernalia of brooms, pails, and other implements. The nest was so placed that every passer by could not fail to perceive it, but the birds sat on their eggs quite unconcernedly.

The song of this bird is not very powerful, but the notes are peculiarly sweet. While singing, it often changes its situation, occasionally singing as it flies.

The nest is placed usually in a hole in a wall, or in a hollow tree. The eggs are five in number, of a greenish blue colour, closely resembling those of the Hedge Accentor. The length of the bird is rather more than five inches. The fourth primary feather is the longest.

RUTICILLA.—(Lat. *sparkling.*)

Phœnicūra (Gr. Φαίνω, I display; οὐρᾶ, a tail), *the Redstart.*

The REDBREAST, or ROBIN REDBREAST, as it is affectionately termed, has, by its fearless conduct, earned itself golden opinions from all kinds of men. Every nation seems to protect it. Even the American Redbreast lives unharmed, possibly on account of its connexion with its English relation, whose oft-told charity towards the Babes in the Wood has turned aside from its posterity even the unsparing hand of the sporting schoolboy.

In the winter, when the berries are gone, insects dead, and the worms hidden under the hard frozen soil, then the Robin flies for refuge to the habitations of man for shelter and food. It is very amusing to see the half trusting, half fearful look with which it hops to the window-sill for the first time. After a while, it becomes bold, and taps at the window, if the expected crumbs are not thrown out. Before very long, it ventures to enter the room, hops about on the table, and quite seems to consider as a right what was first merely a favour. When once established, it is

ERYTHĂCUS.*—(Gr. Ἐρίθακος, a Redbreast.)

Rubecŭla (Lat. *a Redbreast*), *the Redbreast.*

* This word ought properly to be spelled ERITHACUS, but I have taken the etymology of the British Museum Catalogue.

very jealous, and will not suffer a friend to be partaker of the same comforts, but attacks him with the greatest fury; so the unfortunate second comer has to wait shivering outside the window, with his feathers puffed up, and his little bright eye glancing from the depths of the plumage.

About the year 1843, a Robin used to frequent our house. He was so tame as to answer to his name "Bob," and continued his attachment even through the summer. When the rabbits were fed, Bob always came to assist, and usually contrived to perch on the edge of the pan from which the rabbit was eating. Both parties seemed perfectly satisfied, and Bunny and Bob always continued very good friends.

The nest of this bird is built in a crevice of an old ivied wall, in a bank, sheltered by the roots of trees, or in a mass of ivy clinging to an old tree. The eggs are five in number, of a pale grey colour, profusely marked with reddish spots.

Sub-family c. *Accentorinæ.*

ACCENTOR.—(Lat. *a Singer.*)

Modularius (Lat. *warbling*), *the Hedge Accentor.*

The HEDGE ACCENTOR, or HEDGE SPARROW, is one of our commonest English birds, closely resembling the common sparrow in appearance. The nest is built in holes, and contains five blue eggs like those of the Redstart, but stouter in shape, and of a deeper blue.

It is often very bold when engaged in sitting, and will permit a near approach without leaving the nest. I have repeatedly visited the nest of of one of these birds while the female was sitting, and have parted the boughs of the shrub where the nest was placed, in order to get a good view, while the hen bird still sat quietly in the nest anxiously watching every movement, but not attempting to stir.

The TITS.—The birds of the family of the Tits are remarkable for their active habits among the branches of trees. There are few who have not seen these beautiful and interesting little birds twisting round the branches, perfectly unconcerned at the presence of the spectator, sometimes hanging, head downwards, sometimes chasing an unlucky beetle along the bark, and invariably catching it, in spite of its swift limbs and active wings; sometimes twisting off a bud, and pulling it to pieces with marvellous rapidity, in order to secure the lurking caterpillar within; sometimes pecking away at a piece of loose bark, and extracting an unwilling spider by one of its legs left incautiously projecting from its lurking-place. Pity it is that their funny little sharp beaks should ever be put to worse uses; but they lie under a grave imputation of using these very beaks in the slaughter of the defenceless young of other birds.

Sub-family *d. Parinæ.*

PARUS.—(Lat. *a Titmouse.*)

Major (Lat. *greater*), *the Great Titmouse.*

The GREAT TITMOUSE is common in this country, frequenting gardens, orchards, copses, &c. During the spring it is very active in the capture of insects, but in autumn and winter it is forced to content itself with grains and seeds of various descriptions. Gilbert White, in his "Selborne," mentions that he has seen the Great Tit "while it hung with its back downwards, to my no small delight and admiration, draw straws lengthwise from the eaves of thatched houses, in order to pull out the flies that were concealed among them, and that in such numbers that they quite defaced the thatch, and gave it a ragged appearance."

The nest of this bird is built in a hole of a wall, or a decayed tree, and in it are placed six or eight eggs, of a white colour, spotted with reddish brown. The length of the bird is about six inches.

The little BLUE TITMOUSE is so well known as hardly to require any description. It is most amusingly courageous, and from the strenuous resistance it offers to its capturer, has acquired from rustic boys the name of "Billy-biter." The angry hiss of the female has frequently caused an intruding hand to be rapidly withdrawn, for the sound is so exceedingly like the hiss of an irritated snake, and the little beak is so sharp, that few have the courage to proceed with their investigations. A pair of these birds built their nest in the coping of the Great Western Railway, at the Shrivenham station, not two feet from the fiery and noisy engines, which were constantly passing. The men respected the courage of the little birds, and this whole brood was hatched, and suffered to fly at liberty.

The utter contempt which this bird entertains for fire-arms often leads to its destruction, for when the disappointed schoolboy has been wasting

PARUS.

Cœrulĕus (Lat. *blue*), *the Blue Titmouse.*

his powder and shot in attempting to hit larks and such large game, he consoles himself by shooting the unfortunate Titmouse, who will allow him to come so close that few vestiges of it remain except a tuft of blue feathers.

The eggs of the Blue Titmouse are from six to eight in number, white, marked with reddish brown spots. Its length is about four inches and a half.

The LONG-TAILED TITMOUSE is another well-known species of this amusing family. Unlike the other Tits, it does not frequent human habitations during the winter, but may be seen in great numbers twisting and creeping about the branches of hedge-rows and field trees. In the summer they are quite as bold as their relations, and especially favour apple-trees, for the sake of the diseased buds, which they pick off and devour, thereby drawing upon themselves the vengeance of the gardener, who prepares his gun, fires at the supposed depredators, and possibly succeeds in killing them; but he has also succeeded in doing more damage to the healthy buds by his spare shot, than a score of Tits would injure during the entire season.

PARUS.

Caudātus (Lat. *tailed*), *the Long-tailed Titmouse.*

The beautiful and elaborate nest which this bird constructs is one of its chief peculiarities. It is oval in shape, and entirely closed, except one small hole at the side, just large enough to admit the bird. The exterior of the nest is usually covered with lichens, and it is lined with a thick layer of soft feathers. In this warm and elegant habitation are laid from ten to fourteen eggs, which are small and very delicately spotted. The entire length of the bird is about five inches and a half.

The PIED WAGTAIL.—The Wagtails, so named, from the almost incessant vibration of their tails, are exclusively confined to the Old World. The Pied Wagtail is the most common of its race. We often see it pass rapidly, with its peculiar dipping flight; it settles on the ground and wags its tail; it runs a few paces, and wags its tail again; pecks at an insect, and its tail again vibrates. It does not hop, like the warblers, finches, &c., but runs with great rapidity, and altogether looks very like a diminutive magpie. Sand banks by the sides of rivers are the usual resort of these birds, where they may almost always be seen, running about by the water's edge, sometimes snatching at an incautious may-fly, sometimes wading into the water after a caddis-worm or a stray grub, or pecking at an unfortunate little minnow, which has come too near the surface—and then it flies off to another spot to repeat the same manœuvres. This bird also greatly frequents pastures, and may be seen running about among the cows in the most nonchalant manner imaginable, catching the flies that torment those animals in the summer, or flying off to its unfinished nest with a beak full of hairs. Their nests are built near the water, in crevices among stones, or in the hole of a wall. Frequently when stones are piled by a wet quarry, several nests may be found in one heap of stones. The eggs are four or five in number, of a dusky white colour, spotted with ashy brown. The length of the bird is seven inches and a half.

Sub-family *e. Motacillinæ.*

MOTACILLA.—(Lat. *a Wagtail.*)

Yarrellii (Lat. *of Yarrell*), *the Pied Wagtail.*

The YELLOW WAGTAIL* is very similar in habits to the more common Pied Wagtail, but the yellow tints of some of its feathers, somewhat resembling those of the Yellow Hammer, at once distinguish it.

* See page 216.

MOTACILLA.

Flava (Lat. *yellow*), *Yellow Wagtail.*

The Meadow Pipit, more commonly called the Titlark, resembles the true Larks in the long hind claw and peculiar plumage, but is pointed out as distinct, by the different colour of the bill. Like the skylark, it sings while in the air, but sometimes also pours forth its musical strains while settled upon the ground. It feeds principally on slugs, worms, and insects, which it chases with much activity, after the manner of the wagtails, even vibrating its tail like them. Hilly grounds, commons, and meadows are its chief resort in summer, but during September and October flocks of these birds may be seen congregated in turnip fields, and in the winter they seek the protection of the warm hedge-rows.

Anthus.—(Lat. *a Titlark.*)

Pratensis (Lat. *of a Meadow*), *the Meadow Pipit.*

The nest of the Titlark is made on the ground, and concealed by a tuft of grass. There are usually five or six eggs, light brown in colour, spotted with a darker tint. The length of the bird is six inches.

The Water Ouzel, or Dipper, is one of the most interesting of our native birds. It is found principally in hilly places where there are clear and rapid streams, such as in Derbyshire and Yorkshire. There it may be seen to go through its far-famed movements under the water, which have given rise to so much controversy. It dives for considerable distances with apparent ease, and has a habit of dipping and rising repeatedly, from which practice its name has been derived.

The nest is usually built by the water side, and is most carefully concealed. In general appearance it is not unlike that of the wren, being made of intertwined mosses, with an entrance at the side. It lays five largish eggs, of a pure white. The length of this bird is about seven inches.

Family II. . . . Turdĭdæ.—(Lat. *Turdus*, a Thrush. Thrush kind.)

Sub-family *a*. . *Formicarīnæ.* (From Lat. *Formīca*, an ant.)

HYDROBĂTA.—(Gr. Ὕδωρ, water; βαίνω, I go.)

Cinclus (Gr. Κιγκλος), *the Dipper*.

The MISSEL, or MISSELTOE THRUSH, or STORMCOCK, according to Waterton, "surpasses all other thrushes in size, and is decidedly the largest songster of the European birds. He remains with us the whole of the year, and he is one of three birds which charm us with their melody during the dreary months of winter, when the throstle and lark are silent, and all the migratory birds have left us, to sojourn in warmer climates. He appears

Sub-family *b*. *Turdīnæ.*

TURDUS.

Viscivŏrus (Lat. *Viscus*, a Misseltoe; *voro*, I devour), *the Misseltoe Thrush.*

to be gregarious in the months of August and September." "This bird, though usually known by the name of the Misseltoe Thrush in many parts of England, is invariably called the Stormcock by all the lower orders in our neighbourhood: not that it delights in storms more than in fine weather; but that nature has taught it to pour forth its melody at a time of the year when the bleak winds of winter roar through the leafless trees."

It is very fond of the berries of the misseltoe, but when they fail it turns its attention to those of the mountain ash, which are almost certain to attract this beautiful and powerful songster. In the summer it devours all kinds of garden-fruits, especially cherries and raspberries.

During the breeding season it is very pugnacious, attacking and driving away not only small birds, but the crow, the magpie, or even the prowling cat. The nest is very large, almost as large as a "wide-awake" hat, is always built in a tree, and contains about five reddish spotted eggs. The length of the bird is eleven inches.

The Fieldfare is properly a native of the cold regions of Europe, and only visits this country during the cold winter months. From its excellence as an adjunct to the table, it is perseveringly sought after, but

TURDUS.

Pilāris (Lat. *like a ball*), *the Fieldfare.*

is so shy, that unless the bird is very busy satisfying its hunger, there is some difficulty in approaching within gun-range.

It builds in fir or pine-trees, and lays several bluish-green eggs, spotted with brown. Its length is about ten inches.

The SONG-THRUSH, THROSTLE, or MAVIS, is deservedly considered one of our best singing birds. Its powerful and rich notes may be heard even during the month of January, when most of the other singing birds are either silent, or have departed. Its nest is built almost before any other bird has commenced, and may often be seen conspicuously placed in a bush, some time before the leaves have begun to sprout. In order to defend the callow young from the cold winds of the season when they are hatched, the nest is more substantial than birds are accustomed to build, being thickly plastered within with a coating of mud, effectually keeping out the chilly blasts. Were it only for its singing powers, the Thrush would deserve protection; but the services it renders to the gardener in devouring insects, snails, and other destructive creatures, entitle it to a double share of regard.

TURDUS.

Musĭcus (Lat. *musical*), *the Song-Thrush.*

It is very amusing to watch a Thrush listening for the sound of the earthworm working his way through the ground, or the gnawing teeth of the cockchaffer grub. The grub he unearths and devours without further ceremony, but he knows that if he is not cautious, the earth-worm will withdraw itself out of his reach. He therefore gives several hops near the worm, which, fancying that it hears its enemy the mole pursuing it, comes to the surface, and is instantly seized in triumph by the crafty thrush.

It clears the shells from snails by beating them against a stone, and when it has found a convenient place for that purpose, it invariably returns to the same spot with its prey, so that heaps of broken snail-shells may often be found where the thrushes have been at work.

The eggs of the Thrush are five in number, of a bluish-green colour, spotted with a deep reddish brown. Sometimes the spots are altogether absent.

The BLACKBIRD is another delightful songster, whose jetty hue and "orange-tawny bill" are too well known to need description. It is a very shy bird, and if disturbed in a hedge, has a habit of darting through it, and then escaping on the other side, uttering a sharp cry of alarm. The

TURDUS.

Merŭla (Lat. *a Blackbird*), *the Blackbird.*

habits of this bird are not unlike those of the thrush, especially in its zeal for unearthing the cockchaffer-grubs, and possibly for eating cherries when they are ripe.

Its nest is built usually at the foot of a hedge, frequently in the very centre of a holly bush, safe from most enemies, except weasels and schoolboys.

A beautiful King Charles' dog of my acquaintance was accustomed to search for thrush's and blackbird's nests, and bite out the bottom of them, so that the eggs rolled quietly into his mouth, he having discovered that when he tried to take them from the mouth of the nest, he invariably broke them. The eggs are five in number, of a bluish-green colour, profusely spotted with brown.

The MOCKING BIRD, or POLYGLOT THRUSH, is a native of most parts of America. This wonderful bird stands preeminent in powers of song. Not only are its natural notes bold and spirited, but it has the faculty of imitating with deceptive fidelity every sound it hears. To its flexible organs, the harsh setting of a saw, the song of a nightingale, the creaking of a wheel, the whistled tune of a passer-by, the full and mellow notes of the thrush, the barking of a dog, the crowing of a cock, and the savage scream of the bald eagle, are each equally easy of execution, and follow one another with such marvellous rapidity that few can believe that the insignificant brown bird before them is the sole author of these varied sounds. The Virginian nightingale and the canary hear their exquisite modulations performed with such superior execution, that the vanquished songsters are silent from mere mortification, while the triumphant Mocking-bird only redoubles his efforts. Wilson, whose animated description of this bird has never been surpassed, says:—" His expanded wings and tail glistening with white, and the buoyant gaiety of his action arresting the eye, as his song does most irresistibly the ear, he sweeps round with enthusiastic ecstasy, and mounts and descends as his song swells or dies away. He often deceives the sportsman, and sends him in search of birds that are not perhaps within miles of him, but whose notes he exactly imitates: even birds themselves are frequently imposed upon by this admirable mimic, and are decoyed by the fancied calls of their mates, or dive with precipitation into the depth of thickets at the scream of what they suppose to be the sparrow-hawk."

While sitting on its eggs it is an exceedingly courageous bird, attacking without discrimination man, dogs, or any animal who may approach too near the nest. But the black snake is the special object of its vengeance. The snake, who has perhaps just arrived at the vicinity of the nest, and is contemplating a pleasant breakfast on the young or eggs, is violently attacked by the enraged Mocking-bird, who, by repeated blows on the head, generally destroys its enemy, and then mounting upon a bush, pours forth a triumphant song of victory.

ORPHEUS.—(Gr. proper name of a famous musician.)

Polyglottus (Gr. Πολύς, many; γλῶττα, a tongue), *the Mocking Bird.*

Webber's account of this magnificent songster is so pleasing that I insert part of it here:—"I saw it now leaping up from its favourite perch on a tree-top, much in the manner I had observed before; but now it was in a different mood, and seemed to mount, thus spirit-like, upon the wilder ecstasies, and floating, fall on the subsiding cadence of that passionate song it poured into the listening ear of love, for I could see his mate, with fainter bars across her wings, where she sat upon a thornbush near and listened.

"When this magnificent creature commenced to sing, the very air was burdened with a thousand different notes, but his voice rose clear and melodiously loud above them all. As I listened, one song after another ceased suddenly, until in a few minutes, and before I could realize that I was so, I found myself hearkening to that solitary voice. This is a positive fact! I looked around me in astonishment. What! are they cowed? but his song only now grew more exulting, and, as if feeling his triumph, he bounded yet higher with each new gush, and, in swift and quivering raptures, dived, skimmed, and floated round, round, then rose to fall again more boldly on the billowy storm of sound.

"No wonder the other birds were silent to listen, for, one after one, he hurled the notes of each upon its ear, so alchemyzed with splendour that they knew not their own song.

"This curious phenomenon I have witnessed many times since. Even in the morning choir, when every little throat seems strained in emulation, if the mocking-bird breathes forth in one of its mad, bewildered, and bewildering extravaganzas, the other birds pause almost invariably, and remain silent until his song is done. This, I assure you, is no figment of the imagination or illusion of an excited fancy; it is just as substantial a fact as any other one in Natural History. Whether the other birds stop from envy, as has been said, or from awe, cannot be so well ascertained; but I believe it is from the sentiment of awe; for as I certainly have felt it myself in listening to the mocking-bird, I do not know why these inferior creatures should not also. It must be known that these creatures differ from each other as do men and women in their vocal powers, and there is usually *one* bird in a neighbourhood that supremely surpasses all the rest. It is another remarkable fact, that all other mocking-birds retire from the immediate neighbourhood of this acknowledged monarch, to such a distance that you can hear but the faintest note from them in the pauses of his song, and it sounds as if they but prolonged the echo."

The nest is made generally in a bush or apple-tree, frequently close to houses, as the bird is protected by the inhabitants. The Mocking-bird is often kept tame, in which case, so far from its imitative powers showing any decrease, the variety of domestic sounds heard about the house is often very perplexing.

The GOLDEN ORIOLE.*—The genus Oriolus has the beak notched and rather bent at the tip. The Golden Oriole is only an occasional visitor to this country, and has never as yet been seen in Scotland. It is a very shy bird, frequenting the skirts of woods, especially copses that border on larger woods. In the fruit season it leaves the woods for the orchards, and makes no small havoc among the fruit, particularly the figs, grapes, and cherries.

The nest is made of wool and fine hay; it is generally placed on the fork of a bough. The eggs are five in number, of a purplish white spotted with reddish marks. The length of the bird is rather more than ten inches.

* See page 223.

Sub-family *c. Oriolīnæ.*
ORIŎLUS.—(Linnean generic name.)

Galbŭla (Lat.), *the Golden Oriole.*

Family III. . Muscicapĭdæ.—(Gr. *Musca*, a Fly; *capĭo*, I take. Flycatcher kind.)

Sub-family *a. Muscicapīnæ.*

MUSCICĂPA.

Griŏsla, *the Spotted Flycatcher.*

The SPOTTED FLYCATCHER may be considered as the type of the entire family. It may be constantly seen in gardens and orchards, going through the evolutions that have given it the names of Flycatcher, Post-bird, Beam-bird, &c. It takes its station on some elevated spot, such as the overhanging bough of a tree, a post, or a rail, and from thence watches for a passing insect, on seeing which it darts from its post, secures the insect in the air, and returns to the same spot by a short circular flight. It is not a timid bird, and will permit an observer to stand quite close to it, provided that he does not disturb it.

I have seen one of these birds engaged in the pursuit of flies in a garden at Headington. It perched on a balustrade close to a window from which several persons were watching it, and continued its evolutions perfectly undisturbed by their proximity. On another occasion I was keeping watch in a gig in Nuneham Park, and to pass away the time, amused myself with cutting off the heads of the white clover with the lash. While so engaged, a Spotted Flycatcher came and took up its station on a bough close by the gig, from which it made excursions among the flies and other insects that were driven from the grass and flowers by the whip.

It is only a summer visitor to England, arriving in May and departing about the beginning of October. The note of this bird is a weak chirp, and even that is not often heard.

The nest is built usually in holes of trees or walls, or sometimes between a branch of a wall-fruit tree and the wall itself. The eggs are five in number, spotted with reddish brown on a grey ground. The length of the bird is about five inches.

Family IV. . . Ampelĭdæ.
Sub-family *a.* . *Ampelīnæ.*

Ampĕlis.—(Gr. Ἀμπελίς or Ἀμπελίων.)

Garrŭlus (Lat. *chattering*), *the Bohemian Waxwing.*

The Bohemian Waxwing, or Waxen Chatterer, is only occasionally seen in England during severe frosts, at which time flocks of them sometimes arrive. One of these birds was shot near Oxford in the winter of 1846. It is very common in Norway and Russia, and is plentiful in North America. The name of Waxwing is given to it from the singular appendages to the secondary quill feathers, bearing much resemblance to a drop of red sealing-wax pressed on the wing.

Berries of all kinds, especially those of the dog-rose and the hawthorn, form the principal food of this bird; but it is related that when in captivity it rejects scarcely any vegetable substance, but loses at the same time all its vivacity and social habits. The note of the Waxwing is not unlike that of the thrush, but it is very weak and more uncertain than the notes of that beautiful songster. While singing it agitates the crest on its head, but shows scarcely any of that swelling in the throat so perceptible in the canary and other singing birds.

The length of the bird is rather more than eight inches.

Family V. . . Lanidæ.—(Lat. *Lanius*, a Butcher. Butcher-bird kind.)
Sub-family *a. Laninæ.*

The SHRIKES or BUTCHER BIRDS well deserve their name, as they live upon insects and small birds, which they kill and afterwards transfix with a thorn, preparatory to devouring them. They take their prey much after the same manner as the flycatchers, by darting on it from some place of concealment.

LANIUS.—(Lat. *a Butcher.*)

Excubitor (Lat. *a Sentinel,*) *the Great Grey Shrike.*

The GREAT GREY SHRIKE is supposed to be only an occasional visitor to this country. It feeds upon mice, birds, frogs, and other small animals. After pouncing upon its prey, the Shrike, by a few blows on the head from its powerful bill, destroys it. The unfortunate animal is then carried to the nearest hedge, impaled on a thorn, and the Shrike devours it at his leisure. Large insects are treated in the same manner. The object of this impalement is apparently that the creatures thus suspended should become tender or "high," so that the modern epicure, who hangs up his venison until no one with an unsophisticated taste would venture to touch it, has but borrowed his custom from the Shrike. The bird after hanging a lizard or a mouse in this fashion, generally goes off and fetches another, always preferring to eat those which have remained longest on the thorn, and which are as it were cooked in the sun.

There is a strong bodily resemblance between this Shrike and the Mocking-bird, the distinction lying generally in the outline, while the plumage is so similar that many persons have actually confused the two birds, giving to one the habits of the other. Moreover, the resemblance is not merely in outward form. The Grey Shrike can also imitate the notes of other birds, and often does so.

Lest I should be suspected of committing the same mistake, and supposed to have given the song of the mocking-bird to the Shrike without foundation, I have given my authority. Audubon, in his work on the American birds, has this passage:—

"This valiant little warrior possesses the faculty of imitating the notes of other birds, especially such as are indicative of pain. Thus it will often mimic the cries of sparrows and other small birds, so as to make you believe you hear them screaming in the claws of a hawk; and I strongly suspect this is done for the purpose of inducing others to come out from their coverts to the rescue of their suffering brethren. On several occasions I have seen it in the act of screaming in this manner, when it would suddenly dart from its perch into a thicket, from which there would immediately issue the real cries of a bird on which it had seized. On the banks of the Mississippi, I saw one which for several days in succession had regularly taken its stand on the top of a tall tree, where it from time to time imitated the cries of the swamp and song-sparrows, and shortly afterwards would pitch down like a hawk, with its wings close to its body, seldom failing to obtain the object of its pursuit, which it would sometimes follow even through the briars and brambles among which it had sought refuge. When unable to secure its prey, it would reascend to its perch, and emit loud and discordant notes of anger. Whenever I could see it strike its victim, it appeared to alight on its back, and instantly strike its head, which on such occasions I have several times found torn open. If not disturbed, the Shrike would then tear up the body, and swallow in large pieces, not well cleaned of the feathers, every part excepting the wings. It now and then pursues birds that are on the wing to a considerable distance. Thus I saw one follow a turtle-dove, which, on being nearly caught, pitched on the ground, when its skull was bruised in a moment; but the next instant both birds were in my possession."

One of the New Holland Butcher Birds, *Vanga destructor*, has been known to learn to whistle a tune like a bullfinch. The melody in question was the Scotch tune of "Over the water to Charlie."

The name Excubitor or Sentinel is given it from its habit of watching for birds of prey, and chattering loudly directly it perceives them; thereby proving that, like most other tyrants, it has a great objection to suffering any injury itself. The bird-catchers on the Continent take advantage of this peculiarity, to assist them in the capture of the peregrine falcon. The fowler places a small net on the ground, with a pigeon fastened to it by way of bait. A string is attached to the net and brought within a turf hut where the fowler sits. Close to the hut a Shrike is tied to the ground, and two pieces of turf are set up as a shelter for the bird from the weather, and as a refuge from the hawk. The fowler remains within his hut busied with some sedentary occupation, knowing well that his vigilant watchman will not fail to give him notice of the approach of a hawk. Directly a hawk appears in the distance, the Shrike becomes agitated; as it draws nearer, he begins to scream with fright; and just as the hawk pounces on the pigeon, he runs under his turf, which is the signal to the fowler within the hut to pull the string, thereby enclosing the hawk within the folds of the net.

The nest is built on trees, and contains about six eggs, greyish-white, spotted with dark ash on the larger end; the length of the bird is from nine to ten inches.

The RED-BACKED SHRIKE is much more common than the last-mentioned bird, and may be seen in and about hedges, in the spring, when it is occupied in building its nest. It is rather a noisy bird, and the nest is so large as to be easily discovered. It feeds principally on insects, such as bees, beetles, &c., which may frequently be found impaled on thorns.

LANĬUS.

Collurĭo (Gr. Κολλυρίων), *the Red-backed Shrike.*

I have found many insects impaled by this bird, but the insect most commonly found by myself, in this position, was the Staphylinus erythropterus, but I have also found ground beetles and humble bees thus impaled.

These impaled insects form a very good indication as to the locality of the nests, and are probably placed there for food; certainly not, as some authors have stated, for the purpose of decoying other birds to the spot in order to murder and devour them. The nest and eggs much resemble those of the Great Shrike, but are smaller. The length of the bird is seven inches and a half.

The JAY.—The CORVIDÆ are peculiarly remarkable for a kind of preternatural air of sagacity with which they set about any self-imposed task, especially if that task be a mischievous one. The ravens and magpies are most conspicuous in these qualities.

The Jay, so well known for the beautiful blue markings on its wings, is rather a shy bird, preferring to reside in the thickest woods, and seldom coming into the open country. It is easily tamed when young, and is very amusing when domesticated.

This bird possesses, like several others of the same family, considerable talents for mimicry. It has been known to imitate the sound of a saw, the bleat of a lamb, or even the neighing of a horse, with the most perfect accuracy. Nor do its powers cease here, for although its natural voice is harsh and grating, yet it can imitate the sweet notes of singing birds, such as the Greenfinch, with wonderful fidelity. It has also frequently been taught to articulate words.

The name of Glandarius has been given to the Jay, because it feeds on vegetable productions, such as acorns, &c., more than the true Crows. It is also partial to fruits, especially ripe cherries, and is consequently persecuted by the gardener. It is also said to devour eggs and young birds.

Its nest is built about twenty feet from the ground, the upper part of

Tribe IV. . . CONIROSTRES.—(Lat. *Cone-shape beaked.*)
Family I. . . Corvĭdæ.—(Lat. *Corvus*, Crow. Crow kind.)
Sub-family *a. Garrulīnæ.*

GARRŬLUS.—(Lat. *talkative.*)

Glandarĭus (Lat. *of the Acorn*), *the Jay.*

a thick bush being preferred. The eggs are five or six in number, of a yellowish white, thickly speckled with brown. The length of the bird is nearly fourteen inches.

It is somewhat remarkable that in some parts of England a Jäy and a Jaypie should be considered as distinct birds. Such is the case in Wiltshire. A year or two ago I was rather in want of a few good Jay's eggs for my collection, and on happening to mention the fact in the hearing of one of the country boys, he said that Jaypies were common in Wiltshire, and that he would get me some. Of course, I imagined that a Jay and a Jaypie must be the same bird, and merely thought that the suffix of "pie" was meant to mark that the Jay belonged to the Pies, just as the bird in the engraving is called either the nutcracker, or the nutcracker crow. So the boy went off happy in a commission to get some Jaypies' eggs at a certain sum for each. In a day or two, I was told that the boy had brought the Jaypies' eggs.

I thought it somewhat remarkable that the boy should have thought it necessary to bring a friend with him, and that both should be bare-headed, but when they produced their hats, stating them to be full of Jaypies' eggs, I believe that I stared at them in helpless astonishment, supposing that they must have ransacked every Jay's nest in the county at least. On uncovering the hats, they were found to be filled with the eggs of the misselthrush, which bird it appears is called the Jaypie in those parts. So I had to pay for nearly two hundred misseltoe thrushes' eggs, of which I had plenty, and could have obtained as many as I wished by taking some twenty steps into our orchard. After I had paid for them, the boys wanted to know whether they should get some more. I was so cross about the whole matter, that, I regret to say, I put them up as "cockshies," and together with my pupils spent more than an hour in demolishing these ill-omened eggs, by throwing stones at them from a short distance. I have found no real Jay's eggs from that time to this.

Sub-family *b. Corvinæ.*

NUCIFRAGA.—(Lat. *Nux*, a nut; *frango*, I break.)

Caryocatactes (Gr. Καρυοκατάκτης, properly, the Nuthatch), *the Nutcracker Crow.*

The NUTCRACKER CROW, whose true position in the scale of creation has so long bewildered naturalists, is about the size of a jackdaw, but its form is more slender, and the tail is longer. It is seldom found in this country, but is very common in more northern districts. In its habits it displays a singular mixture of the woodpecker and the nuthatch, and exhibits so few of the well-known habits of the Crows, that observers might well be perplexed where to place it.

It is now supposed to be one of the connecting links between the crows and the woodpeckers.

It runs about the branches of trees, using its tail for a support, and pecks away the bark in order to reach the insects beneath, just as the woodpeckers do. It also pecks open the fir-cones, in search of the hidden seed, and breaks nuts by repeated strokes of its bill, like the nuthatch. It is usually seen in flocks, but is not so wary as the crows.

Its eggs are laid at the bottom of a hole in some tree. They are of a greyish yellow colour, diversified with a few dark grey spots.

PICA.—(Lat. *a Magpie.*)

Caudāta (*long-tailed*), *the Magpie.*

The MAGPIE, who seems to rival the Parrot in the proud title of the Monkey of the Birds (the Raven being the ornithological Baboon), is a well-known inhabitant of this country. Its thieving and hiding propensities have been frequently told; but I must still venture to give a few anecdotes of a tame magpie that resided in Wiltshire. This bird found a malicious enjoyment in pecking the unprotected ankles of little boys not yet arrived at manly habiliments, and was such a terror to the female servants that they were forced to pass his lurking-place armed with a broom. One of the servants having neglected this precaution, was actually found sitting down on the stones to protect her ankles, the magpie triumphantly pacing round her, until aid was brought, and the bird driven away. But to little boys and girls the magpie showed no mercy, springing out of its hiding-place and chasing them completely along the garden walk.

It had also a great penchant for tearing and biting to pieces any papers that came in its way, probably because it had perceived that people valued them. One Sunday morning, after the family had returned from church, the rector found his study strewed with pamphlets, torn newspapers, &c., so that until the delinquent was discovered, he really thought that thieves had been in the house. A Magpie never seems to be happy unless it possesses a hiding-place, nor did this one form an exception to the general rule, as it had pecked a hole in the thatch of a barn, wherein to dispose its ill-gotten goods, and displayed great uneasiness if anybody approached it.

Another Magpie gained entrance into the chapel of Wadham College, Oxford, and remained quiet enough until the service had begun, when it gravely walked up the centre, bowing and saying, "Pretty Mag! Pretty Mag!" much to the discomposure of the junior members. A curious story is told respecting the power of the Magpie to count numbers.

"George Le Roy states that a magpie having stolen some game, it was resolved to shoot it. A man hid himself in a hut near its nest for this purpose. The bird flew away when he entered, nor would return. The next day two men entered and one came out. Mag was not to be cheated; she waited till the second left also. Three went in and two came out, with the same result. Four then entered, and three came away. The bird went back, and was shot.—So magpies, says George Le Roy, can count three but not four."

The nest of the Magpie is built on a high tree, and curiously defended with thorns, having only a small hole just large enough to admit the owners, so that the liberal use of a pocket knife is frequently requisite in order to obtain the eggs. The nest is covered with a dome of thorns, respecting which a curiously quaint fable is told.

"The birds," says the historian, "not knowing how to build nests, went in a body to request the magpie to teach them. He willingly undertook the office. 'First,' he said, 'you must look out for a good, strong, forked branch, and begin by laying two sticks crosswise.' 'That's just what I did,' said the rook. 'Next, you must raise the sides a little, and then put in some hay, which you must work well into the sticks.'—'The very thing I have been doing,' said the crow. 'Now, for fear the eggs should be broken or thrown out, you must raise the sides about as high as your head when you sit in the bottom of the nest, and put in some soft wool.' 'Why,' said the thrush, 'I did as far as that before I came here.' 'Oh! then,' replied the magpie, 'as I see that you all know how to make nests, there is no occasion for me to teach you.' And that is the reason why the other birds are only able to build half nests."

The interior of the nest is defended by a coating of mud, worked smooth. The eggs are five in number, of a greenish white, covered with brown markings. The length of the bird is about eighteen inches.

The Raven is very common on the Continent, and most parts of Asia and America, but is now seldom seen in this country except in a domesticated state. It is more frequently found in the Hebrides than in any other part of Great Britain. In those islands it lives principally on carrion of various kinds, such as dead sheep or lambs, whose death the Raven is accused with some justice of hastening, and on fishes or cetaceous animals which have been cast on shore by the waves. In these cases the Raven conducts itself much in the manner of the vulture. It commences by taking out the eye and tongue, and then proceeds to tear open the abdomen, operations for which its sharp and powerful bill seems quite as well fitted as the hooked beak of the rapacious birds. It is a very crafty bird, and can with difficulty be approached; but by laying a

Corvus.—(Lat. *a Crow.*)

Corax (Gr. Κόραξ, a Raven), *the Raven.*

dead carcase near its haunts, and being carefully concealed, it may be seen cautiously approaching: first, perching on an eminence, it looks carefully round; then, advancing with a sidelong step, it examines its expected prey. When fully satisfied, it pecks out the eyes, and proceeds to satiate itself with food. The Raven seems to rear in storms, and to be deterred by no inclemency of weather from seeking its prey.

Although formerly so plentiful in England that innumerable omens were drawn from its appearance, its croaking, or its flight, it has almost become extinct, much to the discomfiture of omen seekers. No incantation and no dance of witches seemed to be considered complete, without a black cat, a toad or two, a bat, and a raven. Certainly the extraordinary gravity which marks the demeanour of the Raven has something almost preternatural in it. The manner in which he sets about a piece of mischief, as if he considered it a moral duty, is most absurd, and the pertinacity with which he prosecutes a great work, such as the feat of Charles Dickens' Raven, who "new pointed the greater part of the garden wall, by digging out the mortar, and tore up and swallowed in splinters the greater part of a wooden staircase of six steps and a landing," is perfectly astounding.

A Raven in our possession used to watch the gardener taking particular pains to prop up and secure a valuable plant. His labour was always in vain, for the Raven, with a sidelong step and an unconcerned air, as if he were thinking of anything but the plant, would sidle by it, when one wrench of his iron bill laid the unfortunate plant on the earth, and the Raven moved off with a most provoking air of innocence. The lady to whom the garden belonged was quite afraid of the bird, and declared that she almost believed that it was possessed by some evil spirit. It used to walk behind her, so that she could never see it; for when she turned round, the Raven hopped round too, and kept himself completely out of her sight. At last it became so very mischievous that it was sent away, much to my regret.

Not long ago, I was visiting a small collection of living birds, among which was a Raven, whose wings were clipped, and who was permitted to have the free range of the yard. He gained considerable benefit from his freedom, for he could steal the provisions of the other birds, unless they were very quick. When I went to his residence, I took the back of a letter, and was reading the address, when I saw the Raven watching my proceedings with great curiosity. The paper was of no consequence, so I let it fall, and walked on as if it had been an accident. The Raven waited until I had left the paper some few paces behind, when he took a sidelong kind of a walk towards it, tore it into scraps, and ran away with the largest piece under a water-butt, where he kept watch over it.

It has a great capacity for imitating sounds, and can be taught to pronounce whose sentences, or sing songs with wonderful accuracy.

In the northern parts of Scotland it makes its nest on high rocks, but not unfrequently builds on the summit of a tall tree. The nest is a large irregular structure of heath, grass, wool and feathers, and sea-weed, if it builds near the sea-shore. It lays from four to seven eggs, of a pale green colour, spotted with greenish brown. The length of the bird is two feet two inches, and the expanse of wing four feet eight inches.

The Rook inhabits almost every part of Europe, and is very common in England, where it lives in a kind of semi-domestication, usually inhabiting a grove of trees near a house, or in a park, where it is protected by the owner, although he makes it pay for this accommodation by shooting the young ones every year. Apparently in consequence of this annual persecution, the Rook has an intense horror of guns, perceiving them at a great distance. While feeding in flocks in the fields, or following the ploughman in his course, and devouring the worms and grubs turned up by the share, the Rook has always a sentinel planted in a neighbouring tree, who instantly gives the alarm at the sight of a gun, or of a suspicious-looking object.

The good which the Rook does by devouring the grubs of the cockchaffer, and the tipulas, or daddy-long-legs, both of which are exceedingly injurious to the crops. more than compensates for the damage it sometimes causes, by pulling up young corn, or newly set potato cuttings; in the latter case more, I believe, to get at the wireworms, which crowd to

CORVUS.

Frugilĕgus (Lat. *Corn-gatherer*), *the Rook*.

the slices of potato, than to eat the vegetable itself. In the fruit season, the Rook, like most other birds, likes to have his share of the cherries, pears, and walnuts, but may be easily kept away by the occasional sight of a gun.

Towards evening, the Rooks may be seen flying in long lines to their resting-place—"The blackening train of crows to their repose." They then perform sundry evolutions in the air, and finally settle to rest.

Round the base of the Rook's beak is a whitish-looking skin, denuded of feathers, the reason or cause of which is not very obvious. A white variety of the Rook is sometimes seen. The gamekeeper at Ashdown had a very fine white Rook, which he kept tame in his garden.

The eggs of this bird are five in number, similar to those of the raven in colour, but much smaller. The length of the bird is nineteen inches.

The JACKDAW is another well-known bird. It does not build in the branches of trees like the rook, to which it is very similar in many respects, but prefers holes in decayed trees or old buildings, particularly frequenting church towers and steeples. The Jackdaw feeds upon almost any substance that it can find. It kills mice with a single blow of its beak, and then devours them piecemeal. Grasshoppers, beetles, &c. are also killed by a squeeze across the thorax, and the head, wings, and legs, are twisted off before the bird begins to eat them. It treats bees, wasps, and other stinged insects, with much more caution. The feathers upon the crown of its head are of a greyish white colour, a peculiarity instantly distinguishing it from the rook. It is frequently kept tame, and is very amusing in captivity.

CORVUS.

Monedŭla (Lat. *a Jackdaw*), *the Jackdaw.*

A tame Jackdaw, in the possession of one of my scholars, used to travel backwards and forwards from the school to his own home. When in the train, his terror at the sight of the trees whizzing past him was most ludicrous. He uttered a sharp cry of fear, twisted round in his basket, and thrust his beak through the interstices on the opposite side. After he had recovered from his fright, he was usually very talkative. He had a great notion of his own consequence, and if any one spoke in a louder tone than usual, thought it his duty to answer. His love of washing was unbounded, and he would have washed twenty times a day, if he could have persuaded us to give him the water, for which he used to ask in a very expressive manner.

The eggs are of a lighter colour than those of the rook, smaller and more sparingly spotted. The length of the bird is fourteen inches.

The CROW, or CARRION CROW, as it is erroneously called, seldom feeds on carrion; for poor indeed would be his meals were he dependent on dead sheep or horses for a livelihood. Possibly the name was given as a distinction between it and the rook. Waterton states that the flesh of the Carrion Crow is just as good as that of the rook, and relates how he once served up a pie of these birds to some friends, who thought them pigeons. It will also eat cherries and walnuts like the rook, and when the supply of insects has failed, it will then turn its attention to the duck-pond and farm-yard, and carry off a young duckling or chicken.

"Sometimes he approaches the farm-house by stealth, in the search of young chickens, which he is in the habit of snatching off, when he can

CORVUS.

Corōne (Gr. Κορώνη), *the Crow.*

elude the vigilance of the mother hen, who often proves too formidable for him. A few days ago, a crow was observed eagerly attempting to seize some young chickens in an orchard, near the room where I write; but these clustering round the hen, she resolutely defended them, and drove the crow into an apple-tree, whither she pursued him with such spirit and intrepidity, that he was glad to make a speedy retreat and abandon his design." *

It also carries off eggs, by pouncing upon them, and driving its bill through the shell. It will be seen, from the following anecdote, that mice and rats are not unaccustomed food.

In a field near a gentleman's house, about a mile from Caernarvon, there are some out-buildings much infested with rats. Four or five traps are set on the premises every night, and it is the business of a servant-man to go to the spot between five and six in the morning. He is always punctually met by a company of crows that station themselves at a little distance, and most narrowly watch all his proceedings. No sooner does he remove his captives from the traps and throw them into the field, than the carnival begins. The crows seize upon their booty, scientifically perforate the integuments, and scoop out and devour every particle of flesh, even in the head. In a very short time the skins are turned inside out, and a few clean picked bones are the only memorials of the banquet.

The nests of this bird are placed on the summit of some tall tree, and

* Waterton.

contain about five eggs, closely resembling those of the rook. The length of the bird is eighteen inches.

The HOODED CROW, otherwise called the Royston Crow or the Grey Crow, is rather a scarce bird in the British Islands, although scattered over nearly every portion of Great Britain, even including Scotland.

It is one of the winter visitors to this country, generally leaving us about April, although it sometimes remains during the summer, and brings up a brood of young. Like most of its congeners, it builds its nest on the tops of very tall trees, such as the pine, but is also known to build on

CORVUS.

Cornix (Lat. *a Crow*), *the Hooded Crow.*

precipitous rocks. It is said to use these rocks in the stead of an oyster-knife, for as it is very fond of oysters, and does not possess a knife to open them with, it must discover some other method of getting at the enclosed animal. To attain this purpose, it is said to seize the oyster in its beak, soar up to a great height in the air, and to let the oyster drop from that elevation upon the hard rock, when the shell is dashed to pieces, and the Crow is enabled to pick out the animal with ease.

There is but little of the usual Corvine black hue about this bird, only the head, throat, wings and tail being so decorated, the remainder of the bird being of an ashy grey. The length of the bird is about twenty-two inches.

The CHOUGH is rather larger than the jackdaw, and is principally distinguished by the red hue of its bill and legs. It inhabits the counties of the western coast of England, and is, perhaps, more common in Corn-

Sub-family c. *Pyrrhocoracinæ.* (Gr. πυῤῥὸς,* flame-coloured; κόραξ, a Crow.;
CORACIA.—(Gr. Κορακίας, like a Raven.)

Gracŭla (Lat. *a Chough*), *the Chough.*

wall than in any other county. When tame, it shows a very inquisitive disposition, examining every novelty with the greatest attention.

It builds its nest in the cavities of high cliffs, and lays four or five eggs of a yellowish white colour, spotted with light brown. The length of the bird is seventeen inches.

The EMERALD BIRD of PARADISE.—This most gorgeous and elegant bird was once the subject of much discussion between naturalists. The natives of New Guinea were accustomed to dry them, having first cut off their legs, and then to offer them for sale. In this footless state they reached Europe, where it was immediately stated that the bird lived always in the air, buoyed up by the lightness of its feathery covering; that the shoulders were used as its nest; that the only rest it took was by suspending itself from a branch by the filamentary feathers of the tail; that its food was the morning dew; together with many other conjectures not less ingenious than amusing.

This bird is about the size of a jay. Its body, breast, and lower parts are of a deep rich brown; the front set close with black feathers shot with green; the throat is of a rich golden green; the head yellow; the sides of the tail are clothed with a splendid plume of long downy feathers, of a soft yellow colour. By these are placed two long filamentous shafts, which extend nearly two feet in length.

Of these beautiful feathers the bird is so proud, that it will not suffer

* In allusion to the colour of its bill and legs.

the least speck of dirt to remain upon them, and it is constantly examining its plumage to see that there are no spots on it. When in its wild state, it always flies and sits with its face to the wind, lest its elegant filmy plumes should be disarranged.

So far from living exclusively on dew, it eats no small amount of insects, such as grasshoppers, which it will not touch if dead, and commences its

Family II. Paradiseïdæ.

PARADISĔA.—(Gr. Παράδεισος, a pleasure-ground.)

Apŏda (Gr. 'Απους, *without feet*), *the Emerald Bird of Paradise.*

"Which, like a bird of Paradise,
Or herald's martlet, has no legs."

---

repast by stripping off the legs and wings. When in confinement, it also eats boiled rice, plantains, and other vegetables; but in the wild state it seems to feed mostly on the seeds of the teak-tree, and a kind of fig.

There are several species of Paradise Birds known, but the one given in the engraving is the most common, and is the one of which the above-mentioned fables were told.

The SATIN BOWER-BIRD.—It is a singular thing to find a bird building a kind of playground, without reference to its nest, but merely for amusement. The Bower-bird has this curious habit. It builds a kind of bower of thin twigs, interwoven so as to meet above, forming a kind of tunnel. The entrance of this bower is decorated with any brilliant article that the bird can find, such as shells, bones, and feathers of several parrots, some feathers being stuck in among the twigs, and others strewn at the entrance. Mr. Gould, who first brought this curious bird before the public, says, "The propensity of these birds to pick up and fly off with any attractive object, is so well known to the natives (of Australia), that they always search the runs for any small missing article, as the bowl of a pipe, &c. that may have been accidentally dropped in the brush. I, myself, found at the entrance of one of these a small neatly worked stone tomahawk, of an inch and a half in length, together with some slips of blue cotton rags, which the birds had doubtless picked up at a deserted encampment of the natives. For what purpose these curious bowers are made, is not yet, perhaps, fully understood: they are certainly not used as a nest, but as a place of resort for many individuals of both sexes, which, when there assembled, run through and around the bower in a sportive and playful manner, and that so frequently, that it is seldom entirely deserted."

Family III. . Sturnĭdæ.—(Lat. *Sturnus*, a Starling. Starling kind.)

Sub-family *a*. *Ptilonorhynchīnæ.*

Ptilonorhynchus.—(Gr. Πτίλον, a feather; ῥύγχος, a beak.)

Sericĕus (Lat. *silky*), *the Satin Bower-Bird.*

The Starlings comprise many genera, among which the Pensile Orioles of America are the most interesting. These birds build, or rather weave, a fabric not unlike loose cloth, composed of hemp or flax.

This nest is of the singular form represented in the engraving,* and the entrance is at the side. In all probability this singular formation is for the purpose of keeping out the Black Snake, who is constantly on the look-out for young birds. The parent Orioles often attack the snake, and compel him to retreat.

* See page 241.

The plumage of the male when full grown is very brilliant. The head, throat, and back are black, the under parts are orange, the breast vermilion. A band of orange passes over the shoulders, and he tail its orange and black. The length of the bird is almost eight inches. This

Sub-family *d. Icterinæ.*

Ictĕrus. — (Gr. Ἴκτερος.)

Baltimōrus (Lat. *belonging to Baltimore*), *the Baltimore Oriole.*

is not the only bird that constructs pensile nests; the Weaver Birds also form these nests, but of a different form. They look like great pistols hung up by the butt, the entrance being at the muzzle, and the nest in the butt.

The Common Starling is a bird well known both for its beauty and the singular method of flight. When a flock of Starlings begin to settle for the night, they wheel round the place selected with great accuracy. Suddenly, as if by word of command, the whole flock turn their sides to the spectator, and with a great whirring of wings, the whole front and shape of the flock is altered. No body of soldiers could be better wheeled or countermarched than are these flocks of Starlings, except, perhaps, an unfortunate few, who are usually thrown out at each change, and whom we must charitably suppose to be recruits.

Sub-family *g. Sturninæ.*

STURNUS. (Lat.)

Vulgāris (Lat. *common*), *the Starling.*

The Starling lives principally among old buildings, and is very fond of gaining admittance into dovecotes, where it is a harmless visitor, and may be suffered to remain without detriment to the pigeons or their eggs. Its nest is made usually in a hole in a wall, sometimes in a decayed tree, and contains five eggs of a very delicate uniformly pale blue.

There is never any difficulty in discovering the nest of the Starling, for if it builds in a hole of a wall it generally leaves several straws sticking out, as if to indicate the locality; and when it goes to take food to its young, both parent and children set up such an outcry, that it may be heard a long way off. Consequently, there are few eggs so prevalent in the string of the country boy as those of the Starling.

Family IV. . Fringillĭdæ.—(Lat. *Fringilla,* a Finch. Finch kind.)

Sub-family *a. Coccothraustīnæ.*

COCCOTHRAUSTES.—(Gr. Κόκκος, a berry; θραύω, I break.)

Vulgāris (Lat. *common*), *the Hawfinch.*

The GROSBEAK or HAWFINCH.—We now arrive at the Finches; a very large and interesting family. None of the species are large, and most of them are excellent songsters. Their beaks are conical, and fitted for the destruction of corn, peas, &c.

The Grosbeak, or Hawfinch, well deserves its generic name of "Berry-breaker," for its beak is capable of breaking the hard kernels of the cherry, and, according to Willoughby, even those of the olive. It is not a very rare bird, although it is but seldom seen. This fact is accounted for by its great shyness and dread of mankind; so that, although it remains in this country throughout the year, it seldoms ventures out of the thick woods in which it delights to dwell.

The nest of this bird is very shallow, and slightly put together, being hardly superior to that of the wood-pigeon. The eggs are from four to six in number, of a greenish white, covered with dark marks and spots. The length of the Grosbeak is seven inches.

The CHAFFINCH or PIEFINCH, as it is often called, is so well known as to need no description. It is chiefly remarkable for the beautiful nest which it constructs. The forks of a thorn or wild crab-tree are favourite places for the nest, which is composed of mosses, hair, wool, and feathers, covered on the exterior with lichens and mosses, so exactly resembling the bough on which the nest is placed, that the eye is often deceived by its appearance. In the nest four or five very pretty eggs are laid: these are of a reddish-brown colour, sparely marked with deep brown spots, especially towards the larger end.

The name Cœlebs or Bachelor, is given to this bird, because the females quit this country about November, leaving large flocks of males behind them.

Sub-family *b. Fringillinæ.*

FRINGILLA.

Cœlebs (Lat. *Bachelor*), *the Chaffinch.*

The GOLDFINCH or THISTLEFINCH, so called on account of its fondness for the down of the thistle, is one of our most beautiful birds. Where thistles abound, small flocks of goldfinches may be seen flying from hedge to hedge, and occasionally pecking the white tops of the thistles. The tufted seed of the dandelion, groundsel, and other plants is also eaten by the Goldfinch.

In captivity it is very tame, and can be trained to perform a multitude of tricks; the most common of which are, drawing its own food and water with a chain and bucket, or firing a gun when commanded. The nest is very beautiful, being mostly made of wool and down from various plants,

FRINGILLA.

Carduēlis (Lat. *a Linnet*), *the Goldfinch.*

and is usually placed on the extremity of a spray. The eggs are small, of a whitish tint, spotted with orange brown.

FRINGILLA.

Cannabīna (Gr. Καννάβινος, *fond of hemp*), *the Linnet.*

The Common Linnet frequents commons and neglected pastures. Its song is very sweet, and many bird-fanciers suppose that the mixed breed of a canary and a linnet has a sweeter song than either bird.

Its nest is usually built in the centre of a large and dense bush. The eggs are five in number, greyish-white speckled with red.

The Siskin is hardly to be considered more than an occasional visitor in England, but in Scotland it sometimes breeds, as may be seen from the following extract:—

"The Siskin is a common bird in all the high parts of Aberdeenshire, which abound in fir-woods. They build generally near the extremities of the branches of tall fir-trees, or near the summit of the tree. Sometimes the nest is found in plantations of young fir-wood. In one instance, I met with a nest not three feet from the ground. I visited it every day until four or five eggs were deposited. During incubation the female showed no fear at my approach. On bringing my hand close to the nest, she showed some inclination to pugnacity, and tried to frighten me away with her open bill, following my hand round and round when I attempted to touch her. At last she would only look anxiously round to my finger, without making any attack on me. The nest was formed of small twigs of birch or heath outside, and neatly lined with hair."

Its eggs are a bluish-white spotted with purplish red.

FRINGILLA.

Spinus (Lat. *a Sloe-tree*), *the Siskin.*

FRINGILLA.

Chloris (Gr. Χλώρος, green), *the Greenfinch.*

The GREENFINCH or GREEN LINNET is larger than the Common Linnet. It frequents gardens, shrubberies, and cultivated lands, and feeds on insects or seeds. The notes of this bird are not peculiarly melodious, nor has it many qualifications to entitle it to notice.

The CANARY.—This pretty little songster is so well known as to need but little description, particularly as there are no opportunities of studying its natural course of life. From the manner in which the

CARDUELIS.—(Lat. *a Linnet.*)

Canaria (Lat. *the Canary*).

Canary is usually reared, it is evident that the bird has but very little opportunity of exhibiting its natural instincts.

The SPARROW.—The courageous, impudent, quarrelsome Sparrow is known to all, and, therefore, will not be described. There are few who have not seen this little bird, when pressed by cold in the winter, come to the window, expecting his donation of crumbs. It is very fond of grain of various kinds, and does some damage to the farmer, but the destruction of caterpillars by the bird more than compensates for the loss of the grain. The little impertinent bird has no scruple in perching on the pig's trough, and partaking of his dinner, or in mixing with fowl and taking its share of their provisions; and on a newly thatched house it absolutely revels. Dozens of sparrows may then be seen pecking and pulling at the straws in high enjoyment. I was once watching a flock of sparrows on a newly-thatched barn, hopping, pecking and scrambling in perfect happiness, when

PASSER.—(Lat. *a Sparrow.*)

Domesticus (Lat. *domestic*), *the House Sparrow.*

suddenly a sharp twitter was heard, and the whole body hastily adjourned to a tree close by, making a prodigious chattering. Presently I saw appear, over the ridge of the house, the head of a cat, who had walked up the thatcher's ladder, hoping to secure a few sparrows in the midst of their meal. The nest of the House Sparrow is usually built in holes of roofs. The eggs are speckled black and white, and very variable.

Sub-family c. *Emberizinæ.*

EMBERĪZA.—(Lat. *a Bunting.*)

Miliarĭa (Lat. *belonging to Millet, or that feeds on Millet*), *the Bunting.*

The BUNTING.—A hard round knob in the roof of the upper mandible, points out the genus Emberīza. This knob is probably used for the purpose of breaking the shells of the small but hard seeds on which the bird feeds.

Large flocks of these birds collect together in the autumn and winter in their attacks upon the farm-yard or the field. Partly in revenge for their depredations, and partly on account of the price that they will fetch in the market, great numbers are annually caught in nets or shot.

The nest is always placed either on or near the ground, and contains four or five darkish eggs, covered very irregularly with deep brown marks.

EMBERĪZA.

Citrinella (Lat. *yellowish*), *the Yellow Bunting.*

The YELLOW-HAMMER or YELLOW BUNTING is a very delicately marked little bird, very common in our hedges, where it flits before the traveller, always keeping about twenty yards in front. It makes its nest on the ground, and lays five eggs curiously scribbled over with dark chocolate lines, just as if a child had been trying to write Arabic on the eggs.

Some say that the name ought to be Yellow-Ammer, the word Ammer being German for Bunting, and the German word for this bird being Goldammer.

The ORTOLAN, a little bird in very great repute for the table, is also an Emberiza. This bird is regularly fed like poultry, in the South of Europe, and soon becomes exceedingly fat, when a guinea is frequently the price of it.

EMBERĪZA.

Hortulāna (Lat. *belonging to a little garden*), *the Ortolan.*

It is received as a British bird, because it has been taken off the coast of Yorkshire, and also in Mary-le-bone Fields.

The LARKS are known by their very long hind toe. The Skylark, which pours forth its animated song while suspended high in the air, is an inhabitant of most parts of Europe, Asia, and North Africa, but is not found in America. A very interesting story is told of a Skylark that was taken out to America by a poor emigrant, and which used to collect crowds of delighted listeners round its cage. An English settler, who happened to be passing by while the bird was singing, was so affected by the reminiscences which its song called up, that he offered his horse and cart for the bird, on the spot. The owner, however, would take no price for it, although most extravagant offers were made, and kept it till his death. The bird afterwards passed into other hands, but refused to sing until its cage was hung up in the open air. After its death, its skin was sent back to its native land, and is now stuffed, seated in its old cage, with a suitable inscription attached.

Sub-family *d.* . . *Alaudīnæ.*

ALAUDA.—(Lat. *a Lark.*)

Arvensis (Lat. *belonging to the fields*), *the Skylark.*

The nest is made on the ground, frequently in the print of a horse's foot, and contains five eggs of a greenish-white, thickly spotted with brown. There are generally two broods in the year; one in May, and the other in July or August. Immense numbers of these birds are caught annually and sent to the London markets. The mode of catching the Larks is generally by means of a number of horsehair nooses attached to a long line. Food is scattered among the nooses, and the larks in reaching the food get their limbs entangled in the horsehair, and either strangle themselves, or are held until the fowler comes to take them out. Dunstable is the most celebrated place for them. It does not at all agree with the sense of justice, that these beautiful birds, who charm us with their voices, should be killed to increase the pleasures of the table.

The WOODLARK is another of our British Larks, but differs in some respects from the skylark. It is smaller, and can perch on trees, a power denied to the skylark. It also sings on the wing, but sometimes prefers to pour forth its notes while perched on the branch of a tree.

Its nest, like that of the skylark, is also placed on the ground; the eggs are darker than those of the skylark.

ALAUDA.

Arborĕa (Lat. *belonging to trees*), *the Woodlark.*

The BULLFINCH is a singular instance of the power of art on the song of birds. The natural note of the Bullfinch is low, and can only be heard at a short distance; but when well trained the bird whistles, or "pipes," as it is called, any melody which has been taught it, in a fine flute-like tone. A good piping Bullfinch sells at a very high price. The method of teaching, is to confine the birds in a dark room, and, before their food is given, to play the air that they have to learn, on an instrument called a bird-organ. The birds soon begin to imitate the notes, and by degrees the whole tune is learned. Some trainers substitute a small clarionet for the bird-organ.

When in captivity the Bullfinch is very sociable, and soon learns to know his owners, and to come to them if called.

A rather singular fate befel one of these birds. A lady, on opening her window, saw a bullfinch quietly seated outside on the sill. To her great surprise, it did not move when the window was opened; but permitted itself to be taken up by the hand and carried into the room. When placed on a table, it gave no indication of any wish to escape, but sat still, as if it were suffering from illness. An examination being instituted as to the cause of this singular circumstance, a seed was found firmly fixed in the bird's throat. The impediment was soon removed with a needle, and the bird became quite lively. This relief, however, proved but a temporary respite; for while a cage was being prepared for its reception, it escaped from the hand of its benefactress; flew hastily to the closed window; and, being unaccustomed to glass, dashed itself with such violence against one of the panes, that it fell dead on the ground.

This anecdote was related to me by an eyewitness of the scene.

Sub-family *e. Pyrrhulīnæ.*

PYRRHŪLA.—(Gr. Πυῤῥούλας,) from Πυῤῥός, flame-coloured.

Rubicilla (Lat. *reddish*), *the Bullfinch*

The nest of this bird is made in thick bushes, or fir-trees. The eggs are of a pale greenish white, spotted with orange brown. The name of Bullfinch is given to it, on account of the large proportionate size of its head and neck. When in captivity, its plumage sometimes turns black, the result of feeding it too profusely with hempseed.

Sub family *f. Loxīnæ.*

LOXĪA.—(Gr. Λοξός, crosswise.)

Curvirostra (Lat. *Curved-bill*), *the Cross-bill.*

The genus LOXIA is instantly known by the crossed points of the beak, and the horny scoop at the tip of the tongue. The Crossbill uses these tools to open the fir-cones, on the seeds of which it feeds. The bird inserts both its mandibles under the scales of the cone, then by separating them the scale is raised up, while the seed is scooped out by the horny tip of the tongue. This singular structure of the beak enables the bird to divide an apple in halves, so as to get at the pips. Although the crossed mandibles appear rather a barrier to picking up small objects, yet the Crossbill can pick up and husk the smallest seeds, or shell almonds; which latter feat is accomplished by picking a hole in them, and then wrenching them open, just as an idle schoolboy opens a nut with his penknife, when he ought to be using that instrument in the more legitimate operation of mending his pen. Mr. Yarrell gives an interesting account of a pair of Crossbills, who amused themselves by twisting out the wires of their cage. They actually succeeded in pulling out a flat-headed nail used to confine the network, but the bird lost the point of his bill in his efforts. They were at last banished, on account of their unceasing destruction of cages.

The nest is built on the branches of a fir-tree, and the eggs are bluish white spotted with red.

Family VII. Bucerotĭdæ.

Bucĕros.—(Gr. Βούκερως, ox-horned.)

Rhinocĕros, *the Rhinoceros Hornbill.*

The Rhinoceros Hornbill.—This singular and almost startling family comprises but few species, which are all natives of India and Africa. The enormous bill, with its incomprehensible appendage, although of course heavy, is really much lighter than it looks; being composed of a kind of light honeycombed structure. The upper protuberance is hollow, and the only conjecture formed of its use, is that it serves as a sounding-board to increase the reverberations of the air, while the bird is uttering its peculiar roaring cry.

In spite of the apparently unwieldy bill, the bird is very active, and hops about the branches of the trees with much ease. The appendage to the upper mandible is small when the bird is young, and only attains its enormous size when the Hornbill has reached its full growth. The bill of the hoopoes presents a somewhat analogous peculiarity, as when the bird is young the bill is short and pointed, and increases with the size of the bird. From this circumstance, together with some other resemblances, some naturalists imagine that there is an affinity between the hornbills and hoopoes.

The Hornbills seem to be omnivorous, fruits, eggs, birds, reptiles, &c., forming their food. The African Hornbills are extremely fond of nutmegs, and are, on that account, said to be peculiarly delicate eating, reminding one of the Barmecide's memorable lamb fed on pistachio nuts.

The Rhinoceros Hornbill is a native of India, and the Indian islands. The length of its bill is usually about ten inches.

The SCANSORES, or CLIMBING BIRDS, now engage our attention. According to Mr. Gray, under this order are placed the Toucans, the Parrots, the Woodpeckers, and the Cuckoos. The feet of these birds have two toes in front and two behind.

The Toucans are all natives of tropical America. Their enormous bill is rendered light in the same way as that of the hornbills, by being chiefly composed of a honeycomb structure. It seems to be very sensitive, and well supplied with nerves, as the bird not only appears to enjoy holding meat or fruits with the tip of its bill, but has been seen to scratch that organ with its foot, plainly proving that there must be sensation. It seems to be omnivorous, but is particularly fond of mice, and small birds, which it kills by a powerful squeeze, then strips and finally pulls to pieces and devours, having previously reduced them to a shapeless mass by repeated lateral wrenches with its enormous and saw-like bill. Waterton in his Wanderings describes the usual haunts of the Toucan.

"Heedless and bankrupt in all curiosity must he be, who can journey on without stopping to take a view of the towering mora. Its topmost branch, when naked with age or dead by accident, is the favourite resort of the Toucan. Many a time has this singular bird felt the shot faintly strike him from the gun of the fowler beneath, and owed his life to the distance betwixt them." In the same interesting and amusing work, he remarks a strange habit of the Toucan, called the Houtou by the natives.

"This bird (the Houtou) seems to suppose that its beauty can be increased by trimming the tail, which undergoes the same operation as our hair in a barber's shop; only with this difference, that it uses its own beak, which is serrated, in lieu of a pair of scissors. As soon as his tail is full grown, he begins about an inch from the extremity of the two longest feathers in it, and cuts away the web on both sides of the shaft, making a gap about an inch long: both male and female adorn their tails in this manner, which gives them a remarkable appearance amongst all other birds."

Order III. . . *SCANSŌRES.*—(Lat. *scando,* I climb. Climbing birds.)
Family I. . . Ramphastĭdæ.
Sub-family *a.* *Rhamphastidīnæ.*

RHAMPHASTOS.—(Gr. Ῥαμφηστής, properly, a Pike.)

Toco (native name), *the Toco Toucan.*

When sleeping, the Toucan takes great care of his bill, packing it away, and covering it carefully with the feathers of its back, and altogether presents the appearance of a large round ball of feathers. The body is about eighteen inches in length. These birds, together with the hoopoes and hornbills, have a habit of throwing their food down their throats with a peculiar jerk of the bill.

Family II. Psittacidæ.—(Gr. Ψιτταкоs, a Parrot.)

MACROCERCUS.—(Gr. Μακρός, long; κέρκος, tail.)

Ararauna (from the Brazilian word *Arara*), *the Blue and Yellow Macaw.*

The MACAWS.—Many naturalists imagine, and with some reason, that the Psittacidæ ought to be formed into an order by themselves. In this family the construction of the bill is very remarkable. As the curved tip of the bill would prevent the bird from opening it wide enough to admit its food, the upper mandible is united to the skull by a kind of hinge joint, of equal strength and flexibility. When climbing among the branches of trees, or about their cages, the Parrots invariably make great use of their hooked bills in assisting themselves both in ascending and descending. The crossbills have been observed to climb much in the same way.

The Parrots are said to be very long lived, some have certainly been known to live upwards of eighty years in captivity, and may be imagined to exceed that period in a wild state.

The Macaws are natives of South America. The blue and yellow Macaw inhabits Brazil, Guiana and Surinam, living principally on the banks of rivers. Of one of the Macaws, the Carolina Parrot, or Parrakeet as Wilson calls it, the following anecdote is told by that enterprising naturalist:—

"Having shot down a number, some of which were only wounded, the whole flock swept repeatedly round their prostrate companions, and again settled on a low tree, within twenty yards of the spot where I stood. At each successive discharge, though showers of them fell, yet the affection of the survivors seemed rather to increase; for, after a few circuits round the place, they again alighted near me, looking down on their slaughtered companions with such manifest symptoms of sympathy and concern, as entirely disarmed me."

Wilson also makes mention of a singular idea, that the brains and intestines of the Carolina Parrot (which lives on cockle-burs) are poisonous to cats. Why the brains should be so is rather incomprehensible, although we can easily understand that the Parrot might take some substance into its stomach injurious to cats. Wilson tried the experiment after being repeatedly disappointed of a patient, but came to no conclusion on the subject.

"Having shut up a cat and her two kittens, the latter only a few days old, in a room with the head, neck, and the whole intestines of the parrakeet, I found on the next morning the whole eaten except a small part of the bill. The cat exhibited no symptom of sickness, and at this moment, three days after the experiment had been made, she and her kittens are in their usual health. Still however the effect might have been different, had the daily food of the bird been cockle-burs instead of Indian corn."

PALÆORNIS.—(Gr. Παλαιός, old; ὄρνις, a bird.)

Torquātus (Lat. *collared*), *the Ringed Parrakeet.*

The RINGED PARRAKEET is frequently seen domesticated in this country, where its pleasing manners and gentle disposition render it a great favourite. It seems to be exceedingly fond of ripe walnuts, divided in halves; and, while it is picking out the kernel, continually utters a short clucking sound indicative of pleasure.

It soon learns to repeat words and short sentences, and to speak with tolerable distinctness. Sometimes, when excited, it utters most ear-piercing screams, and always appears to practise any new accomplishment when it thinks that no one is within hearing. A Ringed Parrakeet belonging to one of my scholars was accustomed to live in the schoolroom. At first it used to become angry that it was not noticed during school hours, and to utter a succession of screams; but after being shut up in a dark closet several times, it learned to behave very demurely,—giving an example worthy of imitation to several of its human playfellows. I am sorry to say, that the bird escaped from its cage, and was shot by an ignorant farmer in the neighbourhood.

The colour of the bird is green, and a rose-coloured band round its neck gives it the name of the Rose-ringed Parrakeet. The bill is red.

CACATŬA.—(From the native name.)

Sulphurĕa (Lat. *Sulphury*), *the Great Sulphur Cockatoo.*

The COCKATOOS are remarkable for the powdery surface of their wings, and the crest on the head, which can be raised or depressed at pleasure. The Sulphur-crested Cockatoo is an inhabitant of New Guinea. Its colour is white, and the crest is of a sulphur yellow. Its white plumage glancing among the dense dark foliage of its native forests, imparts a wonderful beauty to the scene; and as Sir Thomas Mitchell remarks, "amidst the umbrageous foliage, forming dense masses of shade, the white cockatoos sported like spirits of light." This Cockatoo is easily tamed, and is of a very affectionate disposition. When in captivity it has been known to live to the age of 120 years. Its nest is built in hollow trees, and the crevices of rocks. The eggs are white. The length of the bird is about eighteen inches.

The WOODPECKERS, whose name indicates their habits, are widely spread, being found in all quarters of the globe except Australia. They subsist on insects and grubs, which they dig out of trees, or discover under the bark. For this purpose, their whole structure is admirably adapted. The bill is long, sharp, and powerful, and the formation of the feet and legs is such that the bird is able to grasp the tree firmly with the feet, while swinging with the force of his whole body against it. Another most singular point in the Woodpeckers, is the method by which they are enabled to thrust the tongue deep into the crevices, and bring out any insects that may happen to be there. The tongue is connected with two elastic ligaments which are inserted near the juncture of the upper mandible with the skull. From thence they sweep round the back of the head, and passing under the lower mandible, enable the tongue to be thrust out a considerable distance. The tip of the tongue is sharp, and barbed with several filaments; and more firmly to secure the prey, a kind of gummy secretion causes those insects to adhere, that would be too small to be impaled.

It appears to be an erroneous opinion, that these birds injure trees. Their only object in pecking away the wood and bark, is to get at the insects, which they know are hidden within. Now insects seldom or never

Family III. . Picĭdæ.—(Lat. *Picus*, a Woodpecker. Woodpecker kind.)
Sub-family *c. Picīnæ.*

PICUS.

Major (Lat. *greater*), *the Great Spotted Woodpecker.*

bore into healthy wood, but a decayed branch or stump is always full of them, as is well known to the entomologist; so the winged entomologist, when he perceives a decayed branch, or finds an unsound spot in the trunk, immediately sets to work industriously, and is rewarded by finding plenty of insects, which he draws out and demolishes, with more benefit to himself, and possibly more good to others, than many human entomologists can boast.

Although the Woodpecker does not scoop away sound trees, yet it is because it has no motive for doing so—not that the power is wanting. Wilson had an Ivory-billed Woodpecker in his possession, which pecked away lath and plaster in its efforts to escape, and utterly ruined a mahogany table to which it was fastened.

The Great Spotted Woodpecker is an inhabitant of England, but is seldom seen. Large woods are its favourite haunts. Like all its tribe, it feeds on the insects which it procures from decayed trees, and also on berries and fruits. Its eggs are laid in a deep hole excavated in a tree. For this purpose, the Woodpecker usually chooses a place where a branch has broken off, or more commonly the part of the trunk where a certain fungus has grown, causing the tree to decay in that spot, although apparently healthy.

SKULL OF THE WOODPECKER.

The Green Woodpecker is by far the most common in this country, and may be often seen in woods, tapping the trees with wonderful rapidity, the blows following each other something like the sound of a watchman's

PICUS.

Viridis (Lat. *green*), *the Green Woodpecker.*

rattle. It generally runs up the trunk of the tree in a spiral direction, occasionally striking off large pieces of dry bark. When it descends, it still keeps its head uppermost.

I have more than once seen the Green Woodpecker busily employed among the trees of the Christ Church Walks, Oxford, and very frequently in Bagley Wood. I have never seen it on the ground, and but once on the smaller branches of the trees.

A few years ago, I saw one of these birds running up the trunk of a tall elm-tree, and seeing half a brick lying on the ground, I took it up and threw it towards the bird, merely wishing to startle it. As the tree was at some distance, I did not imagine that the brick would have gone so far, much less were there any grounds for supposing that a bird in rapid motion could have been hit by so unwieldy a weapon. To my astonishment, it struck the Woodpecker, but did not bring it down; the poor bird flew off with loud screams, and a drop or two of blood fell on the dry leaves. The brick had come with one of its edges against the legs of the bird, and broken them both off. I lamented for a while about the fate of the unfortunate Woodpecker, which at no distant period must have died of starvation, if not from the effects of the wound. I made an inward resolution at the time never to throw a stone at a bird again, however distant it might be.

Sub-family *g. Yuncinæ.*

YUNX.—(Gr. Ἴυγξ.)

Torquilla (Lat. *twisting*), *the Wryneck.*

The WRYNECK is tolerably common in the southern counties of England, but is scarcely ever seen in the north and west. It principally feeds on ants, which it picks up with great rapidity by means of its long tongue, covered with a glutinous secretion like that of the woodpecker. The rapidity with which the ants are taken is so great, that "an ant's egg, which is of a light colour, and more conspicuous than the tongue, has somewhat the appearance of moving to the mouth by attraction, as a needle does to the magnet." The name Wryneck is given it from its habit of rapidly twisting its head and neck, and hissing like a serpent, if disturbed upon its eggs. The young also hiss if they are molested.

Its eggs are laid on the bare wood in the holes of trees. Like most eggs that are laid in holes, they are of a pure white. The length of the bird is seven inches.

Family IV. . Cucŭlĭdæ.—(Lat. *Cucūlus*, a Cuckoo. Cuckoo kind.)
Sub-family *e. Cuculīnæ.*

CUCŪLUS.

Canōrus (Lat. *musical*), *the Cuckoo.*

The Cuckoo, spring's harbinger, has, at all ages, obtained for itself a name at once pleasing and disreputable; pleasing, because its well-known notes are a sign that the cold winter is gone; and disreputable, because it usurps the nests of other birds, of which the hedge sparrow is the usual victim. In its nest the cuckoo deposits one of its own eggs, which are remarkably small in proportion to the size of the bird. The unsuspecting hedge sparrow hatches the intruder together with her own young. The Cuckoo rapidly increases in size, and monopolizes no small portion of the entire nest, besides taking the lion's share of the provisions. The mother, however, never seems to perceive the difference, but feeds and tends the interloper with quite as much care as her own young.

Dr. Jenner states that the young Cuckoo ejects the former and rightful occupants of the nest, by managing to get the egg or young bird upon its back, clambering up to the edge of the nest, and then throwing it over by a sharp jerk.

At some times of the year, Cuckoos are comparatively tame. I have repeatedly decoyed them by imitating their cry, until they came near enough for me to see the movement of the beak. Once a Cuckoo came voluntarily, and settled on a hurdle close by, uttered his peculiar cry several times, and then leisurely flew off.

The Cuckoo feeds principally on the hairy caterpillars, especially those of the tiger moth (*Arctia caja*), the hairs of which form a kind of lining to its stomach. These hairs are placed so regularly, that it was imagined for some time that they were a growth from the stomach itself. To settle the point, the microscope was brought to bear on the subject; and by its aid, the hairs were found to be exclusively those of the caterpillar above mentioned.

The Cuckoo will also feed on other insects, as is proved by Gilbert White, who saw several Cuckoos engaged in feeding by a large pond. They were chiefly employed in catching the dragon flies, some of which they took while resting on the water plants, and others they caught on the wing.

A tame Cuckoo, that lived for more than a year in captivity, seemed to consider a young mouse an especial treat. The mouse was first beaten against the ground or a hard stone, until it was reduced to a soft mass, after which process it was swallowed. The length of the bird is about fourteen inches.

Order IV. *COLUMBÆ.*
Family I. Columbĭdæ.—(Lat. *Columba,* a Dove. Dove kind.)
Sub-family *b. Columbīnæ.*
COLUMBA.

The Dove—This family is supposed to be more widely distributed than any other. The three pigeons engraved are the only species that live wild in this country.

Palumbus (Lat. *a Pigeon*), *the Ringdove.*

The Ringdove, or Cushat, is the largest of our native pigeons. A black ringlet round the neck, edged with white, gives it the name of Ringdove. It is very common in England; and its nests are usually found to consist of a few sticks, thrown loosely together on a spray of fir or holly. The structure of this platform, for nest it can hardly be called, is so loose, that the white eggs can generally be seen from below through the interstices of the nest.

COLUMBA.

Œnas (Gr. Οἰνάς), *the Stockdove.*

The STOCKDOVE builds its nest in the stocks of trees, (from whence its name,) and has been known to lay its eggs in deserted warrens, without making any nest at all. In former times, when forests of beech-trees used to cover the country, enormous flocks of these birds frequented them, in order to feed on the beech-mast. Now they are not so common, although they still assemble in considerable numbers.

TURTUR.—(Lat. *a Turtle-dove.*)

Aurĭtus (Lat. *eared*), *the Turtle-dove.*

The TURTLE-DOVE, a bird much revered by poets for its constancy, is only a spring visitor to our shores, arriving towards May, and leaving us about September. The nest is a mere platform of twigs, on which the eggs are laid. The constancy and affection of this bird for its mate has been deservedly celebrated in all ages; though it is not easy to understand why other birds, such as the raven, whose constancy is quite as remarkable, should be deprived of the meed of praise due to them.

The PASSENGER PIGEON.—This extraordinary bird, whose powers of flight are almost incredible, is a native of America, and overspreads the country in countless myriads during the breeding season. It is well that their power of wing is so great, for were the enormous flocks to be confined to one place, they would devour the whole of the grain. Pigeons have been killed in New York with Carolina rice still in their crops. As their digestion is remarkably rapid, these birds must have flown between three and four hundred miles in six hours, giving an average speed of a mile per minute.

ECTOPISTES.—(Gr. Ἐκτοπίζω, to migrate.)

Migratoria (Lat. *migratory*), *the Passenger Pigeon.*

At the breeding season, the overwhelming multitudes of Pigeons that settle on one spot are almost incredible. Wilson, who was present at one of these breeding places, gives the following account:—

"Not far from Shelbyville, in the state of Kentucky, about five years ago, there was one of these breeding places, which, stretched through the woods in nearly a north and south direction, was several miles in breadth, and was said to be *upwards of forty miles in extent!* In this tract almost every tree was furnished with nests wherever the branches could accommodate them. The pigeons made their first appearance there about the 10th of April, and left it altogether, with their young, before the 25th of May.

"As soon as the young were fully grown, and before they left the nests, numerous parties of the inhabitants, from all parts of the adjacent country, came with waggons, axes, beds, cooking utensils, many of them accompanied by the greater part of their families and encamped for several days at this immense nursery. Several of them informed me that the noise in the woods was so great as to terrify their horses, and that it was difficult for one person to hear another speak, without bawling in his ear. The ground was strewed with broken limbs of trees, eggs, and young squab pigeons, which had been precipitated from above, and on which herds of hogs were fattening. Hawks, buzzards, aud eagles, were sailing about in great numbers, and seizing the squabs from their nests at pleasure; while from twenty feet upwards to the tops of the trees, the view through the woods presented a perpetual tumult of crowding and fluttering multitudes of pigeons, their wings roaring like thunder, mingled with the frequent crash of falling timber; for now the axe men were at work cutting down those trees that seemed to be most crowded with nests, and contrived to fell them in such a manner, that in their descent they might bring down several others, by which means the falling of one large tree sometimes produced two hundred squabs, little inferior in size to the old ones, and almost one mass of fat.

"All accounts agree in stating that each nest contains only one young squab. These are so extremely fat, that the Indians and many of the whites are accustomed to melt down the fat for domestic purposes, as a substitute for butter and lard."

A few observations on the mode of flight of these birds must not be omitted. "A column, eight or ten miles in length, would appear from Kentucky, high in air, steering across to Indiana. The leaders of this great body would sometimes gradually vary their course, until it formed a large bend of more than a mile in diameter, those behind tracing the exact route of their predecessors. This would continue sometimes long after both extremities were beyond the reach of sight, so that the whole with its glittering undulations, marked a space on the face of the heavens resembling the windings of a vast and majestic river. . . . Sometimes a hawk would make a sweep on a particular part of the column, from a great height, when, almost as quick as lightning, that part shot downwards out of the common track, but soon rising again, continued advancing at the same height as before; this inflection was continued by those behind, who, on arriving at this point, dived down almost perpendicularly to a great depth, and rising, followed the exact path of those that went before."

The following group comprises the most conspicuous varieties of the Domestic Pigeon. All these birds, except the Carrier, the Pouter, and and Tumbler, are very similar in their habits, and need no description.

The TUMBLER is a very little pigeon, and derives its name from its singular habit of falling backwards when on the wing. Pigeon fanciers assert that a flight of twelve Tumblers may be covered with a handkerchief.

The POUTER is a large pigeon. It stands particularly erect, and seems exceedingly vain of the swollen crop which gives it the name of Pouter. The bird is enabled to inflate its crop with air, until the head is almost hidden behind it. This inflation sometimes causes the bird to lose its balance, and fall down chimneys, on which it is fond of standing, thereby illustrating the proverb that "Pride will have a fall."

The CARRIER PIGEON is the bird that was so largely employed to take messages, before the invention of the Electric Telegraph rendered even the speed of the wind too slow for the present day. The most valuable Carriers were trained to carry to and from their residence. A letter was written on a small piece of paper, and fastened under the wing of the

DOMESTIC PIGEONS.

pigeon, or to its feet. The feet were then bathed in vinegar to keep them cool, lest the bird should stop on the way to bathe. When the pigeon was set free, it rose high in the air, made one or two circular flights, and then darted off like an arrow in the proper direction. One of these birds has been known to fly nearly one hundred and fifty miles in one hour.

The PEACOCK.—This magnificent bird is not a native of this country, but has been domesticated in England for many years. Some suppose that it was first brought from India by Alexander, and by him introduced into Europe. The gorgeous plumes that adorn the Peacock do not compose the tail, as many suppose, but are only the tail-coverts. The tail feathers themselves are short and rigid, and serve to keep the train spread, as may be seen when the bird walks about in all the majesty of his expanded plumage.

Although pea-fowl seek their food on the ground, they invariably roost on some elevated situation, such as a high branch, or the roof of a barn

Order V. . . . *GALLĪNÆ.*—(Lat. *Gallus*, a Domestic Fowl.)
Family III. . Phasianĭdæ.—(Gr. *Φασιανός*, a Pheasant.* Pheasant-kind.)
Sub-family *a.* *Pavonīnæ.*

Pavo (Lat. *a Peacock*).

Cristātus (Lat. *crested*), *the Peacock.*

or haystack. When the bird is perched on the roof, its train lies along the thatch, and is quite invisible in the dusk.

We have almost dismissed pea-fowl from our entertainments in these

* *I.e.* a bird from the river Phasis in Colchis.

days, but in the times of chivalry, a roasted peacock, still clothed in its plumage, and with its train displayed, formed one of the chief ornaments of the regal board. The nest of this bird is made of sticks and leaves rudely thrown together, and contains from twelve to fifteen eggs. The young do not attain their full plumage until the third year, and only the males possess the vivid tints and lengthened train, the female being a comparatively ordinary bird. A white variety of the Peacock is not uncommon. In this case, the eyes of the train feathers are slightly marked with a kind of neutral tint.

The voice of the Peacock is as unpleasant and unmusical as its external appearance is attractive. There are some Peacocks living at a farm just at the outskirts of Oxford, which frequently startle those who pass by their haunts at night. They are fond of roosting on the roof of a small barn, where they are quite imperceptible in the dusk of the evening. When any one passes down the lonely lane which borders the farmyard, one or other of the birds is tolerably certain to give its jarring scream, much to the alarm of the passer-by.

Sub-family *b. Phasianinæ.*

ARGUS.—(Gr. proper name.)

Gigantēus (Lat. *gigantic*), *the Argus Pheasant.*

The ARGUS PHEASANT is found in Sumatra and the south-eastern parts of Asia. The magnificently marked secondary quill feathers render it a most conspicuous bird. The primary feathers are comparatively short. No living specimen has yet been brought to Europe, as it is said to pine in captivity. In its native haunts it is very shy, avoiding the proximity of human abodes, and living in the solitary depths of woods.

The name Argus is given to the bird because the numerous eye-like spots that cover the whole of its plumage are considered as bearing an allusion to Argus, the hundred-eyed shepherd, who was set by Juno to watch Io.

The COMMON PHEASANT was originally brought from Georgia, and has completely naturalised itself in this country. It is a hardy bird, and bears the cold months very well. Although it can be tamed, and will come to be fed with the poultry, yet an innate timidity prevents it from being thoroughly domesticated. Young pheasants that have been hatched

PHASIĀNUS.

Colchĭcus (Lat. *Colchian*), *the Pheasant.*

under a hen, scamper off in terror if an unexpected intruder makes his appearance among them, although the remainder of the poultry remain perfectly unconcerned.

This bird loves to perch at night on trees, especially on the spreading branches of the larch. Poachers are so well aware of this habit that they always visit the larches first, while on their marauding excursions.

A few spruce-firs surrounded by dense and tall holly hedges form an excellent place of refuge for the birds, who can bid the poacher defiance from their stronghold; while a few dozen wooden pheasants nailed on the branches of the unguarded trees, are admirably adapted for trying the patience and wasting the ammunition of the nocturnal plunderer.

A white variety of the Pheasant sometimes occurs, but seems never to be propagated. The nest of the bird is made on the ground, and contains from ten to eighteen eggs of an uniform dun colour.

The DOMESTIC FOWLS are too well known to need much description. There are many varieties, the most conspicuous of which are the Cochin-China, Crested, and Bantam. The Game Fowl was formerly in great request for the cruel sport of cock-fighting, an amusement which, although happily now almost extinct, was in great vogue but a few years since. The Java Fowl, of which the enormous Cochin-China bird is a

Sub-family *c.* *Gallinæ.*

GALLUS.—(Lat. *a Cock.*)

Domesticus (Lat. *domestic*), *the Domestic Fowl.*

variety, is supposed to be the origin of the Barn-door fowl. The cock has been long celebrated for his warlike propensities, and his habit of greeting the approach of morn by his "shrill clarion."

A young hen of the Cochin-China breed, when introduced among the other poultry of a farmyard, was shamefully persecuted by its companions. It was very absurd to see the poor creature pecking up a stray crumb or two outside the general circle, and flying in terror before a little game hen, if it ventured to approach too close. The principal advantage of this bird seems to be that the chickens, from their superior size, are ready for the market at an earlier age than those of the ordinary fowl.

The Bantam is a very little bird indeed, but exceedingly courageous, and does not hesitate to attack a turkey or such large bird with most amusing pompousness of manner. Some Bantams have their legs thickly feathered down to the very toes. The hackles, or long neck feathers of this and the preceding bird, are much used by anglers for making artificial flies.

The celebrated Jungle Fowl of India belongs to this race, and is by many supposed to be the origin of our domestic game fowl. The Chinese, who are greatly addicted to the sport of cock-fighting, prefer this bird for their cruel amusement.

The Dorking Fowl is a large and delicate species. The chief peculiarity in this bird is the double hind toe, so that it has five toes instead of four.

The TURKEY is an inhabitant of America, and appears to have been imported into Europe about the year 1600. Its habits in a state of

Sub-family *d.* *Meleagrīnæ.*

MELEĀGRIS.—(Gr. Μελεαγρίς, a Guinea-fowl.)

Gallopāvo (Lat. *the Turkey*).

domestication need no description, but when wild in its native woods are rather interesting. It is partly migratory in its habits, moving from the parts about Ohio, Kentucky and Indiana, towards the Ohio and

Mississippi. The march is usually performed on foot in large flocks, the birds seldom using their wings except when attacked, or in order to cross a river. The powerful birds can easily cross a river of a mile in breadth, but the weaker frequently fall into the water, and then paddle to shore with some rapidity. This migration is performed about the end of October. Bonaparte, in his splendid work on the American Ornithology, gives the following account of the ingenious way in which the turkeys escape the insidious attacks of their enemies.

"These birds are guardians of each other, and the first who sees a hawk or eagle gives a note of alarm, on which all within hearing lie close to the ground. As they usually roost in flocks, perched on the naked branches of trees, they are easily discovered by the large owls, and when attacked by these prowling birds, often escape by a somewhat remarkable manœuvre. The owl sails round the spot to select his prey, but notwithstanding the almost inaudible action of his pinions, the quick ear of one of the slumberers perceives the danger, which is immediately announced to the whole party by a *chuck:* thus alarmed, they rise on their legs, and watch the motions of the owl, who, darting like an arrow, would inevitably secure the individual at which he aimed, did not the latter suddenly drop his head, squat, and spread his tail over his back; the owl then glances over without inflicting any injury, at the very instant that the turkey suffers himself to fall headlong towards the earth, where he is secure from his dreaded enemy."

The Guinea-fowl or Pintado was originally brought from Africa, and was anciently confounded with the turkey. From its peculiar cry it has gained the name of "Come-back." In its wild state it is gregarious, assembling in large flocks in some marshy situation. At night the birds roost on the trees in company, like the turkey. It is of a restless, wandering disposition, which does not leave it in captivity, the bird frequently wandering for several miles from its home.

Numĭda.*—(Lat.)

Meleăgris, *the Guinea-fowl.*

It is rather a tyrant to the other inhabitants of the farmyard, and does not hesitate to bite them severely. I say bite, because it does not peck like other birds, but grasps its opponent with its mouth, and holds

* So called on account of its Numidian origin.

on like a bull-dog. One of my friends has one of these birds, which has struck up a great friendship with a barn-door cock. In spite of her friendship, however, she keeps him in good order, and if he presumes to help himself before she is satisfied, she plunges at him tumultuously, and grasps him tight with her beak, while she runs screaming round the yard. When the fowls are fed, she secures all the food to herself by a very ingenious manœuvre. She runs in a circle round the place where the food is lying, so as to include it all, and if she sees any lying beyond, she makes a larger circle. Any one of the other fowls intruding into this space is immediately bitten. So they hang about outside, while she pecks away at the food at her leisure, and when she is satisfied she graciously permits the others to get what they can.

Like the turkey, the Pintado lays its eggs in the closest concealment it can find. The eggs are rather smaller than those of the hen; the shell is very thick, and the colour is a yellowish red profusely spotted with dark brown.

This is the bird that was called Meleagris by the ancients. The sisters of Meleager were said to have been metamorphosed into birds, whose feathers were sprinkled with the tears shed for his death.

Family IV. Tetraonĭdæ.—(Lat. Tetrăo, a Bustard.)

Sub-family *a. Perdicīnæ.*

Perdix.—(Gr. Πέρδιξ, a Partridge.)

Cinerĕa (Lat. *ashy*), *the Partridge.*

The Partridge, an inhabitant of this country, is well known as one of the birds included in the designation of "game." It lays from fifteen to twenty eggs in a rude nest placed on the ground, and displays great attachment to them, and no small ingenuity in decoying an intruder away. Mr. Jesse mentions that a gentleman who was overlooking his ploughman, saw a partridge run from her nest, almost crushed by the horses' hoofs. Being certain that the next furrow must bury the eggs and nest, he watched for the return of the plough, when to his great astonishment the nest, previously containing twenty-one eggs, was vacant. After a search, he found the bird sitting upon the eggs under a hedge, nearly forty yards from the nest, to which place she and her mate had removed the whole number in less than twenty minutes. In some parts of England the Partridge is very plentiful—one sportsman having shot in two days one hundred and sixty-eight brace on one manor.

The length of the bird is twelve inches and a half; the wing is short and rounded, causing the peculiar whirring sound when in motion; the third and fourth primary feathers are the longest.

The QUAIL is a tolerably common little bird, visiting England in the summer. Countless flocks of them are spread over the whole of Southern Europe, and multitudes are taken and sent to the London markets; thirty-six thousand having been purchased during one season by the London poulterers.

Temminck states that hundreds of thousands arrive in Naples and Provence, and are so fatigued that for several days they suffer themselves to be taken by hand. We are here reminded of the flight of Quails with which the Israelites were fed, the sacred narrative even preserving the nocturnal flight of these birds. "And it came to pass, that *at even* the Quails came up and covered the camp." Probably the instinct to fly by night is implanted in them for the purpose of avoiding the birds of prey that would attack them by day. The female lays from seven to twelve eggs in a rude nest on the ground.

COTURNIX.—(Lat.)

Commūnis (Lat. *common*), *the Quail.*

The length of the bird is seven inches; the second primary feather is the longest.

The CAPERCAILLIE, or COCK OF THE WOOD, is common in most parts of Northern Europe, and was once to be found in Scotland and Ireland. The male is a large bird, almost equalling a Turkey in size, but the female is considerably smaller. In the early spring, before the snow has left the ground, this singular bird commences his celebrated "play." This play is confined to the males, and intended to give notice of their presence to the females who are in the neighbourhood. "During the play," says Lloyd, "the neck of the Capercaillie is stretched out, his tail is raised and spread like a fan, his wings droop, his feathers are ruffled up, and in short he much resembles in appearance an angry turkey-cock. He begins his play with a call something resembling *peller, peller, peller;* these sounds he repeats at some little intervals, but as he proceeds, they increase in rapidity, until at last, and after perhaps the lapse of a minute or so, he makes a sort of gulp in his throat, and finishes with sucking in, as it were, his breath.

"During the continuance of this latter process, which only lasts a few seconds, the head of the Capercaillie is thrown up, his eyes are partially

Sub-family *b.* *Tetraoninæ.*

Tetrăo. — (Lat. *a Bustard.*)

Urogallus * (Latinised from the Danish *Ur*, primitive), *the Capercaillie.*

closed, and his whole appearance would denote that he is worked up into an agony of passion. At this time, his faculties are much absorbed, and it is not difficult to approach him."

The nest is made on the ground, and contains from six to twelve eggs.

The Black Grouse or Black Cock † is still found on the moors of Scotland and some parts of England, and, together with the red grouse, tempts innumerable sportsmen annually to spend their leisure months on the moors.

* Agassiz, in his "Nomenclator Zoologicus," derives the name from Urus, a buffalo; but the derivation given above is evidently the correct one. The German name for the bird is exactly similar, Auerhahn, literally, meadow-fowl, Aue and Ur being cognate words.

† See page 275.

TETRĂO.

Tetrix (Gr. Τέτριξ or τέτραξ, *a grouse*), *the Black Grouse.*

The Red Grouse has never been found wild on the Continent, but seems to confine itself exclusively to the heaths of Scotland, Wales, and Ireland. In these places it is very numerous, associating in flocks or "packs;" and together with the black grouse is eagerly pursued by sportsmen, who are frequently baffled by the shy and wary habits of the birds.

Lagōpus.—(Gr. Λαγώς, a Hare; πούς, a foot the Ptarmigan.)

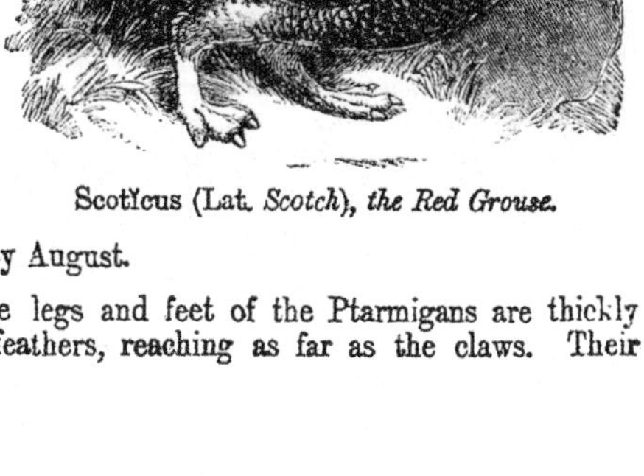

Scotĭcus (Lat. *Scotch*), *the Red Grouse.*

It derives its name of Lagopus or Hare-footed, from the covering of its legs, which are feathered down to the toes, much as the foot of the hare is covered with fur. The nest of the Red Grouse is formed of heath and grass, carelessly heaped together on the ground under the shelter of some low shrub. The young are fully fledged by August.

The Ptarmigan.—The legs and feet of the Ptarmigans are thickly covered with hair-like feathers, reaching as far as the claws. Their plumage bears a singular analogy to the fur of the ermine and some other quadrupeds, as it changes in winter from a rich tortoiseshell colour to a pure white. The common Ptarmigan inhabits the northern parts of Europe and America, and is also found in the north of Scotland, principally among the mountains. The colour of the bird is so similar to that of the mossy and lichen-covered rocks among which it dwells, that a whole covey easily eludes an unpractised eye.

LAGŌPUS.

Albus (Lat. *white*), *the Ptarmigan.*

Enormous numbers of Ptarmigans are annually imported from the north of Europe, especially from Norway and Sweden, to the London market. One poulterer has purchased fifteen thousand of these birds; and twenty-four thousand have been exported in one ship from one place.

Like that of the grouse, the Ptarmigan's nest is a loosely-constructed heap of twigs and grass, and contains from ten to fourteen eggs, of a reddish white spotted with brown.

The Brush Turkey.—The Megapodidæ, deriving their name from the enormous size of their feet, are inhabitants of Australia and the Papuan Islands. In the habits of these birds there is a peculiarity hardly less singular than surprising. Instead of hatching their eggs by the warmth of the body, as most birds do, not excepting the ostrich, the Megapodes bury their eggs in a decaying heap of grass and leaves, trusting to the heat furnished by the fermentation to hatch the eggs.

The Brush Turkey is principally found in the thick brushwood of New South Wales. Mr. Gould, who first brought it before the public, gives this curious account of their nests:—"The mode in which the materials composing these mounds are accumulated is equally singular, the bird never using its bill, but always grasping a quantity in its foot, throwing it backwards to one common centre, and thus clearing the surface of the ground for a considerable distance so completely that scarcely a leaf or a blade of grass is left. The heap being accumulated, and time allowed for a sufficient heat to be engendered, the eggs are deposited, not side by side as is ordinarily the case, but planted at the distance of nine or twelve inches from each other, and buried at nearly an arm's depth, perfectly upright, with the large end upwards. They are covered up as they are laid, and allowed to remain until hatched. I am credibly informed, both by natives and settlers living near their haunts, that it is not an unusual event to obtain nearly a bushel of eggs at one time from a single heap; and as they are delicious eating they are eagerly sought after."

Family V. Megapodĭdæ.—(Gr. Μέγας, great; πούς, a foot. The great-footed kind.)

TALEGALLUS.—(Latinised from native name.)

Lathāmi (Lat. *of Latham*), *the Brush Turkey, or Tallegalla.*

When the Brush Turkey is disturbed, it either runs through the tangled underwood with singular rapidity, or springs upon a low branch of some tree, and reaches the summit by a succession of leaps from branch to branch. This latter peculiarity renders it an easy prey to the sportsman.

The MOUND-MAKING MEGAPODE inhabits the dense thickets bordering on the sea-shore, and is never found far inland. Like the Brush Turkey, it deposits many eggs in one mound, but instead of placing them at intervals in the mound, the bird makes deep holes, from five to six feet, at the bottom of which the eggs are deposited. The natives obtain the eggs by scratching up the earth with their fingers, until they have traced the hole to the bottom; a very laborious task, as the holes seldom run straight, and often turn off at right angles to avoid a stone or root. The mounds are enormously large. Mr. Gilbert was told by the residents

MEGAPODIUS.

Tumulus (Lat. *a Mound*), *the Mound-making Megapode.*

that they were the tombs of the aborigines, nor was it until after some time that their real nature was made known. The height of one mound was fifteen feet, and its circumference at the base sixty feet.

The OSTRICH.—The Struthionidæ include the Ostrich, Emu, Cassowary, and Apteryx. The birds of this family are all remarkable for the shortness of their wings, which are weak and unable to raise them from the ground, but appear to assist them in running. On this account Cuvier called the family Brevipennes, *i.e.* short-winged birds.

The Ostrich is the largest bird as yet known to exist, its height being from six to eight feet. It is an inhabitant of Africa, and from thence the elegant plumes are brought. These plumes are mostly obtained from the wings of the bird, and not from the tail, as is generally imagined.

An immense number of eggs are laid by the Ostriches in one spot, several birds belonging to each nest. The eggs are very large and strong, and are in general use by the Bosjesmans for holding water. By means of these eggs, which they bury at intervals in the sand, after filling them with water, they are enabled to make inroads across the desert and retreat with security, as none can follow them for want of water. Each egg holds rather more than five pints. An excellent omelet is made by the natives, by burying the fresh egg in hot ashes, and stirring round the contents with a stick through a hole in the upper end, until thoroughly cooked.

The principal strength of the Ostrich tribe lies in the legs. These limbs are so powerful that a swift horse has great difficulty in overtaking the bird. As the Ostrich mostly runs in large curves, the hunters cut across and intercept the bird, which would in all probability escape if followed in its exact course.

Order VI. . . *STRUTHIŌNES.*

Family I. . . Struthionĭdæ.—(Gr. Στρουθός, an Ostrich. Ostrich kind.)

Sub-Family *a. Struthionīnæ.*

STRUTHĬO.

Camēlus (Gr. Κάμηλος, *a Camel*), *the Ostrich.*

The Ostrich is easily tamed, as those who have been pursued by the magnificent birds in the Zoological Gardens can testify. These frequently astonish the visitor by suddenly snatching out of his hand a bun or cake which he had intended for his own especial benefit, their long necks enabling them to reach to a surprising distance. Many of my readers have doubtless seen the tame Ostriches at the Hippodrome, who ran races bearing riders on their backs, and really seemed to enjoy the sport as much as any of the spectators. The interesting narrative of Captain Cumming contains some useful remarks on the habits of the Ostrich, and the method in which it is destroyed by the Bosjesmans.

"While encamped at this vley we fell in with several nests of ostriches; and here I first ascertained a singular propensity peculiar to these birds. If a person discovers the nest, and does not at once remove the eggs, on returning he will most probably find them all smashed. This the old birds almost invariably do, even when the intruder has not handled the eggs, or so much as ridden within five yards of them. The nest is merely a hollow scooped in the sandy soil, generally amongst heath or other low bushes; its diameter is about seven feet; it is believed that two hens often lay in one nest. The hatching of the eggs is not left, as is generally believed, to the heat of the sun, but, on the contrary, the cock relieves the hen in the incubation. These eggs form a considerable item in the Bushman's cuisine, and the shells are converted into water flasks, cups, and dishes. I have often seen Bush-girls and Bakalahari women, who belong to the wandering Bechuana tribes of the Kalahari desert, come down to the fountains from their remote habitations, sometimes situated at an amazing distance, each carrying on her back a kaross, or a net-work containing from twelve to fifteen ostrich egg-shells, which had been emptied by a small aperture at one end; these they fill with water, and cork up the hole with grass.

"A favourite method adopted by the wild Bushman for approaching the Ostrich and other varieties of game, is to clothe himself in the skin of one of these birds, in which, taking care of the wind, he stalks about the plain, cunningly imitating the gait and motions of the Ostrich, until within range, when, with a well-directed poisoned arrow from his tiny bow, he can generally seal the fate of any of the ordinary varieties of game. Their insignificant looking arrows are about two feet six inches in length; they consist of a slender reed, with a sharp bone head, thoroughly poisoned with a composition, of which the principal ingredients are obtained sometimes from a succulent herb, having thick leaves, yielding a poisonous milky juice, and sometimes from the jaws of snakes. The bow barely exceeds three feet in length; its string is of twisted sinews. When a Bushman finds an ostrich's nest he ensconces himself in it, and there awaits the return of the old birds, by which means he generally secures the pair. It is by means of these little arrows that the majority of the fine plumes are obtained which grace the heads of the fair throughout the civilized world."

The food of the Ostrich is vegetable, and it swallows many stones, &c. to assist it in grinding its food. When in confinement it picks up anything, glass, nails, &c., from the effects of which it sometimes dies. I have assisted at the dissection of an ostrich, and have seen an astonishing amount of pebbles and other hard materials taken from its stomach, among which were a tolerably large piece of deal, and a considerable portion of a brickbat.

Capt. Cumming remarks a fact not generally known, viz. the care that the Ostrich takes of its young. It has generally been supposed, that after the eggs are laid the female leaves them to be hatched in the sun, and takes no more care for them. The following anecdote would do honour to the far-famed Lapwing. "I fell in with a troop of about twelve young ostriches, which were not much larger than Guinea-fowls. I was amused to see the mother endeavour to lead us away, exactly like a wild duck, spreading out and drooping her wings, and throwing herself down on the ground before us as if wounded, while the cock bird cunningly led the brood away in an opposite direction."

The Rhea, or American Ostrich, is abundant on the banks of the river La Plata, and is chased by the Gauchos, who pursue it on horseback, and kill it by throwing the celebrated "bolas." These curious weapons are made of a long leathern thong, having a heavy stone or leaden ball attached to each end. The Gaucho can throw it so as either to stun his prey with a blow from the ball, or strangle it by causing the thong to twist round its neck.

It is known that the Rhea can swim well, and it has been seen to cross rivers several hundred feet in width, a power which the ostrich and the cassowary are not ascertained to possess. There are two species of this bird, one, the Darwin's Rhea, has been but lately introduced to science.

CASUARIUS.—(Latinised form of Cassowary.)

Casŏar, *the Cassowary.*

The CASSOWARY is a native of the eastern parts of Asia. Like the ostrich, it cannot fly, but runs with great swiftness, and if attacked by dogs kicks with extreme force and rapidity. The feathers of this bird are remarkable for being composed of two long, thread-like feathers, sprouting from the same root. The wing feathers are round, black, and strong, and resemble the quills of the porcupine. At the end of the last joint of the wing is a sort of claw or spur. The crest upon its head is composed of a cellular bony substance.

The food of the bird consists of vegetable substances, and it will frequently swallow a tolerably large apple entire, trusting to the pebbles, &c. in its stomach to bruise it.

The EMU is a native of New Holland, and nearly equals the ostrich in bulk, its height being between five and six feet. Its feathers lie loosely on the body, and its wings are small and hardly to be distinguished. The skin of the Emu furnishes a bright and clear oil, on which account it is eagerly sought after. Mr. Bennet gives the following account of the habits of this bird:—

"In its manners the Emu bears a close resemblance to the ostrich. . .

DROMAIUS.—(Gr. Δρομαῖος, running swiftly.)

Novæ-Hollandiæ (Lat. *of New Holland*), *the Emu.*

Its food appears to be wholly vegetable, consisting chiefly of fruits, roots and herbage, and it is consequently, notwithstanding its great strength, perfectly inoffensive. The length of its legs and the muscularity of its thighs enable it to run with great swiftness; and as it is exceedingly shy, it is not easily overtaken or brought within gun-shot. Captain Currie states that it affords excellent coursing, equalling if not surpassing the same sport with the hare in England; but Mr. Cunningham says that dogs will seldom attack it, both on account of some peculiar odour in its flesh which they dislike, and because the injuries inflicted upon them by striking out with its feet are frequently very severe. The settlers even assert that the Emu will break the small bone of a man's leg by this sort of kick; to avoid which, the well-trained dogs run up abreast, and make a sudden spring at their neck, whereby they are quickly dispatched.

"Its flesh has been compared to coarse beef, which it resembles both in appearance and taste. There is but little fit for culinary use upon any part of the Emu except the hind quarters."

The voice of the Emu is a kind of low booming sound. The eggs are six or seven in number, of a dark green colour, and are much esteemed by the natives as food. When the natives take an Emu, they break its wings, a curious custom of no perceptible utility. Young men and boys are not permitted to eat the flesh of this bird.

The APTERYX.—This extraordinary bird, whose name is derived from the apparent absence of wings, those members being merely rudimentary, inhabits the islands of New Zealand. It conceals itself among the densest fern, and when hunted by dogs, it hastens to seek a refuge among rocks and in the chambers which it excavates in the earth. In these chambers its nest is made and the eggs laid. The natives hunt it with great eagerness, as the skin is used for the dresses of chiefs, who are so tenacious of them that they can hardly be persuaded to part with a single skin. The feathers are employed to make artificial flies. When attacked it defends itself by rapid and vigorous strokes with its powerful feet.

Sub-family *b*. *Apterygīnæ*.

APTĔRYX.—(Gr. α, priv.; πτέρυξ, a wing wingless.)

Austrālis (Lat. *Australian*), *the Apteryx.*

Dr. Shaw first brought this bird before the notice of the public, but for many years naturalists considered it an extinct species. Latterly the question has been set at rest, not only by the researches of Gould and other naturalists, but by the arrival in this country of several skins and one living specimen, now in the Zoological Gardens. This bird has a singular habit of resting with the tip of its bill placed on the ground. The nostrils of the Apteryx are placed almost at the very extremity of the bill. The aborigines of New Zealand give it the name of Kiwi Kiwi. The food of the bird consists of snails, insects and worms, which latter creatures it obtains by striking the ground with its feet, and seizing them on their appearance at the surface.

A small but well preserved skin is mounted in the Ashmolean Museum, Oxford, in which the rudimentary wings are very well shown. An entire skeleton is in the museum of the College of Surgeons, and other specimens are to be seen in various collections.

Sub-family *c*. *Didinæ*.

Didus.—(Latinised form of Dodo.)

Ineptus (Lat. *stupid*), *the Dodo*.

The Dodo.—This singular bird, which is supposed to be extinct, was discovered at the Mauritius by the earlier voyagers. For many years their accounts of the Dodars were supposed to be mere flights of fancy. Lately, however, the discovery of several relics of this bird in various countries has set the question of its existence at rest, but not the question of the proper position of the bird. Some think it belongs to the pigeons, and some to the ostriches. In the Ashmolean Museum at Oxford are a head and foot of the Dodo, sole remnants of a perfect specimen known to have existed in 1700; and in the same place, in the year 1847, during the meeting of the British Association, were gathered together the whole of the existing remains from every country.

In the travels of Sir T. Hubert, in the year 1627, are several accounts. From the work of this traveller, whose amusement it was to re-write his travels, each time completely changing the language but retaining the matter, an extract is taken.

"The Dodo, a bird the Dutch call Walghvogel, or Dod Eersen; her body is round and fat, which occasions the slow pace, or that her corpulencie, and so great as few of them weigh less than fifty pound: meat it is with some, but better to the eye than stomach, such as only a strong appetite can vanquish. . . It is of a melancholy visage, as sensible of nature's injury in framing so massie a body to be directed by complimental wings, such, indeed, as are unable to hoise her from the ground, serving only to rank her among birds. Her traine, three small plumes, short and improportionable, her legs suiting to her body, her pounces sharpe, her appetite strong and greedy. Stones and iron are digested; which description will better be conceived in her representation." The "representation" here alluded to is that of a globular-shaped bird, perfectly naked, with the exception of three separate feathers on the tail, and a few feathers on the wing. The expression of lugubrious wisdom on the countenance is irresistibly ludicrous.

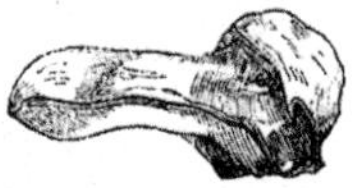
BEAK OF THE DODO.

It is still within the range of possibility that this bird may again be discovered, as at present but little of Madagascar has been searched, and in that island, if anywhere, it will be found.

Another bird, the gigantic Dinornis, has been extirpated from the face of the earth by man. This enormous bird, whose leg is rather larger than that of a fossil elk, and whose head could not have been less than ten feet and a half from the ground, was at one time an inhabitant of New Zealand, but has been extirpated for many years, a fate likely to befal the defenceless Apteryx. In the Anatomical Museum at Oxford is a cast of the leg of the Dinornis, standing side by side with that of an ostrich. The leg of the ostrich is quite insignificant by the side of the enormous cast.

The Great Bustard, our English representative of the Otidæ, is now scarcely ever seen in this country, although formerly it was tolerably common. It runs with great swiftness, and will never rise on the wing until forced, so that instances have been known of Bustards being captured by greyhounds. It is exceedingly wary, and can hardly be approached within gun-shot, except by adopting some disguise, as a labourer with the gun in his wheelbarrow, or by driving a cart or a carriage by the spot where it is feeding.

The male Bustard possesses a membranous pouch on the fore part of the neck, capable of holding six or seven pints of water. There is an opening to this pouch under the tongue, and its use is possibly, like that of the pelican, to carry water for the use of the young; but this is not ascertained. The length of the bird is rather more than three feet. Its

Sub-family *d. Otinæ.*—(Gr. Ὠτίς, a Bustard.)

OTUS.

Tarda (Lat. *slow*), *the Great Bustard.*

nest is a loose heap of straw on the ground, and contains two pale brown eggs, spotted with brown, rather larger than those of the turkey.

The Plovers are known by their long legs, short toes, and long powerful wings. Many are inhabitants of England, of which the Lapwing and Golden Plover are the most common.

The Stone Curlew, or Thick-knee, or Norfolk Plover, is common in England, and is to be found on open plains. White gives an accurate description of the bird in his Natural History of Selborne. "The history of the Stone Curlew is as follows. It lays its eggs, usually two, never more than three, on the bare ground, without any nest in the field, so that the countryman in stirring his fallows often destroys them. The young run immediately from the egg, like partridges, &c., and are withdrawn to some flinty field by the dam, where they skulk among the stones, which are their best security; for their feathers are so exactly of the colour of our grey spotted flints, that the most exact observer, unless he catches the eye of the young bird, may be eluded. The eggs are short and round, of a dirty white, spotted with dark bloody blotches. Though I might not be able just when I pleased to procure you a bird, yet I could show you them almost any day; and any evening you may hear them round the village, for they make a clamour which may be heard a mile. *Œdicnemus* is a most apt and expressive name for them, since their legs seem swollen, like those of a gouty man. After harvest I have shot them before the pointers in turnip fields."

Order VII. . *GRALLÆ.*—(Lat. *Stilted Birds.*)
Family I. . Charadrīdæ.—(Gr. *Χαραδρίος*, a Lapwing.)
Sub-family *a*. *Œdicnemīnæ.*—(Gr. *οἰδέω*, I swell; *κνήμη*, the leg.)

ŒDICNĒMUS.

Crepĭtans (Lat. *crackling*), *the Stone Curlew.*

The Stone Curlew is one of the migratory birds, coming in March or April, and leaving our shores about the autumn, very few remaining until October. It is a very shy bird, and not very easily approached.

The LAPWING, or PEEWIT, is very common in most parts of England, and is well known for its plaintive cry, and the stratagems it employs to decoy intruders away from its nest, or rather eggs, for nest it has none. Frequently, however, the attempts of the bird only draw the attention of the passer-by to the evident vicinity of the eggs. These eggs are dark brown, blotched with black, and are hardly to be distinguished from the soil where they are laid. If an intruder approach them, the bird glides before him, and flutters along, drooping her wings, as if wounded, invariably endeavouring to lead him away from her

Sub-Family *d*. *Charadrīnæ.*

Vanellus.—(Linnean generic name.)

Cristātus (Lat. *crested*), *the Lapwing or Peewit.*

nest. When it has succeeded in decoying away the intruder, it suddenly mounts in the air, uttering its cry of pee-weet, leaving its pursuer to gaze with astonishment at the escaping bird. The eggs are sold in great numbers, under the title of "Plovers' eggs," and are considered great delicacies. When flying, the black and white colours of its plumage make it very conspicuous. On the head of the bird is a kind of crest.

Family II. . . . Ardeĭdæ.—(Lat. *Ardĕa,* a Heron. Heron kind.)

Sub-Family *b*. . *Gruīnæ.*

Grus.—(Lat. *a Crane.*)

Cinerĕa (Lat. *ashy*), *the Crane.*

The Common Crane is now but rarely seen on our shores, although formerly as common as was the bustard. It flies at so great a height, that although its hoarse cry is audible, the bird itself is far out of the reach of sight. It generally feeds on snails, frogs, and worms, but is not by any means averse to newly sown grain. The nest is made among reeds and rushes, and contains two bluish green eggs, marked with brown. The length of the bird is nearly four feet.

The HERON, or HERNE, is a bird renowned in the noble science of falconry, and respecting which much curious knowledge is to be gained from the work of Dame Juliana Berners, a book of most amusingly quaint language.

The Common Heron generally breeds in company, like the rook; indeed, these two birds frequently inhabit contiguous trees, but never interfere with each other. In the dawn of the early morning, or while the moon casts an uncertain light, the Heron may be seen standing in the shallow water, stiff and motionless, and by the faint light may be mistaken for a stump of a tree. But his eye is keenly directed on the water, and no sooner does a fish approach, than a dart of his unerring bill secures it, and the Heron soars exultingly to his nest, bearing his prey with him. The fixed patience that the Heron displays has caused it to be chosen as the emblem of Solitude.

Sub-family c. *Ardeinæ.*

ARDĔA.

Cinerĕa (Lat. *ashy*), *the Heron.*

The plumes of the Heron were formerly considered as ornaments only to be worn by the noble. It is not an uncommon sight to see this splendid bird slowly winnowing his way through the air, when suddenly a magpie, or a crow, gives the alarm, and the poor bird is instantly beset by its annoying enemies, especially the crows, who resent the Heron's approach to their own residence, and frequently drive him away.

There was, and may be still, a belief that the legs of the Heron exhaled some scent that was attractive to the fish, and caused them to approach as the bird stood with its feet in the water. For that reason, anglers were accustomed to make some kind of a preparation from the skin taken from the legs of a Heron, a sort of Heron-skin tea, and with this to prepare their bait, which they thought was rendered quite irresistible by such a proceeding.

If the Heron is fired at, the gunner should aim at the head, for the bird is capable of carrying away without damage, or resisting by the thickness of its plumage, a goodly charge of shot. I have myself lodged a charge of shot against a Heron's ribs without materially injuring the bird, although I waited until it had spread its wings, and aimed at its side. The Heron certainly appeared startled, but although the shot came against its feathers with an audible sound, the charge did not appear to have penetrated, for the bird flew away very unconcernedly.

Perhaps it was as well that it did fly away, for I was not particularly cautious, and as the bird would hardly have been killed on the spot, I might probably have lost an eye or so in attempting to secure it; for the Heron is a very cunning bird in these matters. When it is wounded, it will be perfectly quiet as if dead, and when the fowler stoops to pick it up, it will dart its beak at his head with such unerring aim, and such unexpected rapidity, that he may think himself fortunate if his eyes escape. Not long since, a gentleman who had shot a Heron, nearly lost one of his eyes when he approached the wounded bird, who darted its bill at his face with sufficient force to dash in pieces the glass of a pair of spectacles which he was wearing at the time.

The Heron sometimes killed the falcon in its stoop by throwing its head back, whether purposely or not is not known, and receiving its enemy on the point of its sharp beak, by which the falcon was transfixed as if on a bayonet.

It has been lately ascertained that the Heron can swim in deep water, and does so when it sees any prey that cannot be reached by wading, such as a nice nestfull of young moor-hens, or a water-rat engaged at his dinner.

The nest of the Heron is a flat mass of sticks, laid on the highest branches of a tree, and contains five bluish green eggs. The length of the bird is about three feet. An old name of this bird was the Herne, or Hernshaw, from which was derived the saying, "He does not know a Hawk from a Hernshaw." The last word has been corrupted into "handsaw," and of course renders the proverb most unmeaning.

The Demoiselle Crane.*—This bird is chiefly remarkable for the considerable idea that it appears to have respecting the beauty of its own person. Its deportment is very singular, and at times even ludicrous. Whenever it takes it into its head to be ridiculous, it does so most effectually, and affectedly also. It moves about with a consequential air, hanging its head first on one side and then on the other. It then will run some twenty or thirty yards, treading only on the tips of its toes, as if it wore white satin shoes, and were trying to pick its way over a very dirty road. Then it will have a little dance all to itself, and suddenly stand still again quite grave and composed, as if it had been doing nothing at all. From these habits, cynical naturalists have named it the Demoiselle.

It is rather a tall bird, being between three and four feet in height.

* See p. 291.

ANTHROPOÏDES.—(Gr. 'Ανθρωποειδής, like a man.)

Virgo (Lat. *a virgin*), *the Demoiselle Crane.*

The BITTERN.*—The beautiful Bittern has been almost banished from this country, although it was formerly a common bird. It frequents morasses, and dense beds of reeds, where it lies concealed until the evening, when it leaves its rushy bed and soars to a vast height, continually uttering its sepulchral booming cry. This singular sound is not unlike the bellowing of a bull, and is most startling in its effect.

In olden times the Bittern was one of the birds chiefly sought after in falconry, as the stout defence it makes against its enemies, by darting its sharp and powerful beak at them, and beating violently with its feet, renders it by no means an easy prey. For this reason the falconer's first care, on reaching the Bittern when brought to the ground by his falcon, was to secure its head, and by fixing its bill deep in the earth, to save his eyes from the rapid and well-aimed blows of the wounded bird. The falcon also was in danger of being transfixed by the sharp beak of his victim.

The plumage of this beautiful bird is a rich reddish-yellow ground, boldly variegated with various black marks, which are most conspicuous in the loose, long feathers that decorate its neck. In size, it is a little less than the heron. It feeds principally on small reptiles, field mice,

* See page 292.

BOTAURUS.—(Lat. *boo*, I bellow; *taurus*, a bull.)

Stellāris (Lat. *starry*), *the Bittern.*

and fish. Its nest is built on some slight elevation in a morass, and contains five bluish green eggs.

The WHITE SPOONBILL.*—The Common Spoonbill is found in Europe, Asia, and Africa, and frequents Holland, together with the stork. The strange shape of the tip of its beak has gained it the name of Spoonbill. It has rarely been taken in this country. It feeds on worms, snails, and water plants, searching for the latter by agitating the water with its broad beak.

The nest of the White Spoonbill is sometimes placed in trees, and sometimes amid rushes. It contains three whitish eggs, slightly spotted with red. The length of the bird is not quite three feet.

The STORK † is extensively found throughout Europe, Asia, and Africa. In Holland storks are very abundant, and are encouraged by the Dutch to build in their towns. Among the ruins of Persepolis they are very common, scarcely one pillar being without a stork's nest at the summit.

* See page 293. † See page 294.

PLATALĔA.—(Lat.)

Leucorodĭa (Gr. Λευκόροδον, *a white rose*), *the White Spoonbill.*

In Holland a kind of false chimney is built by the inhabitants for these birds to make their nests in. When the Stork cannot find a building on which to make its nest, it chooses the flat spreading branches of a cedar or pine, and there collects a large mass of sticks and twigs, on which it lays from three to five whitish eggs. When disturbed, the birds make a great clattering with their bills.

A recent visitor to Constantinople remarks that the very Storks seemed to have become Ottoman, for they sat on the tops of the houses, looking staid and solemn, as becomes the Oriental character, and managed their beaks just as if they were pipes. It is true that they wore no turbans, but each of them appeared to have left a turban of preposterous dimensions, viz. his nest, on the roof of a house close by.

The draining of our morasses seems to have driven the Stork completely out of this country, where it was formerly tolerably common. The food of this bird consists of rats, mice, frogs, &c., and it is for the benefits it confers upon man by devouring these vermin that it is so carefully protected and encouraged, especially in the East, where the inhabitants do not trouble themselves by removing carrion or offal, but leave that office to the vultures, hyænas, and other scavengers of nature. The height of the Stork is nearly four feet.

Sub-family *d. Ciconinæ.*

CICONIA.—(Lat. *a Stork.*)

Alba (Lat. *white*), *the Stork.*

The ADJUTANT.*—This very remarkable bird is a native of various parts of India, and must not be confounded with the Marabou, which belongs to the same genus, but lives in the African tropics.

The Adjutant is one of the largest birds in the world, standing five feet in height, and measuring seven feet and a half from the tip of the bill to the claws, while its expanse of wing is rather above fourteen feet. On the front of the breast there hangs a pouch of skin, into which the bird sometimes appears to withdraw its neck altogether, looking on such occasions as if it had no neck at all. Its bill, as will be seen from the cut, is enormously large, and capable of receiving morsels of considerable magnitude, inasmuch as in the crop of one of these birds were found a land tortoise, ten inches in length, and a large male black cat, which the Adjutant had snapped up entire. It has also been known to swallow entire a small leg of mutton, a hare, and a small fox, so that there is no reason to complain that it does not make use of the "terrors of its beak." But its beak only affords terror to those who are afraid of it, for the Adjutant is an arrant coward, and will not venture even to oppose a hen when she is defending her chickens which the Adjutant has been trying to catch unobserved. When opposed, it makes a great demonstration

* See page 295.

LEPTOPTILOS.- (Gr. Λεπτός, slender; πτίλον, a feather.)

Argala (*native name*), *the Adjutant.*

with opened beak, and roars with a proportionately loud voice, but if the opponent is undismayed and continues to attack, the Adjutant shuts his big beak and runs away. It is related that the great bill of this bird was once exercised in a very singular manner. A cooly, or native servant, was coming round a corner, when an Adjutant dashed round the same corner in the opposite direction, and literally spitted the poor cooly on its beak.

The plumage of this bird is of an ashy grey fading into white on the under parts. The generic name is given to the Adjutant on account of the delicate plumes of the slender tail coverts.

A bird that will swallow a cat or a fox whole is likely to be rather voracious, and to require a considerable amount of food to appease its hunger. This object is attained by assisting the vultures, dogs and hyænas at their scavengering repasts, while it fills up its leisure hours by swallowing all the reptiles and vermin of all kinds that it can find. For this reason it is in high favour among the natives, who protect it with jealous guardianship. It has often been tamed, and becomes rather troublesome in its familiarity.

The SACRED IBIS inhabits Egypt, but does not seem to breed there. This is the bird so frequently depicted in the hieroglyphics as playing a conspicuous part in religious ceremonies. Their mummies are constantly

Sub-family *f.* *Tantalīnæ.*—(From Gr. Τάνταλος, a proper name.)

IBIS.—(Gr. Ἶβις.)

Religiōsa (Lat. *sacred,*) *the Sacred Ibis.*

found in the tombs, and in one of these mummies Cuvier discovered remnants of skin, and scales of snakes. It is a migratory bird, appearing simultaneously with the rise of the Nile, and departing as the inundation subsides. The Sacred Ibis is about the size of an ordinary fowl.

The CURLEW,* or WHAUP, is often found in the northern parts of England and Scotland, and is spread over the whole of the Old World, from South Africa to the polar regions. In winter it collects in large flocks at the muddy shores of the sea, where its long curved bill can easily penetrate in search of food. It is an exceedingly shy bird, and cannot easily be approached within gun-shot.

Its nest is composed of grass and rushes, collected under the shelter of a tuft of heath or grass, and contains four greenish olive eggs blotched with brown. The length of the bird slightly exceeds two feet.

* See page 297.

Family III. Scolopacĭdæ.—(Gr. Σκολόπαξ, a Woodcock. Woodcock-kind.)
Sub-family *a. Limosīnæ.*—(From Gr. λειμών, a Meadow.)

CRACTICORNIS.—(Gr. Κρακτικός, *clamorous ; ὄρνις, a bird.*)

Arquatus (Lat. *arcuea*), *the Curlew.*

Sub-family *b. Totaninæ.*

TRINGOĪDES.—(Like a Sandpiper.)

Hypoleuca (Gr. ὑπόλευκος, whitish), *the Common Sandpiper.*

The COMMON SANDPIPER, or SUMMER SNIPE, comes to England in April, and leaves in September. It is very common in Wales, and is spread throughout most parts of England. It derives its name from the low piping sound which it utters while running along the sandy banks of the rivers where it finds its food. While it runs, it keeps its head and tail constantly on the move, something like those toy birds whose heads and tails are moved by two strings and a weight.

It can dive and swim well, although the shape of the feet and legs do not seem very applicable to such a purpose. Even the young birds only an hour or two out of the egg will run away if alarmed, and take to the water as boldly as if they had been accustomed to it for years.

The nest of the Sandpiper is built in a hole in a bank near fresh water, and is generally shaded by a tuft of grass or sedge. The eggs are four in number, of a reddish white colour, spotted with brown.

Sub-family *c.* *Recurvirostrinæ.*

RECURVIROSTRA.—(Lat. *with bill curved upwards.*)

Avocetta, *the Avocet.*

The AVOCET.—The bill in the genus Recurvirostra is exactly the reverse of that in the genus Cracticornis, the curve being upwards instead of downwards. The common Avocet is spread throughout the warmer regions of Europe, and is also found in some parts of Africa. It is very common in Holland, and is frequently seen on the eastern coasts of England, but seldom visits Scotland. It frequents marshes and the mouths of rivers, where it finds in the mud myriads of the small worms and insects on which it feeds, and which it obtains by scooping them up from the mud with its curiously curved bill. It is a good swimmer, but seldom has recourse to that art except when it wades unexpectedly out of its depth.

The eggs of the Avocet are laid on the ground, in a depression sheltered by a tuft of herbage. Their colour is a bluish green, spotted with black. The birds when disturbed at their nests feign lameness, like the lapwing, in order to draw the intruder to a distance. The length of the bird is eighteen inches.

The WOODCOCK is a native of the northern parts of Europe and Asia, and is common in this country, but rarely has been known to breed here. It generally reaches England at the beginning of October, and leaves us in March or April, at which time its flesh loses the delicacy that characterises it, and becomes coarse and valueless. The Woodcock frequents dense thickets during the day, but at night it leaves these retreats, and visits the swamps and flooded meadows, where it finds a sufficiency of worms and insects.

The nest of this bird is a loose mass of grass and leaves, gathered together in some sheltered depression. The eggs are four in number, of a yellowish brown, blotched with dark brown and grey.

The SNIPE is too well known to need description. In its habits it much resembles the woodcock, excepting that it breeds plentifully in several counties of England, Scotland, and Ireland. Its flight is very singular, rendering it a difficult mark. The Jack Snipe confines itself to one spot, and cannot be induced to leave it even when fired upon. Its flight is fully as perplexing as that of the common Snipe. Stanley, in his History of Birds, mentions "a gentleman, a very bad shot, who having at length succeeded in killing a Jack Snipe, deeply lamented the loss of a bird which, as he was always sure of finding it in the same place, had afforded him constant amusement during a whole winter."

Sub-family d. *Scolopacinæ.*

SCOLOPAX.

Rusticŏla (Lat. *fond of the country*), *the Woodcock.*

NUMENĪUS.—(Gr. Νουμήνιος.)

Scolopacīnus (Lat. *like a Woodcock*), *the Snipe.*

The RUFF is celebrated for its pugnacious habits and the singular change of its plumage at certain seasons of the year. Towards the breeding season a beautiful frill of long feathers is formed round the neck. It is a singular fact, that in hardly any two of these birds is the frill of the same colour; and more remarkable, that the frill of the same bird is of different colours at different seasons. At the same time that the frill forms, the male birds choose each for themselves a small spot, on which

Sub-family *e. Tringīnæ.*

PHILOMĂCHUS.—(Gr. Φιλόμαχος, a lover of battles.)

Pugnax (Lat. *quarrelsome*), *the Ruff.*

no other bird is permitted to intrude without a severe battle taking place. The females, called Reeves, now arrive, and their approach is the signal for a general melée; and the ground is soon denuded of grass by the constant battles.

The nest of the Reeve is merely a slight depression in a tuft of grass. The eggs are four in number, of a greenish white blotched with reddish brown.

Great numbers of these birds are annually sent to the London markets. Various precautions are taken to prevent their destructive quarrels from taking place, as captivity in no way diminishes their pugnacity.

The JACANAS * are found in Asia, Africa, and America. Their light bodies and widely extended claws enable them to walk on the leaves of aquatic plants with equal ease and safety. As their weight is just suffi-

* See page 301.

cient to sink the leaf a little below the surface, they have quite the appearance of walking on the water itself. The Common Jacana inhabits the hotter parts of South America, and is abundant in Brazil and Guiana. It possesses large and sharp spurs on the wing. It is not a very large bird, barely exceeding a pigeon in bulk.

Family IV. . Palamedeïdæ.—(Gr. Παλάμη, the palm of the hand.)

Sub-Family a. *Parrinæ.*

PARRA.—(Lat.)

Jacāna, *the Jacana.*

The CORNCRAKE * or LANDRAIL is very common in England. It reaches us at the beginning of April, and leaves us at the end of October, after hatching its eggs. During the early part of the summer months its harsh cry may be heard in almost every field, but the bird itself is very seldom seen, as it threads its way among the long grass with marvellous rapidity. Its cry can be so exactly imitated by drawing a quill sharply across the teeth of a comb, that the bird may be decoyed by the sound until quite close to the operator. The Corncrake is so averse to rising on the wing, that a dog is frequently employed to hunt it. The young when taken feign death with admirable accuracy, nor do they move until they imagine that the intruder is safely out of the way.

The nest of the Corncrake is by no means uncommon. It is formed of hay, collected and worked into some depression in the ground, and contains from eight to twelve eggs, of a greyish yellow, covered with dark brown spots. The length of the bird is about nine inches.

The WATER-HEN,† or MOOR-HEN, is very common along the reedy banks of rivers and ponds. It is very widely distributed, being found in almost all parts of the Old World. It swims very gracefully, constantly nodding its head, and dives with great skill and rapidity, particularly when alarmed, in which case it generally dives under some floating herbage, and remains there with merely its beak above the water until

* See page 302. † See page 302.

Family V.. Rallĭdæ.—(Latinised by Linnæus from the word Rail.)

Sub-family *a. Rallīnæ.*

Ortygometra.—(Gr. 'Ορτυγομήτρα, migrating with the Quails; the Landrail.)

Crex (Gr. Κρεξ, a Crake; derived from its cry), *the Corncrake or Landrail.*

Sub-family *b. Gallinulīnæ.*

Gallinŭla.—(Lat.)

Chlorŏpus (Gr. Χλωρός, green; πούς, a foot), *the Water-hen.*

the danger is passed. On account of this habit, it is almost useless to shoot this bird unless the sportsman is accompanied by a dog, for if it is not shot dead it instantly dives, and nothing but a dog can discover its retreat. It runs on land with considerable activity, constantly flirting up its tail, so as to show the white feathers beneath, and when alarmed, instantly makes for the water.

The nest of the Water-hen is built among sedges and reeds at the water side, and contains from five to eight or nine eggs, of a cream yellow spotted with dark brown. When the Water-hen leaves her nest, she covers the eggs with dried grass and reeds, so as completely to conceal them, apparently lest the rats should discover them. The young when hatched look like round tufts of black down. They swim and dive well, following their parent with great address. The pike is their chief enemy, and destroys numbers by darting at them from under the cover of water-lilies or other plants.

The Coot much resembles the water-hen in its habits. It is usually found in large sheets of water, particularly if sheltered by trees. The nest is a huge mass of flags, reeds, and grass, usually at the water's edge, but sometimes actually in the water. In 1849 I took five Coots' eggs from a nest situated at the Reservoir near Swindon. The nest was nearly fifty yards from

FULICA.—(Lat.)

Atra (Lat. *black*), *the Coot.*

the bank, and was made on a very small sunken hillock, in three feet water. In the nest are from seven to ten greenish white eggs, spotted with brown.

The FLAMINGO * is an inhabitant of the warmer parts of Europe, and is common in Asia and the coasts of Africa. The singularly shaped beak of this splendid bird is peculiarly adapted to its long and flexible neck. When the bird wishes to feed, it merely stoops its head to the water; the upper mandible is then lowest, and is well fitted to receive the nutritive substances which are entangled in a filter placed on the edges of the beak, much resembling the analogous apparatus of the whale.

The Flamingo frequents marshes, lakes, and mouths of rivers, bidding defiance to the pestilent exhalations that drive man far from their haunts. The colour of their plumage is a deep brilliant scarlet, except the quill feathers, which are black. When a number of these birds stand ranged in a line, according to their custom, they present the appearance of a small and well-drilled body of soldiers, but are far more dangerous to approach than the most formidable army, for the miasma of the marshes has a more deadly aim than the rifle, and its breath is more certainly fatal than the bullet.

The nest of the Flamingo is a curious conical structure of mud, with a cavity at the summit, in which are placed two or three whitish eggs. When the female bird sits on the nest, her feet rest on the ground, or hang into the water. The height of the bird is between five and six feet.

* See page 304.

Order VIII. . *ANSĔRES.*—(Lat. *Geese.*)

Family I. . . Anatĭdæ.—(Lat. *Anas*, a Duck. Duck-kind.)

Sub-family *a. Phœnicopterīnæ.*

PHŒNICOPTĔROS.—(Gr. φοινικόπτερος, red-winged.)

Rubra (Lat. *red*), *the Flamingo.*

The BERNICLE GOOSE * inhabits the northern parts of Europe and America, but during the winter it resorts to our shores in great numbers. It is an extremely shy bird, and cannot be approached without the greatest caution and skill. Of the origin of this bird most absurd tales have been told. All agreed that it was produced from a tree, but the latest and most approved account was that of Gerard, who in 1636 wrote as follows:—"But what our eyes have seen, and hands have touched, we shall declare. There is a small island in Lancashire called the Pile or Foulders, wherein are found the broken pieces of old and bruised ships, some whereof have been cast thither by shipwracke, and also the trunks and bodies with the branches of old and rotten trees, cast up there likewise; wherein is found a certain spume or froth, that in time breedeth into certaine shels, in shape like those of the muskle, but sharper pointed, and of a whitish colour; one end whereof is fastened into the inside of the shell, even as the fish of oisters and muskles, the other end is made fast into the belly of a rude masse or lumpe, which in time commeth to the shape and form of a bird: when it is perfectly formed the shell gapeth open, and the first thing that appeareth is the foresaid lace or string; next come the legs of the bird hanging out, and as it groweth greater it openeth the shell by degrees,

* See page 305.

till at length it is all come forth and hangeth only by the bill: in short space after it commeth to full maturitie, and falleth into the sea, where it gathereth feathers, and groweth to a fowle."

Sub-family *c.* *Anserinæ.*—(Lat. *Anser*, a goose.)

BERNICLA.

Leucopsis (Gr. λευκός, white; ὤψ, a face), *the Bernicle Goose.*

Of the TAME GOOSE, *Anser ferus*, nothing need be said, except that enormous flocks are bred in Lincolnshire, containing from two to ten thousand birds each. The birds are periodically subjected to the operation of plucking out the quill-feathers, in order to supply the vast demand for pens, &c.

The MUTE or TAME SWAN, a well-known ornament to our lakes and rivers, is not an in-

Sub-family *d.* *Cygninæ.*
CYGNUS.—(Lat. a *Swan.*)

Olor (Lat. *a Swan*), *the Mute Swan.*

habitant of England, but was introduced from Eastern Europe and Asia, several hundred years back. All are familiar with the graceful deportment of this bird while sailing on the surface of the water. Unfortunately, its progress on land by no means corresponds with its aquatic grace, being confined to an awkward waddle.

The female Swan makes its nest of a great mass of dry reeds, placed among osiers or rushes near the water, and lays six or eight large white eggs. During the time of incubation, and while the young are still small, the parent birds defend them with great assiduity and courage.

Several large Swanneries are still in existence. The Crown, and the Dyers' and Vintners' Companies own the greater part of the Swans on the Thames, and their Swans are annually marked on the bills by men termed Swan-uppers or hoppers. The mark of the Vintners' Company is a notch or nick at each side of the bill, from which arose the term, "Swans with two nicks," corrupted into "necks."

The spelling books always say that a Swan can break a man's leg with a blow of its wing. Whether they can break the leg of a man or not, I cannot say with certainty, but I have had ocular witness that they cannot break that of a boy. I have repeatedly seen a boy chase a swan into a corner, catch it by the neck, and drag it out, in spite of all the flapping of its wings.

CYGNUS.

Ferus (Lat. *wild*), *the Whistling Swan.*

The WHISTLING SWAN, or HOOPER, (*Cygnus ferus,*) resides during summer within the Arctic circle, but in winter visits the northern parts of Europe, including England. It migrates in flocks of various numbers, arranged in the form of a wedge. The down of this bird is very valuable, and is sought after by the Icelanders, who choose the time when the birds have shed their quill-feathers and are unable to fly, to chase them with dogs. The Whistling Swan wants the grace that characterizes the Tame Swan; its neck being generally carried upright, without the elegant arch domesticated relative. It is not quite so large as the Tame Swan; the expanse of its wings is about eight feet.

"Like a BLACK SWAN," was formerly a well-known proverb, analogous to the Horse Marines of the present day; unfortunately for the proverb, a Swan has been discovered in Australia, the whole of whose plumage is

CHENŌPIS—(Gr. Χήν, a Goose; ὤψ, a face.)

Atrāta (Lat. *blackened*), *the Black Swan.*

a jetty black, with the exception of the quill feathers, which are white. It has been domesticated in this country, and may be seen in St. James' Park, eagerly seeking after the crumbs offered by juvenile hands. It is rather smaller than the Whistling Swan.

The WIDGEON is one of our autumn visitors, arriving in this country about the beginning of October. Large flocks of these birds generally arrive at the same time, and are eagerly sought after, as their flesh is very

Sub-family *e.* *Anatīnæ.*

MARECA.

Penelope (Gr. Proper name), *the Widgeon.*

delicate, the widgeon living almost exclusively on a vegetable diet.

The nest of this bird is usually composed of decayed grass, leaves, or rushes, lined with the soft feathers plucked from the body of the parent bird; and is placed near water, usually among the substances of which it is composed.

The MALLARD or WILD DUCK is the origin of our domestic bird, and is widely spread over the northern parts of Europe, Asia, and America. In the winter it migrates in countless flocks, many reaching this country In Lincolnshire incredible numbers of these birds are taken in a very ingenious trap, called a decoy. It is a perfect edifice of poles and nets, and is built in the form of a tube, very wide at the mouth, and very narrow at the extremity. The ducks are induced to enter the "pipe" by the antics of a dog, and by some hemp-seed previously strewn on the water. They are then driven onwards to the smaller end, where they are caught and killed.

ANAS.—(Lat *a Duck.*)

Boschas (Gr. Βόσκας, a Mallard; from βόσκη, pasture *), *the Mallard.*

Wilson, in his American Ornithology, gives the following account of the method of catching wild ducks practised in America.

"In some ponds frequented by these birds, five or six wooden figures, cut and painted so as to represent ducks, and sunk by pieces of lead nailed on their bottoms so as to float at the usual depth on the surface, are anchored in a favourable position for

* The keeper of a decoy in Lincolnshire mentioned that "the Mallard, Pintail, and Teal, frequent rich flooded lands, *swittering* with their *nebs* (beaks) in the soil, and sucking out all its *strength*; but the Wigeon is *an amazing fowl to graze, and a strange eater of grass*.'—RICHARDSON.

being raked from a concealment of brush, &c. on shore. The appearance of these usually attracts passing flocks, which alight, and are shot down. Sometimes eight or ten of these painted wooden ducks are fixed on a frame in various swimming postures, and secured to the bow of the gunner's skiff, projecting before it in such a way that the weight of the frame sinks the figures to their proper depth; the skiff is then dressed with sedge or coarse grass, in an artful manner, as low as the water's edge; and under cover of this, which appears like a party of ducks swimming by a small island, the gunner floats down sometimes to the very skirt of a whole congregated multitude, and pours in a destructive and repeated fire of shot among them. In winter, when detached pieces of ice are occasionally floating in the river, some of the gunners on the Delaware paint their whole skiff or canoe white, and laying themselves flat at the bottom, with their hand over the side silently managing a small paddle, direct it imperceptibly into or near a flock before the ducks have distinguished it from a floating mass of ice, and generally do great execution among them. A whole flock has sometimes been thus surprised asleep with their heads under their wings."

The Tame Duck is so well known as to need no description. The manner in which it fights the cock is highly amusing and but little known. It frequently happens while the fowls are being fed, that the duck runs among them, and by the help of his larger beak, gobbles up an undue share of the provisions. This the cock resents by giving him a peck. The Duck takes no notice, but gets behind the cock, deals him a hard peck, and looks innocent. The cock jumps round, but sees nothing. Presently another hard peck comes, and he gets very angry. A third peck; but this time the cock sees his enemy, and rushes at him furiously. Down flops the duck on the ground, and lets the cock pass over him. After running over him once or twice, and then jumping on him, the cock is persuaded that his enemy is quite dead, and walks off on the tips of his toes. Presently the duck first opens one eye and then the other, gets up, and quietly pecks the cock again. The same manœuvres are repeated, until at last the duck wins, like Fabius, by delay, and drives his antagonist fairly off the field.

The Teal* is the smallest of our ducks. It frequently breeds in England, mostly choosing the northern lakes for that purpose. Its flight is exceedingly rapid, soon carrying it out of the reach of gun-shot. It chooses night for its feeding-time, and during the day conceals itself under the herbs that fringe the banks of the water where it has chosen its habitation. Its nest is also carefully placed among dense herbage, and contains from eight to twelve whitish eggs.

The delicacy of its flesh is well known, and it therefore figures on most well-furnished tables.

* See page 310.

QUERQUEDŬLA.—(Lat.)

Crecca, *the Teal.*

The EIDER DUCK.—The Eider Duck furnishes the celebrated down in such request for pillows and beds. It is a singular fact, that the down must be plucked from the bird when living, as it seems to lose its peculiar elasticity and softness when taken from the bird after its death. The down is plucked by the bird itself from its breast, for the purpose of lining its nest, which is then repeatedly robbed until the Eider is reduced to laying its eggs on the down from the male bird. These eggs are generally permitted to be hatched, or the birds would forsake the spot, and never return again. So completely does the poor bird denude itself, that one female will furnish half-a-pound's weight of down.

Sub-family *f. Fuligulīnæ.*—(From *Fuligŭla,* Lat. dim. of *Fūlix, a Fen-Duck.*)

SOMATERĬA.—(Gr. Σῶμα, the body; τείρω, to wear away.)

Mollissĭma (Lat. *very soft*), *the Eider Duck.*

The Colymbidæ are remarkable for their powers of diving. The legs are placed very far behind, and the toes are so arranged as to fold up when returning from the stroke.

The foot of the Grebes is not webbed like that of most water birds, but each toe is separate and flattened, so as to serve as a separate paddle. The Grebes dive so instantaneously that it is difficult to shoot them, as they dive at the flash, and do not reappear for nearly two hundred yards, and then they merely raise their head above water for a second, and again disappear.

Family II. . . Colymbĭdæ.—(Gr. Κόλυμβος, a Diver.)
Sub-family *a*. . *Colymbīnæ.*

COLYMBUS.

Glaciālis (Lat. *icy*), *the Great Northern Diver.*

All the Grebes feed upon fishes and the various water insects, but their stomachs are almost invariably found to contain a mass of their own feathers. This circumstance presents a singular analogy to those masses of compacted hair which are often found in the stomachs of cows. In all probability the reason for their presence is the same, that the feathers and hairs are accidentally conveyed to the stomach after the creature has been making its toilet.

Of the three British species of Divers, the GREAT NORTHERN DIVER* is the largest. It is generally found on the shores of the Orkneys and Shetland.

This bird justly deserves its name of Diver, as it can pursue fish under water with the greatest ease and certainty, and can remain under water without inconvenience for a considerable time.

The nest of this bird is a tolerably large flattened mass of dead herbage, and is placed near the water's edge, in some place where the bird imagines that the reeds and flags, among which it is laid, will guard it from discovery. But unfortunately, the bird dislikes flying, and prefers to walk to and from its nest, thereby leaving a very evident track, by which it is often discovered.

The eggs are usually two in number, although three have been found in one nest. Their colour is dark olive brown, sparingly marked with dark spots.

Sub-family *b.* *Podicepīnæ.*

PODĬCEPS. — (*A hybrid word.*)

Cristātus (Lat. *crested*), *the Crested Grebe.*

The CRESTED GREBE is found in some of the fens of the Midland counties of England, and also inhabits parts of Scotland. This bird, together with the other Grebes, builds its nest of a mass of roots and reeds, among sedges. The female, like the water hen, covers up her eggs when she leaves her nest, which, unlike the nests of most of the aquatic birds, floats upon the surface of the water.

The DABCHICK, or LITTLE GREBE, is very common in most parts of England, and is spread over Europe and Asia. It is easily alarmed, and instantly dives, after which it is of little use to look for the bird. Even in a small pond where tame Dabchicks are kept, if they are startled at anything, they all disappear as if by magic, and reappear in the same mysterious manner. Even when confined in the limited space of a small pond, the sharpest eye cannot detect them as they hide under floating herbage, or are sheltered by an overhanging bank.

* See page 311.

It is easily domesticated, and is often seen placed as an ornament in lakes, or even in ponds, where it swims about very merrily, and seems to enjoy playing a game at hide-and-seek with any observer who is attempting to watch its movements.

PODĬCEPS.

Minor (Lat. *lesser*), *the Little Grebe or Dab-chick.*

The PUFFIN is common at the Needles and the western islands of England. It forms deep burrows in the soil, in which one egg is deposited, or usurps the burrow of a rabbit. The hole is generally from three to four feet in depth, when the Puffin is forced to labour for itself; it usually takes a winding course; and the inhabitant is secured from surprise by forming two entrances, in order that if one entrance is attacked, it may escape by the other. The egg is always deposited at the furthest extremity of the hole, and is not easy to be obtained, on account of the vigorous resistance made by the parent bird. It is an excellent diver, plunging fearlessly from a lofty cliff into the sea, and speedily returning with its beak full of fish, usually sprats, which are secured by their heads, and lie in a row along the bill of the Puffin, forming a kind of piscatorial fringe. Its enormous and sharp-edged bill renders it a formidable antagonist to intruders. The length of the bird is thirteen inches.

It is said that the Raven and the Puffin have occasional conflicts, the object of dispute being generally the egg or young of the Auk, for which the Raven has a great

Family III. . Alcīdæ.—(Lat. *Alca*, an Auk.)

Sub-family *a.* *Alcīnæ.*

FRATERCŬLA.—(Lat.)

Arctĭca (Lat. *Arctic*), *the Puffin.*

predilection. The issue of the combat d pends principally on its position, each bird trying to keep to its own peculiar element. If the Puffin can drag the Raven over the rocks into the sea, it is speedily victorious, as it drowns its sable adversary without much trouble, but if, on the contrary, the Raven can keep to shore, its superior size and strength gain the dominion.

ALCA.—(Lat. *an Auk.*)

Impennis (Lat. *wingless*), *the Great Auk.*

The ALCIDÆ or AUKS are never seen inland, but exclusively inhabit the sea-shores. In this family the wings are small, and in some genera useless for flight. The feet being required more for diving than walking, are placed so far back that the birds, when sitting, assume an erect attitude, and their walk is nothing but an undignified (I was going to say *awkward*) shuffle.

The GREAT AUK is an inhabitant of the Arctic Circle, but is sometimes seen in the northern islands of Scotland. The wings of this bird are incapable of raising it into the air, but serve admirably as paddles when diving. It breeds principally on the shores of Iceland and Spitzbergen, laying one large egg on a cleft of a high rock. The eggs are extremely scarce, and fetch a very high price among collectors, a circumstance which has caused some most ingenious impositions. In one case two of these eggs were offered for sale at a shop where natural curiosities are bought and sold. They were offered, I believe, at £5 each, which being a very low price, excited the suspicions of the buyer, who asked the seller to leave them while he examined them. He examined them accordingly, and although he doubted, yet they looked very genuine indeed. They had the peculiar smell of the Auk's eggs, the hole through which the contents were extracted was perfectly natural, the lining membrane of the egg being still in its place. Just as the price was about to be paid, a visitor happened to enter the shop, who recognised the seller as a man who had sold many of these eggs of late at the same price, but who manufactured all the eggs himself. They were, in fact, nothing but models, exquisitely copied, and accurate in every particular, but yet only a composition of plaster of Paris with other ingredients.

The length of the bird is nearly three feet.

Sub-family c. *Spheniscinæ.*

SPHENISCUS.—(Gr. *σφηνίσκος, a little wedge.*)

Demersus (Lat. *submerged*), *the Cape Penguin.*

The CAPE PENGUIN is very common at the Cape of Good Hope and the Falkland Islands. From the extraordinary sound it produces while on shore, it is called the Jackass Penguin. Darwin gives the following interesting account of this bird:—"In diving, its little plumeless wings are used as fins, but on the land, *as front legs*. When crawling (it may be said on four legs) through the tussocks, or on the side of a grassy cliff, it moved so very quickly that it might readily have been mistaken for a quadruped. When at sea and fishing, it comes to the surface, for the purpose of breathing, with such a spring, and dives again so instantaneously, that I defy any one at first sight to be sure that it is not a fish leaping for sport."

These birds feed their young in a very singular manner. The parent bird gets on a hillock, and apparently delivers a very impassioned speech for a few minutes, at the end of which, it lowers its head and opens its beak. The young one, who has been a patient auditor, thrusts its head into the open beak of the mother, and seems to suck its subsistence from the throat of the parent bird. Another speech is immediately made, and the same process repeated, until the young is satisfied.

This Penguin is very courageous, but utterly destitute of the better part of courage—discretion; for it will boldly charge at a man just as Don Quixote charged the windmills, and with the same success, as a few blows from a stick is sufficient to lay a dozen birds prostrate.

Sub-family *d. Urinæ.*

Uria.—(Gr. Οὔρια, a Diver.)

Trollē, *the Guillemot.*

The Common Guillemot makes its appearance on our coasts in the beginning of spring, and inhabits the cliffs overhanging the sea. Each female deposits one egg on a naked ledge of rock, and sits upon it with great perseverance, even suffering itself to be taken by hand. The egg is usually a pale green, streaked and blotched with brown, but is very variable both in colour and markings. The length of the bird is fifteen inches.

The Fulmar Petrel is an inhabitant of the Arctic circle, but breeds abundantly in St. Kilda and the Orkneys. The inhabitants of those islands consider the Fulmar as one of their principal means of subsistence, and to obtain the birds they expose themselves to the greatest dangers. The feathers of the Fulmar Petrel are used for their beds, its flesh they eat, its oil is delicate and gives an excellent light when used in a lamp, besides which it is considered a good remedy for wounds. To obtain the birds, the inhabitants wait until they are nearly fledged, when they lower themselves down the face of the most fearful precipices, saved from destruction merely by a rope. This rope is one of the principal items of the property of the people who live in the Orkneys. It is sometimes made of hide, but the best ropes are woven of hair, and are found to be less liable to fray against the rocks than if they were made of any other material. There are many stories of the dangers encountered by the daring cragsman, but there is no space for their insertion.

The Fulmar Petrel lays one white egg, large and brittle, which is imbued with the peculiar oily odour that characterises the bird. The food of the Fulmar consists of the flesh and blubber of dead whales and

Family IV. Procellarīdæ.

Sub-family a. *Procellarīnæ.*

PROCELLARĪA.—(Lat. *stormy.*)

Glaciālis (Lat. *icy*), *the Fulmar Petrel.*

other cetacea, and also of molluscs and crustacea. The length of the bird is sixteen inches.

THALASSIDRŌMA.—(Gr. Θάλασσα, the sea; δρόμος, a race.)

Pelagĭca (Lat. *belonging to the sea*), *the Stormy Petrel.*

The STORMY PETREL is, under the name of Mother Carey's chicken, the terror of the sailor, who always considers the bird as the precursor of a storm. It is the smallest of the web-footed birds. Few storms are violent enough to keep this curious little bird from wandering over the waves in search of the food that the disturbed water casts to the surface. Like the Fulmar, the Stormy Petrel is so exceedingly oily in texture, that the inhabitants of the Feroe Islands draw a wick through its body and use it as a lamp. Wilson gives the following account of its habits while following a ship under sail :—

"It is indeed an interesting sight to observe these little birds in a galē, coursing over the waves, down the declivities, up the ascents of the foaming surf that threatens to bend over their heads; sweeping along the hollow troughs of the sea, as in a sheltered valley, and again mounting with the rising billow, and just above its surface, occasionally dropping its feet, which, striking the water, throws it up again with additional force; sometimes leaping with both legs parallel, on the surface of the roughest waves for several yards at a time. Meanwhile it continues coursing from side to side of the ship's wake, making excursions far and wide, to the right and to the left, now a great way ahead, and now shooting astern for several hundred yards, returning again to the ship, as if she were all the time stationary, though perhaps running at the rate of ten knots an hour! But the most singular peculiarity of this bird is its faculty of standing and even running on the surface of the water, which it performs with apparent facility. When any greasy matter is thrown overboard, these birds instantly collect round it, and face to windward, with their long wings expanded and their webbed feet patting the water, the lightness of their bodies and the action of the wind on their wings enable them to do this with ease. In calm weather they perform the same manœuvre by keeping their wings just so much in action as to prevent their feet from sinking below the surface. According to Buffon, it is from this singular habit that the whole genus have obtained the name Petrel, from the apostle Peter, who, as Scripture informs us, also walked on the water."

The WANDERING ALBATROS, the largest of the genus, is a well-known bird in the southern seas, following ships for many miles in hopes of obtaining the refuse thrown overboard. So voracious is the Albatros, that it will swallow entire a fish of four or five pounds' weight. The flight of this bird is peculiarly majestic. Its extreme length of wing prevents it from rising at once from the ground, but when once launched into the air, it seems to float and direct its course without effort. Gould in describing the flight of this bird says:—

"The powers of flight of the Wandering Albatros are much greater than those of any other bird that has come under my observation. Although during calm or moderate weather it sometimes rests on the surface of the water, it is almost constantly on the wing, and is equally at ease while passing over the glassy surface during the stillest calm, or sweeping with arrow-like swiftness before the most furious gale; and the way in which it just tops the raging billows, and sweeps between the gulfy waves, has a hundred times called forth my wonder and admiration. Although a vessel running before the wind frequently sails more than 200 miles in the twenty-four hours, and that for days together, still the Albatros has not the slightest difficulty in keeping up with the ship, but also performs circles of many miles in extent, returning again to hunt up the wake of the vessel for any substances thrown overboard."

The voracity of the Albatros renders it an easy prey. A hook is baited with a piece of blubber, fastened firmly to a string, and suffered to tow astern. The bird immediately sweeps down to seize its prey, and is arrested by the hook, by means of which it is drawn into the ship. It seems rather remarkable that a bird that lives in or over the sea during its whole life, should prove a landsman when taken on board. Yet, when the Albatros is caught and placed on deck, it begins to stagger about, and soon becomes as thoroughly sea sick as the most inexperienced cockney. The best description of the nidification of the Wandering Albatros is that given by Mr. Earl, quoted by Gould.

Mr. Earl, after climbing a fearfully dangerous precipice in the Island of Tristan d'Acunha, arrived at a large plain of dark grey lava, on the summit of which the nests of the Albatros were made. "A death-like

DIOMEDĒA.—(*Proper name.*)

Exŭlans (Lat. *banished*), *the Wandering Albatros.*

stillness prevailed in these high regions, and to my ear our voices had a strange unnatural echo, and I fancied our forms appeared gigantic, whilst the air was piercing cold. The prospect was altogether sublime, and filled the mind with awe. The huge Albatros here appeared to dread no interloper or enemy; for their young were on the ground completely uncovered, and the old ones were stalking around them. They lay but one egg, on the ground, where they make a kind of nest by scraping the earth around it; the young is entirely white, and covered with a woolly down, which is very beautiful. As we approached, they snapped their beaks with a very quick motion, making a great noise; this and the throwing up the contents of the stomach are the only means of offence and defence they seem to possess. I again visited the mountain about five months afterwards, when I found the young albatroses still sitting on their nests, and they had never moved away from them." The expanse of wing in the Wandering Albatros is from eleven to fourteen feet.

Family V. . . Larĭdæ.

Sub-family *b*. *Larīnæ.*

Larus.—(Lat. *a Gull.*)

Marīnus (Lat. *belonging to the sea*), *the Black-backed Gull.*

The Black-backed Gull is a common bird on our coasts. During the winter it seeks the warmer coasts of southern Europe. It breeds in great numbers on the shores of the Bristol Channel, the Orkneys, and other coasts of Great Britain. Its nest is composed of grass, rushes, and other materials, and contains three or four eggs, of an olive green marked with very dark brown. Neither the gulls nor the terns dive, but snatch up their prey when at or near the surface.

The Terns or Sea-Swallows are possessed of great power and endurance of flight, their long forked tails and pointed wings indicating strength and swiftness.

The Common Tern is found in plenty along the southern shores of Europe, in many parts of Asia and Africa. It is frequently seen on the southern shores of England, and has been found in North America. It preys on fish, which it snatches from the surface with unerring aim, as it skims over the waves with astonishing velocity.

Sub-family *c. Sterninæ.*

STERNA.—(Lat. Sterno, *I strew.*)

The nest of this bird is made on the sand above high-water mark, and contains two or three eggs, on which the female usually sits by night. The length of the Common Tern is about fourteen inches.

The Noddy, so frequently celebrated by travellers who have passed the equator, is a species of Tern.

Hirundo (Lat. *a Swallow*), *the Common Tern.*

The TROPIC BIRD, as its name imports, is seldom seen many degrees beyond the tropics, although a storm occasionally drives it from its accustomed habitiat.

Family VI. . . Pelecanĭdæ.—(Gr. Πελεκάν, a Pelican. Pelican-kind.)
Sub-family *b.* . *Phaetonīnæ.*

PHAĔTON.—(Gr. Φαέθω; proper name.)

Ætherĕus (Lat. *belonging to the sky*), *the Tropic Bird.*

Its rapid flight seems to be accomplished almost without the aid of wings. It preys extensively on the flying-fish, who frequently escapes his airy foe but to fall into the jaws of some rapacious rover of the deep. It has been known to continue on the wing for whole days and nights, but sometimes rests on the back of a turtle sleeping at the surface of the water. The length of the Tropic Bird is about eighteen inches.

Sub-family c. *Pelecaninæ.*

SULA.

Bassanĕa (Lat. *belonging to the Bass Rock*), *the Gannet or Solan Goose.*

The Gannet, or Solan Goose, is common on some of our shores, especially at the Bass Rock at the entrance of the Frith of Forth. This rock is literally covered with Gannets, and is rented at a high price from the proprietor, who makes over to the tenant the vast flocks of birds that take up their residence on the rock. Great numbers of Gannets breed at St. Kilda, and many are sent to Edinburgh and other markets.

The Gannet feeds almost entirely on herrings, which it seizes by plunging with extraordinary force from a considerable height. This method of procuring food has led to an ingenious device for capturing the bird. A herring is fastened to a board, and suffered to float on the surface of the water. The Gannet, seeing the fish apparently sporting on the surface, plunges at it with such force that it is instantly killed by the blow. A Gannet was once taken when the board was sunk to the depth of six feet, yet even at that depth the bird's neck was dislocated, and its bill firmly stuck into the wood. The length of the Gannet is about two feet eight inches.

The Booby is a species of Gannet. Sailors have given it this rather inelegant name on account of the stupidity it displays in suffering itself to be knocked down with a stick, or even taken up by hand.

The CORMORANT is found in abundance on our coasts, and is widely spread over many parts of the world. It is exceedingly voracious, and devours an almost incredible amount of fish. It is an excellent diver, and chases the fish actually under the water, seldom if ever returning without having secured its prey. Like the otter, when engaged in chase, it occasionally rises to take breath, and then resumes the pursuit with renewed vigour. Waterton gives the following amusing description of the proceedings of a Cormorant:—"First raising his body nearly perpendicular, down he plunges into the deep, and after staying there a considerable time he is sure to bring up a fish, which he invariably swallows head foremost. Sometimes, half an hour elapses before he can manage to accommodate a large eel quietly in his stomach. You see him straining violently with repeated efforts to gulp it; and when you fancy that the slippery mouthful is successfully disposed of, all of a sudden the eel retrogrades upwards from its dismal sepulchre, struggling violently to escape. The Cormorant swallows it again, and up again it comes, and shows its tail a foot or more out of its destroyer's mouth. At length, worn out with perpetual writhings and slidings, the eel is gulped down into the Cormorant's stomach for the last time, there to meet its dreaded and inevitable fate. This gormandising exhibition was witnessed here by several individuals, both ladies and gentlemen, on Nov. 26, 1832, through an excellent eight-and-twenty guinea telescope, the Cormorant being at that time not more than a hundred yards distant from the observers. I was of the party."

PHALACROCŎRAX.—(Gr. φαλακρός, bald, κόραξ, a Raven.)

Carbo (Lat. *a Coal*), *the Cormorant.*

The Cormorant has the power of perching on trees, an accomplishment which we should hardly suspect a web-footed bird of possessing. Milton, in a well-known passage in his "Paradise Lost," alludes to this habit. Speaking of Satan under the disguise of the Cormorant, he tells that he,

"——— on the tree of life,
The middle tree, and highest there that grew,
Sat."

The Cormorant is easily tamed, and its fishing propensities can be turned to good account. The Chinese, at the present day, employ a kind of Cormorant for that purpose, having previously placed a ring round the bird's neck, to prevent it from swallowing the fish. The eggs of this bird are usually laid on the rock, but sometimes in the branches of trees. A thick coat of chalk envelopes the eggs, and can be easily scraped off with a knife. The length of the bird is about three feet.

PELECĀNUS.—(Latinised from Gr. Πελεκάν.)

Onocrotălus (Gr. Ὀνοκρόταλος; derived from ὄνος, an ass, and κρόταλον, a rattle), *the White Pelican.*

The WHITE PELICAN inhabits Africa, India, and great part of the south-eastern portions of Europe. It is a very conspicuous bird, its singular membranous pouch offering a distinction perfectly unmistakeable. The pouch, when distended, holds two gallons of water, but the bird has the power of contracting it so that it is scarcely to be discerned. The pouch also serves as a net in which to scoop up the fish on which the Pelican feeds.* Another most important use of the pouch is to convey food to the young. The parent Pelican presses the pouch against its breast, in order to enable the young to obtain the fish; which action, in all probability, gave rise to the fable of the Pelican feeding its young with its own blood. The red tip of the bill probably aided the deception.

Although a web-footed bird, the Pelican, like the cormorant, can perch on trees, although it prefers sitting on rocks. The colour of this bird is a pure white, with a very slight tinge of rose-colour, and the pouch is yellow. The length of the bird is nearly six feet.

The FRIGATE PELICAN, or MAN-OF-WAR BIRD, is usually found between the tropics. Although when stripped of its feathers it is hardly

* The beautiful Pelicans in the Zoological Gardens exhibit this pouch and its uses admirably.

larger than a pigeon, yet no man can touch at the same time the tips of its extended wings. The long wing bones are exceedingly light, and the whole apparatus of air-cells is extremely developed, so that its real weight is very trifling. It flies at a great height above the water, and from that elevation pounces down on fish, especially preferring the poor persecuted flying-fish for its prey. According to some authors, the name of Man-of-War Bird was given to it because its appearance was said to foretel the coming of a ship; probably because the Frigate Pelican and

Fregāta.—(From the Spanish Fragáta, a ship.)*

Aquĭla (Lat. *an Eagle*), *the Frigate Pelican.*

real frigates are equally averse to storms, and both like to come into harbour if the weather threatens. Under the throat of the Frigate Pelican is a large pouch, of a deep red colour, which can be distended with air at the pleasure of the bird. The pouch is larger and of a more brilliant red in the male than in his consort, and the general plumage of the female is not so bright as that of the male.

Although its swiftness of wing and general activity enable it to snatch a fish from the surface of the water, or to pounce upon the flying-fish before it can again seek the protection of its native element, yet it too often uses its powers in robbing other birds of their lawful prey. It is enabled, in some mysterious way, to find its way home by night, even though it may be four or five hundred miles from land. The length of the male bird is three feet, and the expanse of wing eight feet.

* Possibly both from Gr. ναῦς καταφράκτης, a decked vessel. The word "freight" is also derived from the same root.

Class III. . . REPTILĬA.—(Lat. *Creeping things.*)
Order I. . . *SAURA.*—(Gr. Σαύρα, a Lizard.)
Sub-order I. Leptoglossæ.—(Gr. Λεπτός, slender; γλῶσσα, the tongue.)
Tribe I. . . CYCLOSAURA.—(Gr. Κύκλος, a circle; σαύρα.)
Family IV. Lacertinĭdæ.—(Lat. *Lacerta*, a Lizard. Lizard-kind.)

Zootŏca.—(Gr. Ζωός, living; τίκτω, to bring forth.)

Vivipăra (Lat. *viviparous*), *the Common Lizard.*

We now arrive at the singular Class of Reptiles. The animals of this class vary exceedingly in their forms, sizes, and habits, but the peculiar formation of the circulatory system, together with many other anatomical distinctions, plainly mark them out as a distinct class.

The Lizards are usually active, bright-eyed little creatures, delighting to bask in the sun, near some safe retreat, to which they dart with astonishing celerity upon the slightest alarm. Two species of Lizards inhabit this country, the Common Lizard, and the Sand Lizard. The latter animal is considerably larger than the Common Lizard, as it sometimes measures a foot in length. It frequents sandy heaths, and in the sand its eggs are deposited, fourteen or fifteen in number. The eggs are hatched by the heat of the sun, and the young immediately lead an independent life. During the winter this as well as the Common Lizard hybernates in a burrow usually made under the roots of a tree, nor does it again make its appearance until the spring.

The Common Lizard is only six inches in length. It is more active than the Sand Lizard, disappearing like magic on being alarmed. When seized, its tail frequently snaps off like glass. Both British Lizards feed on insects.

Tribe II. GEISSOSAURA.—(Gr. Γεῖσσον, a cornice; σαύρα.)

Family XV. Scincidæ.—(Gr. Σκίγκος, a kind of Lizard.)

ANGUIS.—(Lat. *a Snake.*)

Fragilis (Lat. *fragile*), *the Blind-worm or Slow-worm.*

The BLIND-WORM is not a snake, as generally supposed, but a legless lizard of the Skink family. It is perfectly harmless; its small mouth and very minute teeth precluding all attempts to injure, even if it had the will. When alarmed, it snaps asunder at the slightest blow, like the tail of the Common Lizard, and from that peculiarity has derived its name of "fragilis." It feeds almost entirely on small slugs, its jaws not being capable of admitting any larger prey. It is very common in most parts of England, and may be seen basking in the sun in hedgerows or under old walls. Its eyes are very small, but brilliant.

The GECKOS* are nocturnal lizards, remaining hidden in crevices during the day, but wandering forth at night in search of their insect prey. They run about on the smooth walls and ceilings with the greatest ease, as their feet are furnished with an apparatus exactly resembling a boy's sucker, by means of which they are able to adhere to the wall, or even to the roof. They labour in their country under precisely the same imputations that the toad does in England, namely, of being venomous creatures, producing horrible diseases when touched, together with many similar tales. Geckos are spread over every quarter of the globe, but are most numerous in Southern Asia. The species represented is common in India.

The IGUANA family is a very large one, containing 150 species. The Common Iguana is a native of Brazil, Cayenne, Jamaica, &c. In spite of its repulsive appearance, it is with many people a favourite article of food, and is said somewhat to resemble chicken. It is very fierce when attacked, and snaps at its enemies in a most determined manner, often scaring away an intruder by the ferocity of its aspect.† It is generally

* See page 328. † See page 329.

Sub-order II. PACHYGLOSSÆ.—(Gr. Παχύς, thick; γλῶσσα, the tongue.)

Tribe III. . . NYCTISAURA.—(Gr. Νύξ, night; σαύρα, a Lizard.)

Family XXII. Geckotidæ.—(*Geckos.*)

GECKO.

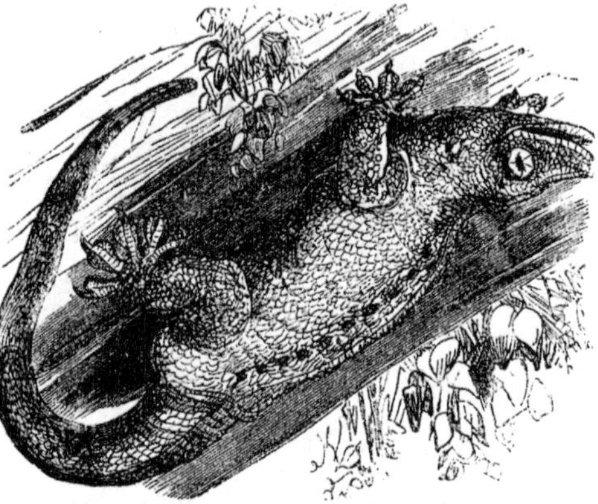

Verus (Lat. *true*), *the Gecko.*

taken by throwing a noose over its head, and dragging it from the branches by main force. It is then immediately killed, as its sharp notched teeth can inflict a very disagreeable wound. Sometimes it is hunted with dogs trained to the sport. It attains a considerable size, frequently reaching the length of six feet. It feeds usually on vegetable substances, such as leaves, fruit, and fungi; but Iguanas have been seen in the Island of Isabella, that feed on eggs, insects, and even the intestines of fowls. An enormous fossil Iguana has been discovered by Dr. Mantell, whose length must have been nearly seventy feet.

The terrible name of FLYING DRAGON belongs to a harmless little lizard, bearing small resemblance to the terrific animal so graphically depicted by Retsch. This curious little lizard lives on trees, and feeds on insects instead of devouring pilgrims bound to the Gnadenbilde. The peculiar structure of its body bears a singular resemblance to that of the flying squirrel. The first six false ribs are greatly elongated, and support a wing-like expansion of skin, which when stretched serves to bear them up as they skim through the air from one tree to another. While running about on the branches, the so-called wings are folded to

Tribe IV. . . . STROBILOSAURA.—(Gr. Στρόβιλος, anything twisted, a Fir-cone; σαύρα, a Lizard.)

Family XXIII. Iguanĭdæ.—(*Iguanas.*)

IGUANA.—(*Native name.*)

Tuberculāta (Lat. *covered with pimples*), *the Iguana.*

the side, but when it wishes to throw itself from the tree, the ribs are

Family XXIV. Agamĭdæ.—(Gr. ἄγαμος, unmarried.)

DRACO.—(Lat. *a Dragon.*)

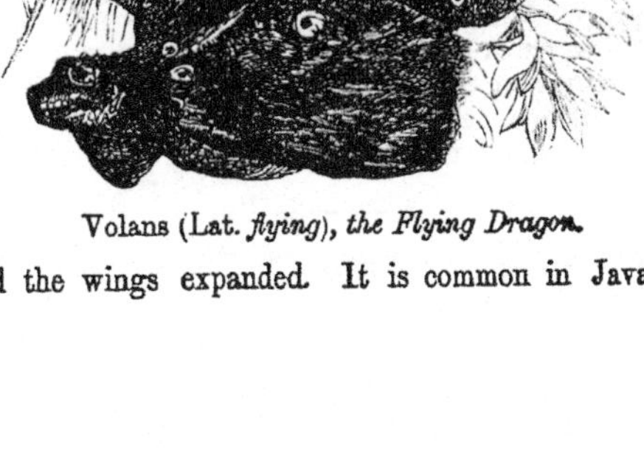

Volans (Lat. *flying*), *the Flying Dragon.*

raised, and the wings expanded. It is common in Java, India, and Borneo.

The COMMON CHAMELEON is plentifully found in northern Africa, the south of Spain, and Sicily. It lives on trees, but exhibits none of the activity usually found in arboreal reptiles. On the contrary, its movements are absurdly grave and solemn. The whole activity of the animal seems to be centered in its tongue, by means of which organ it secures flies and other insects with such marvellous rapidity, that the ancients may be well pardoned for their assertion that the air formed the only food of the Chameleon.

Tribe V. . DENDROSAURA.—(Gr. Δένδρον, a tree; σαύρα, a Lizard.)

Family XXV. Chameleonĭdæ.—(Gr. Χαμαιλέων, a Chameleon. Chameleon kind.)

CHAMELĔON.

Vulgāris (Lat. *common*), *the Chameleon.*

Highly exaggerated descriptions have been given of the changes of colour in this animal. The changes are by no means so complete, nor are the colours so bright, as generally supposed.

> "——And then its hue,
> Who ever saw so fine a blue?"

The poetic moralist further recounts its changes to green, black, and *white*. The umpire referred to in the poem is recorded to have asserted,

> "If you don't find him black, I'll eat him;"

but every one who has watched a Chameleon for any time, will be equally ready to eat him the moment that he turns white.

The power of the Chameleon to move its eyes in different directions at the same time, gives it a most singular aspect. Its enormously long tongue can be withdrawn into the mouth when not in use; but when the creature sees a fly within reach, the tongue is instantly darted forth, and by means of a gummy secretion at the tip secures the fly. The whole movement is so quick as almost to elude the eye.

THE peculiar gliding movements of the Snakes render them excellent types of the Reptiles; a word derived from the Latin *repo*, I creep. The extraordinary flexibility of their bodies is caused by the structure of their vertebræ, each one of which fits into the one behind it by a ball-and-socket joint, thus allowing freedom of motion in every direction.

The RATTLE-SNAKE is a native of America. Its name is derived from the loose bony structure at the extremity of its tail, called the rattle, and which by the sound of its movements gives timely intimation of the

Order II. . . *OPHIDĬA.*—(Gr. Ὄφις, a Serpent.)
Sub-order I. VIPERĪNA.—(Lat. *Vipĕra*, a Viper.)
Family I. . Crotalĭdæ.—(Gr. Κρόταλον, a Rattle.)

UROPSOPHUS.—(Gr. Οὐρά, the tail; ψόφος, a noise.)

Durissus (Lat. durus, *harsh*), *the Rattle-snake.*

vicinity of this terrible reptile. Fortunately, its disposition is exceedingly sluggish, and it invariably sounds its rattle when irritated or disturbed. Its bite is inevitably mortal, and death always ensues within a few hours after the wound has been inflicted.

The deadly weapons with which the venomous serpents are armed, are two long curved fangs belonging to the upper jaw, and moving on a hinge, by which they lie flat in the mouth when not wanted. An aperture exists in the point of the fang, by which a poisonous fluid, secreted in a gland at the base of the tooth, is poured into the wound, and, mixing with the blood, rapidly carries its deadly influence throughout the entire system. A short time since, an American physician was exhibiting a caged rattle-snake to his friends; he approached his hand too near the irritated reptile, who instantaneously inflicted a wound; and, although every precaution was taken, the bite proved fatal in a few hours.

Waterton gives some useful hints respecting the snakes, especially those of the venomous kind :—

"When a man is ranging a forest, and sees a serpent gliding towards him, (which is a very rare occurrence,) he has only to tack off in a side direction, and he may be perfectly assured that it will not follow him. Should the man, however, stand still, and should the snake be one of those overgrown monsters capable of making a meal of a man,—in these cases the snake would pursue its course; and when it got sufficiently near to the place where the man was standing, would raise the forepart of its body in a retiring attitude, and then dart at him and seize him. A man may pass within a yard of Rattle-snakes with safety, provided he goes quietly; but should he irritate a Rattle-snake, or tread incautiously upon it, he would infallibly receive a wound from its fang,—though, by the bye, with the point of that fang curved downwards, not upwards."

This latter passage refers to a plate in Audubon's Ornithology, where a Rattle-snake is represented attacking a mocking-bird's nest, and threatening the birds with two fangs curved *upwards* at the points—a mistake which Waterton never loses sight of.

The same author was nearly falling a victim to a Rattle-snake. He saw what he thought was a green locust struggling in the grass. On stooping down to examine it, he was considerably alarmed at discovering it to be the tail of a rattle-snake.

The inhabitants of those countries where the Rattle-snake lives are not very much afraid of it, as they know that it will be sure to run away directly it hears the approach of human footsteps. It appears that when a man is cutting wood or otherwise engaged in a forest, and hears a Rattle-snake near him, he has no fear, as long as he can keep its rattle going, but directly the sound ceases, the man is rather in dread, not knowing where the animal may turn up next: so he keeps the snake in a constant state of alarm, by throwing bits of wood or sticks at the place where the reptile is lying, and on again hearing the sound of the rattle, he continues his work in confidence, until the snake is silent, when some more missiles are sent in the same direction.

An American told me that, even when these snakes are ready for a spring, they can be avoided by smartly clapping the hands together, or striking the ground with a stick. The snake has the whole powers of its mind bent upon its fatal stroke, and, on hearing such an unexpected sound, it is startled, like a man suddenly waked from sleep, and falls down in its coil again, giving time for its intended victim to escape before it has made up its mind to another assault.

Some years ago a number of these snakes were wanted for certain menageries, and were caught in the following manner. There are some places where the Rattle-snakes abound, and may be seen lying in their holes. A party of sailors were despatched to one of these haunts, furnished with baskets, ropes, poles, and various other implements. The sailors thought it great fun, and laid their plans as follows. They gave three men to each snake, two of them having a rope, and the third a pole and basket. They commenced by making one of those slip-knots so common among sailors, in the centre of the cord. This was laid over the hole, and drawn together until it was just large enough to surround it. The third man, then, either threw stones at the entrance of the den, or poked about it with the pole, until the snake put out its head to see what was the matter. Directly the reptile's head and neck were fairly outside, the two men drew the rope tight, and carried it between them to the open basket, into which they dropped it, while the third man shut down the lid with his pole, and then fastened it. The cord being slackened, the snake soon wriggled itself out of the noose, and the men set off after another victim.

The length of this snake has seldom been known to exceed seven feet.

The PUFF ADDER is an inhabitant of Southern Africa. It is a short, thick, flattish snake, of a most sinister and malignant aspect. The following alarming adventure occurred to Mr. Cole, a resident in the Cape.

"I was going quietly to bed one evening, wearied by a long day's hunting, when, close to my feet and by my bed-side, some glittering substance caught my eye. I stooped to pick it up; but, ere my hand had quite reached it, the truth flashed across me—it was a snake! Had I followed my first natural impulse, I should have sprung away, but not being able clearly to see in what position the reptile was lying, or which way his head was pointed, I controlled myself, and remained rooted breathless to the spot. Straining my eyes, but moving not an inch, I at length clearly distinguished a huge Puff Adder,—the most deadly snake in the colony, whose bite would have sent me to the other world in an hour

Family II. Viperidæ.—(Lat. *Vipĕra*, a Viper.)

CLOTHO.*—(Gr. proper name.)

Ariĕtans (Lat. *butting like a Ram*), *the Puff Adder.*

* This is the name of one of the three Fates; viz. Clotho, Lachesis, and Atropos. All three names are used as genera of venomous serpents.

or two. I watched him in silent horror; his head was from me; so much the worse—for this snake, unlike any other, always rises and strikes back. He did not move, he was asleep. Not daring to shuffle my feet, lest he should awake and spring upon me, I took a jump backwards, that would have done honour to a gymnastic master, and thus darted outside the door of the room; with a thick stick I then returned and settled his worship."

The same author remarks in his "Five Years' Residence in South Africa," that its (the Puff Adder's) bite will kill occasionally within an hour.

It is the more dangerous, because it has a way of flattening itself upon the ground; so that, when it is lying thus concealed upon the sand, an incautious pedestrian is very likely to tread upon it.

"One of my friends lost a favourite and valuable horse by its bite in less than two hours after the attack. It is a sluggish reptile, and therefore more dangerous, for instead of rushing away like its fellows, at the sound of approaching footsteps, it half raises its head and hisses. Often have I come to a sudden pull up on foot or on horseback, on hearing their dreaded warning."

CERASTES.—(Gr. Κεράστης, horned.)

Hasselquisti (Lat. *of Hasselquist*), *the Cerastes.*

The CERASTES is a well-known snake in Egypt, and derives its name from the horny scale over each eyebrow. Bruce mentions that the Cerastes can spring several feet in any direction; but his description of the stratagems employed by it, "to surprise any one who is too far from it," is probably more fanciful than correct, as snakes do not attack unless suddenly surprised or irritated. The size of the Cerastes is by no means great, as its average length is only eighteen inches. The snake-charmers of Egypt employ these reptiles precisely as their brethren of India employ the Cobra de Capello.

The COMMON VIPER, or ADDER, is the only venomous reptile inhabiting England, nor is its bite nearly so dangerous in its consequences as has been reported. Seldom has the bite of the Viper proved mortal; and in all probability, had proper precautions been taken, no case would have been fatal. Viper-catchers employ olive oil as a remedy against the bite, and, from all accounts, it appears to be a certain preservative against all evil effects. The oil should be heated to produce its full efficacy.

PELÏAS.—(Gr. proper name.)

Berus, *the Viper.*

It is asserted that, when danger threatens, the female viper opens her mouth and permits her brood to hide themselves, but it is by no means an ascertained fact.

Frogs, lizards, mice, and other small animals, form the food of this reptile, but sometimes it falls a victim to its own voracity. In the Magazine of Natural History, a Viper is mentioned which had swallowed a lizard nearly as large as itself, and one of whose legs was protruding from its side.

In former times, preparations from Vipers, and especially viper-broth, were in great request as medicines.

The BOA-CONSTRICTOR.—The enormous Boa-constrictor inhabits tropical America. It is not venomous, but is not the less dangerous, as the tremendous power of its muscles enables it to crush its prey in the coils of its huge body. In order to procure its food, the Boa-constrictor lies in wait by the side of some river or pool, where animals of all kinds are likely to come to quench their thirst. It patiently waits until some animal draws within reach, when, with one spring, the Boa fixes its teeth in the creature's head, coils its body round its victim, and crushes it to death. After the unfortunate animal has been reduced almost to a shapeless mass by the pressure of the snake, its destroyer makes preparations for swallowing it entire, a task which it accomplishes, although the slaughtered animal is usually very much larger than the dimensions of the serpent. At last, the snake succeeds in swallowing its prey, and then lies torpid for nearly a month, until its enormous meal is digested, when it again sallies forth in search of another.

Sub-order II. . COLUBRĪNA.—(Lat. *Coluber*, a Snake.)
Family IV. . . Boïdæ.

BOA.—(Native name.)

Constrictor (Lat. *a binder*), *the Boa.*

Even the buffalo has been known to fall a victim to this fearful serpent, whose length frequently exceeds twenty-five feet.

The COBRA DE CAPELLO is a native of India. It must not be confounded with several other hooded snakes, such as the Haje of Egypt, the snake so frequently depicted on the hieroglyphical monuments.

The serpent-charmers invariably use this formidable reptile for their performances. The exhibitors possess several Cobras shut up in baskets, and when commencing their performances, the lid of the basket is opened, and the snake creeps out. Its course is arrested by the sound of the rude fife that the charmer always carries, and it immediately expands its beautiful though threatening hood, erects its neck, and commences a series of undulating movements, which are continued until the sound of the fife ceases, when the snake instantly drops, and is replaced in its basket by its master. The charmers appear to be able to discover snakes, and to induce them to leave their retreats. Indeed it is rather a singular fact, that those travellers who most strongly insist that the snakes thus caught are tame and divested of their fangs, appear to forget that even in that case the creatures must have been previously caught in order to deprive them of their weapons. The length of this snake is about five or six feet.

Family V. Colubrĭdæ.—(Lat. *Coluber*, a Snake.)

NAJA.—(Native name.)

Tripudĭans (Lat. *dancing*), *the Cobra de Capello.*

A Cobra in the Zoological Gardens was a long time in learning caution. It was accustomed to lie coiled up at the bottom of the cage until a spectator came close, when it invariably darted at him, of course striking its nose against the glass with no small violence. On my first visit to the Reptile House after its arrival, it made its customary attack, and after the space of a week, it again struck at me. On my next visit, several months afterwards, it laid very quietly at the bottom of its cage, and contented itself with a hiss.

The Egyptian Asp, or Haje, is supposed to be the asp by whose bite

Cleopatra died, and is in all probability the deaf adder alluded to in the Scriptures, "which stoppeth her ears, and refuseth to hear the voice of the charmer, charm he never so wisely."

The Common Ringed or Grass Snake is a harmless inhabitant of this country, and may be frequently seen or heard gliding along the hedge-banks in search of food. It is easily tamed, and soon learns to know its master. It lives principally on frogs, mice, young birds, newts, &c. It is an excellent swimmer, and from the peculiar structure of its lungs can remain under water for some time. It seems very fond of the water, and is most commonly found on marshy land, or in hedges planted over a wet ditch. The viper, on the contrary prefers dry sandy situations.

Natrix.—(Lat. *a Water Snake.*)

Torquāta (Lat. *collared*), *the Ringed Snake.*

Several snakes kept tame at a village in Wiltshire were fed with frogs and small newts, which latter animals the snake was in duced to swallow, by the simple process of opening the snake's mouth and pushing the newt down its throat. This plan, although apparently rather rude, seemed to cause the snakes no inconvenience.

Like all other serpents, the Ringed Snake sheds its skin several times during the year. The entire skin comes off, even the covering of the eyes. A rent opens in the neck, and the snake, by entangling itself in the thick grass or bushes, actually creeps out of its skin, turning it inside out in the effort.

The Tortoise.—The whole of this order is characterised by the complete suit of bony armour with which the animals are protected. The so-called "shell" is in fact a development of various bones, and not a mere horny appendage, like the coverings of the armadillo and manis. The upper shield is called the "carapace," and is united to the under shield, or "plastron," by certain bones, leaving orifices for the protrusion of the head and limbs. Most species are able to withdraw their head and limbs completely within the shell, and in some few the orifices are closed by a kind of hinge joint. The tortoiseshell of commerce is a series of horny plates that cover the exterior of the shield, and is in great request on account of the beautiful wavy markings that are so familiar to our eyes.

Order III. . *CHELONĬA.*—(Gr. Χελώνη, a Tortoise)
Family I. . Testudinĭdæ.—(Lat. *Testūdo*, a Tortoise.)

Græca (Lat. *Greek*), *the Tortoise.*

The Tortoises and Turtles possess no teeth, but the sides of their jaws are very hard and sharp, enabling them to crop vegetable substances, or to inflict a severe bite.

The family is divided into Land Tortoises, Marsh Tortoises, River Tortoises, and Marine Tortoises, or Turtles.

The Common Land Tortoise is found in abundance in the south of Europe. It is often kept in captivity in this country, and is very long lived, individuals being known to have exceeded two hundred years. Its movements are very slow, but it can excavate a burrow with unexpected rapidity. Secure in an impenetrable covering, it bids defiance to any ordinary enemy, except, as Sydney Smith wittily observes, "man, and the boa-constrictor. Man, however, takes him home and roasts him, and the boa-constrictor swallows him whole, shell and all, and consumes him slowly in the interior, as the Court of Chancery does a great estate."

I had a common Land Tortoise for a few months, part of whose life is described in the following passage, which has already appeared as a note to White's "Natural History of Selborne."

Some time since, a man arrived in Oxford, bringing with him tortoises for sale. They passed their existence in a basket, where they were packed close, like so many bricks, standing on their tails, and their heads looking out of the basket. When I purchased one of them, the man emptied out his whole basketful upon the table, and then turned out the contents of four large pockets, until a large table was entirely covered with them.

The tortoise which I purchased was a very small one, and was tolerably lively, walking about the room, and always settling on the hearthrug. It had a great genius for climbing, and would sometimes spend nearly an hour in endeavouring to scale the fender, probably attracted by the heat. Unfit as the form of the creature may seem for such a purpose, it did contrive to scramble upon a footstool which was placed by the fender. Its method of attaining this elevation was as follows:—First it reared up against the footstool in the angle formed by it and the fender, and after several ineffectual attempts, succeeded in hitching the claws of one of its hind feet into the open work of the fender. On this it raised itself, and held on to the top of the stool by its fore-feet while it gained another step on the fender, and so managed to raise itself to such a height, that it only had to fall flat on the top of the footstool. When once there, it could hardly be induced to leave the elevation which it had gained with such difficulty.

Its food consisted of bread and milk, which it ate several times a-day, drinking the milk by scooping up some of it in its lower jaw, and then, by throwing its head back, the milk ran down its throat. Tortoises are generally long-lived, but this animal died within a few months after it came into my possession, in all probability because, for some days, its food was placed in a brass vessel.

Several days before its death it was very restless, and went about the room mewing like a young kitten, and made such a noise, that it had to be ejected during working hours. I could not for some time believe that the mewing could proceed from the tortoise, as the resemblance to that of a kitten was most exact.

The Common Green Turtle.—The feet of the Marine Tortoises, or Turtles, are modified into fins or flippers just as are the feet of the seals, and consequently, although the Turtles are active in the water, on land their walk is nothing but an awkward shuffle. The flippers, however, are admirable instruments for scooping out the sand, in which the eggs are laid, and afterwards covered over. Nearly two hundred eggs are laid in one nest. The eggs are held in great estimation, but the albumen, or "white," does not become hard by boiling.

The Common Green Turtle, whose flesh is considered such a luxury, is common in Jamaica, and most of the islands of the East and West Indies. The turtles are captured by turning them on their backs; for the carapace is so flat, and their legs are so short, that they are forced to lie helpless until their captors have leisure to drag them away.

Family V. Cheloniădæ.
CHELONĬA.—(Gr. Χελώνη, a tortoise.)

Virĭdis (Lat. *green*), *the Turtle.*

The Green Turtle has been known to reach the weight of five or six hundred pounds. The tortoiseshell of commerce is almost entirely obtained from the Hawksbill Turtle.

The CROCODILE.—These animals are separated from the Lizards on account of the peculiar horny covering with which they are protected.

The Crocodile is an inhabitant of the Old World, the Alligator of the New, and the two animals are best distinguished by the construction of the jaws. In the Crocodiles the lower canine teeth fit into a *notch* in the edge of the upper jaw, and there is in consequence a contraction of the muzzle just behind the nostrils. The lower canine teeth of the Alligators fit into a *pit* in the edge of the upper jaw, and in consequence no contraction is needed. At the back of the throat is a valve completely shutting out water, but leaving the passage to the nostrils free, so that the Crocodile can keep his mouth open when beneath the surface, without swallowing the water, or can hold his prey to drown under the water while he breathes at ease with his nostrils at the surface. There is no true tongue.

The Common Crocodile inhabits many African rivers, and is, probably, the reptile infesting the Ganges. The Nile, however, is the best known haunt of this terrible creature.

The Crocodile feeds on fish, floating carrion, and dogs or other animals, which it is enabled to surprise as they come to drink at the water's edge, but man frequently falls a victim to its voracity. In revenge for this treatment, all nations persecuted with this pest have devised various methods of killing it. The Negroes of some parts of Africa are sufficiently bold and skilful to attack the Crocodile in his own element. They fearlessly plunge into the water, and diving beneath the Crocodile plunge the dagger with which they are armed into the creature's belly, which is not protected by the coat of mail that guards the other parts of its body. The usual plan is to lie in wait near the spot where the Crocodile

Order IV. . *EMYDOSAURI.*—(Gr. 'Εμύς, the Water-Tortoise; *σαύρα.*)
Family I. . Crocodilĭdæ.—(Gr. Κροκόδειλος. Crocodile kind.)*

CROCODĪLUS.

Vulgāris (Lat. *common*), *the Crocodile.*

is accustomed to repose. This is usually a sandy bank, and the hunter digs a hole in the sand, and, armed with a sharp harpoon, patiently awaits the coming of his expected prey. The Crocodile comes to its accustomed spot, and is soon asleep, when it is suddenly roused by the harpoon, which penetrates completely through its scaly covering. The hunter immediately retreats to a canoe, and hauls at the line attached to the harpoon until he drags the Crocodile to the surface, when he darts a second harpoon. The struggling animal is soon wearied out, dragged to shore, and dispatched by dividing the spinal cord. In order to prevent

* The word Crocodile literally signifies, "one afraid of saffron."

the infuriated reptile from biting the cord asunder, it is composed of about thirty small lines, not twisted, but only bound together at intervals of two feet.

When on land it is not difficult to escape the Crocodile, as certain projections on the vertebræ of the neck prevent it from turning its head to any great extent.

The eggs of this creature are very small, hardly exceeding those of a goose; numbers are annually destroyed by birds of prey and quadrupeds, especially the Ichneumon.

The ALLIGATOR, or CAYMAN, is an inhabitant of the New World, and is unpleasantly common in the rivers of North America. It pursues fish with exceeding dexterity, by driving a shoal of them into a creek, and then plunging amid the terrified mass, and devouring its victims at its pleasure. It also catches pigs, dogs, and other animals that venture too close to the river. In that case, as the animal is too large to be swallowed entire, the Alligator conceals it in some hole in the bank until it begins to putrefy, when it is dragged out, and devoured under the concealment of the rank herbage fringing the river.

The usual method of taking this creature is by baiting a most formidable four-pointed hook, composed of wooden spikes artistically arranged, and suffering it to float in the river. When an alligator has swallowed it, he is hauled on shore by the rope, and slaughtered. Waterton gives a very amusing account of catching a cayman. The reptile had swallowed the hook, and was being towed ashore. Waterton was waiting for him, armed with the mast of the boat, to force it down the throat of the cayman should he prove restive. "By this time the cayman was within two yards of me; I saw he was in a state of fear and perturbation. I instantly dropped the mast, sprang up, and jumped on his back, turning half round as I vaulted, so that I gained my seat with my face in a right position. I immediately seized his fore legs, and by main force twisted them on his back; thus they served me for a bridle.

"He now seemed to have recovered from his surprise, and probably fancying himself in hostile company, he began to plunge furiously, and lashed the sand with his long and powerful tail. I was out of reach of the strokes of it, by being near his head. He continued to plunge and strike, and made my seat very uncomfortable."

In Audubon's American Ornithology is an account of a wounded ibis chased by the alligators. A white ibis had been shot, and had fallen into the water with a broken wing. "The exertions which it made to reach the shore seemed to awake the half torpid alligators that lay in the deep mud at the bottom of the pool. One showed his head above the water, then a second and third. All gave chase to the wounded bird, which, on seeing its dreaded and deadly foes, made double speed towards the very spot where we stood. I was surprised to see how much faster the bird swam than the reptiles, who, with jaws widely opened, urged their heavy bodies through the water. The ibis was now within a few yards of us. It was the alligator's last chance. Springing forwards, as it were, he raised his body almost out of the water; his jaws nearly touched the terrified bird, when, by pulling three triggers at once, we lodged the contents of our guns in the throat of the monster. Threshing furiously

Family II. Alligatorĭdæ.—(Alligator kind.)

ALLIGĀTOR.—(Spanish, *El Lagato*, the Lizard.)

Mississipensis (Lat. *of the Mississipi*), *the Alligator.*

with his tail, and rolling his body in agony, the alligator at last sank to the mud; and the ibis, as if in gratitude, walked to our very feet, and then lying down, surrendered himself to us."

Like the Crocodile, the Alligator lays its eggs in the sandy bank of the river. Fortunately, but few of the young ever reach maturity, as their ranks are thinned by various birds and beasts of prey before the eggs are hatched, and by the attacks of large fishes, and even their own species, when they have reached the water.

Class IV. . . AMPHIBĬA.—(Gr. 'Αμφίβιος, leading a double life, *i.e.* on land and on water.)

Order I. . . . *BATRACHĬA*.—(Gr. Βάτραχος, a Frog.)

Sub-order I. SALIENTĬA.—(Lat. *Leaping animals.*)

RANA.—(Lat. *a Frog.*)

Temporarĭa (Lat. *temporary*), *the Common Frog.*

The FROG.—The appearance and habits of the Frog and the Toad are so familiar as to require but little description. A short account, however, is necessary, of the peculiarities common to both Frogs and Toads.

In the early stage of their existence, these animals are termed tadpoles. They at first appear to be nothing but head and tail, but after several days have passed, four legs are observed to become developed. These rapidly increase, and the little creature closely resembles a small eft. In due time, however, the tail is lost, and the creature becomes a perfect frog. Another important change also takes place. In its tadpole state the creature was essentially a water animal, but after its change has taken place it is not able to exist under water for any great length of time, and is forced to come to the surface to breathe.

The tongue of the Frog is curiously fixed almost at the entrance of the mouth, and when at rest points backwards down the throat. When, however, the Frog comes within reach of a slug or insect, the tongue is darted out with exceeding rapidity, the slug secured, carried to the back of the throat, and swallowed.

Both frogs and toads hybernate, the former congregating in multitudes in the mud at the bottoms of ponds and marshes, while the latter choose a hole in the ground, frequently at the roots of a tree, and pass the winter in solitary dignity. In February 1852, two frogs were dug out of the gravelled play-ground of Magdalen School, Oxford. They were about a foot from the surface of the ground, and their habitation was quite smooth. Both were sitting with their mouths pointed upwards, but I could not ascertain if there had been any communication with the open air.

The skin of these animals has the property of imbibing water, so that if an apparently emaciated frog is placed in a damp place, it will soon look quite plump.

The Common Frog is a well-known frequenter of marshy places and the banks of rivers. It is an admirable swimmer, and from the peculiar construction of its lungs can remain for some time under water, but is forced periodically to come to the surface for the purpose of breathing.

The Bull-Frog is an inhabitant of North America. It is very voracious, feeding upon fishes, molluscs, and even young fowl. Its powers of leaping are so great, that an Indian was not able to overtake an irritated bull-frog after it had sprung three hops in advance. It is very large, measuring about seven inches in length.

The Tree Frogs are very peculiar animals. The construction of their feet, something resembling that of the geckos, enables them to traverse the branches, and even to hang on the under surface of a pendent leaf, which it so resembles in colour that the unwary insect passes by and is instantly seized by the watchful frog. The Green Tree Frog is the most common, and is plentifully found in southern Europe and northern Africa. There are several specimens in the Zoological Gardens, which present a most absurd appearance as they stick against the pane of glass forming the front of their cage.

BUFO.—(Lat. *a Toad.*)

Vulgāris (Lat. *common*), *the Common Toad.*

The COMMON TOAD has had its full share of marvellous tales. Its poisonous properties are celebrated in many an ancient chronicle, as are also the virtues of the jewel contained in its head.

Its skin certainly does secrete an acrid humour, which at all events defends it from dogs, who can seldom be induced to bite a toad a second time; but of course such absurd notions as the romantic story of the death of a young lady and her lover, who each ate a leaf of a shrub at the root of which a toad had made its habitation, need no refutation.

The Toad is easily tamed. A correspondent from the country has kindly sent an account of a tame toad, that had lived in the family for several years, and which was accustomed to sup on a lump of sugar.

The well-known instances of imprisoned toads who must have spent many years in their narrow habitations, are apparently explained by the supposition that some aperture or fissure existed, through which air and minute insects could pass, sufficient for their nourishment while in a semi-torpid condition. Certainly those experimented on by Dr. Buckland in 1825, and from whom all air was cut off, died before a year's imprisonment. The Toad casts its skin at certain times, but we never find the slough as we do that of the snake, as the toad invariably swallows its former covering.

Sub-order II. GRADIENTĬA.—(Lat. *walking animals.*)
Family I. . . Salamandrĭdæ.—(Gr. Σαλάμανδρα, a Salamander.)
TRITON.—(Gr. Τρίτων, a Sea-god.)

Cristātus (Lat. *crested*), *the Common Newt.*

The NEWTS are separated from the lizards on account of their changes while young. Like the frogs, they are first tadpoles, and do not assume their perfect shape until six weeks after their exclusion from the eggs.

The Common Newt is a beautiful inhabitant of the ponds, ditches, and still waters. It feeds principally on tadpoles and worms, which it eats with a peculiar rapid snap. I have frequently seen it attack the smaller newt with great perseverance, but I was never fortunate enough to see it kill its prey.

I kept some newts for some time in a large glass vessel, and noticed that when a new inhabitant was added, it always cast its skin within two or three days. The skin came off in pieces, the covering of the feet slipping off like a glove; but I could never see how the creature contrived to pull these glove-like relics off.

It is constantly in the habit of rising to the surface of the water in order to breathe.

Many country people have great horror of these beautiful and harmless little animals. In a little village in Wiltshire there is a current anecdote of a girl who was bitten in the arm by an *effet*, who spit fire into the wound. The girl consequently lost her arm. Some of these newts or efts were placed in a trough where the cows were accustomed to drink. After a few days a calf died, and nothing would convince the rustics that the effets were not the cause of the untimely decease of the calf, although it had never come near the trough, but was safely fastened in the cow-house. The Newt has received the name of Cristatus, or crested, on account of the beautiful crimson-tipped wavy crest of loose skin, that extends along the whole course of the back and tail, and which, together with the rich orange-coloured belly, makes it a most beautiful creature. The female has a singular habit of laying her eggs upon long leaves of water-plants, and actually tying them in the leaf by a regular knot.

THE FEMALE NEWT.

The first of the two cuts represents the male Newt. It is known by the beautiful ridge of undulating skin that runs down its back. While on land, this ridge falls flat on the back, and is hardly distinguishable, but in the water it remains erect, and waves about with every movement of the animal. The female is destitute of this crest, and is not so beautiful in its appearance as the male.

The PROTEUS is an extraordinary animal, which has been found in dark subterranean lakes, many hundred feet below the surface of the earth, where no ray of light can possibly enter. The eyes of this singular creature are mere points covered with skin, and useless for vision; indeed when in captivity it always chooses the darkest parts of the vessel in which it is confined.

I have seen seven specimens of this strange creature, which have lived for several years in a glass vessel covered with green baize in order to keep them in the dark. They have not been known to take any nourishment whatever during the time of their captivity, except the very trifling amount of nutrition that might have been obtained by changing the water.

Order V. . *MEANTĬA.*—(Lat. *gliding animals.*)

Family I. . Proteïdæ.

PROTEUS.—(*Proper name.*)

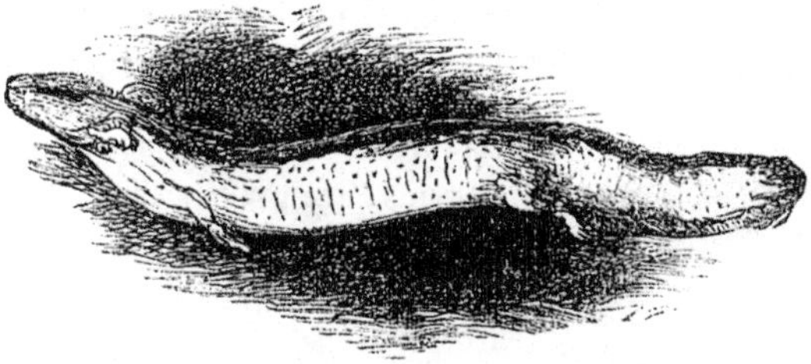

Anguīnus (Lat. *like a snake*), *the Proteus.*

The Proteus breathes in two ways—by lungs and by gills, the latter organs appearing in the form of two tufts, one on each side of the neck, just above the fore limbs. The circulation of the blood in these branchial tufts can easily be seen with a microscope of moderate power. These tufts are of a rather deeper pink tinge than the remainder of the body, which is of a very pale flesh-colour. Exposure to light darkens the tints both of gills and body. It bears some resemblance to the young of the newts, which are furnished with branchial tufts, which they lose upon attaining maturity, and was therefore for some time thought to be the young of some unknown reptile. It has, however, been proved to be a perfect animal, ana has been found of all sizes.

The blood disks of this animal are exceedingly large; so large, indeed, as almost to be distinguished by the naked eye. When in captivity, its movements are slow and eel-like, nor does it seem to make much use of its almost rudimentary limbs.

It has usually been found on the soft mud of a small lake in the grotto of Maddalena. It is not always present, and has been conjectured to be the inhabitant of some unknown subterranean body of water, and to have been forced through the crevices of the rocks. Besides the grotto of Maddalena at Adelsburg, they have also been found at Sittich, thirty miles distant, thrown up from a subterranean cavity.

As the FISHES live exclusively in the water, it is necessary that their organs of respiration should be differently formed from those of the animals breathing atmospheric air. Instead of the purification of the blood being accomplished by the contact of atmospheric air in the apparatus called lungs, that office is performed by the water, which passes into the mouth of the fish, and from thence out at the gill-covers, on its way being strained through the singular structure called the "gills." These gills are able to extract from the water sufficient oxygen to purify the blood of the fish. If the oxygen has already been extracted, the fish instantly dies. The same effect is produced if the fish be so held as to prevent the water from flowing in the proper direction, so that it is perfectly possible to drown a fish; although L'Estrange may doubt the fact:—

> "And like those sages that would drown a fish,
> I am condemn'd to suffer—what I wish."

Most anglers are perfectly aware of the power obtained by keeping the head of a hooked fish *down* the stream.

The elongated form of fishes, and their smooth covering, affording but little resistance to the water, beautifully show their perfect adaptation for the element in which they reside.

Their rapid movements through the water are principally performed by means of a lateral vibration of the tail, just as a boat is sculled along by a single oar at the stern, or by a constant vibration of the rudder. The dead and mangled carcase of a flensed whale has been frequently known to swim for a considerable distance by the mere force of the muscular movements of the tail after death. The fins serve principally as balancers.

Most fish possess a singular organ called the "swimming-bladder." This is a membranous pouch, varying exceedingly in size and shape, situated close under the spine, and filled by some means with gas, mostly found to be nitrogen, but in deep-sea fishes, an excess of oxygen is discovered to exist. The fish seems to be able to rise or sink by means of compressing or expanding this pouch, without being forced to make use of its tail or fins.

The smooth scaly covering with which most fish are furnished, is admirably fitted both for defence against the water, and for enabling the fish to glide easily through places where a rough covering would have held it prisoner. Many valuable characteristics are derived from the shape of the scales in different fish. There are four principal varieties, called, 1. Placoid, or flat scales; 2. Ganoid, or polished scales; 3. Ctenoid, or toothed scales; and 4. Cycloid, or circular scales. These names are derived from, 1. πλακοῦς, a flat cake; 2. γανόω, I polish; 3. κτείς, κτενός, a comb; 4. κύκλος, a circle. The scales of the 1. Dogfish; 2. Sturgeon; 3. Perch; and 4. Carp, are excellent instances of the four kinds of scales.

The Acanthopterygii are so called from their spinous fin rays. The Perch is an excellent example of this order.

The Red Gurnard, or Cuckoo Gurnard, as it is sometimes called, from the sound it utters when taken out of the water, is very common on the English coast. It is rather a small fish, rarely exceeding fourteen inches in length. The colours of its body when living are very beautiful, the upper part being bright red, and the under parts silvery white.

Class V. . PISCES.—(Lat. *Fishes.*)
Sub-class I. PISCES OSSĔI—(Lat. *bony fishes.*)
Order I. . *ACANTHOPTERYGĬI*—(Gr. Ἄκανθος, a thorn; πτερύγιον, a fin.)
Sub-order I. Dactylophŏri.—(Gr. Δάκτυλος, a finger; φέρω, I bear.)
Family I. . Triglĭdæ.—(Gr. Τρίγλα, a Mullet.)

TRIGLA.

Cucūlus (Lat. *a Cuckoo*), *the Gurnard.*

There are nine species of Gurnard known to frequent the coasts of England, some, as the Sapphirine and the Mailed Gurnards, being most extraordinary in form.

The Flying Gurnard is common in the Indian seas. Its pectoral fins are so much enlarged, that when it springs out of the water, when pursued by the dolphin or bonito, the wide quivering fins are able to sustain it in the air for a limited period.

This fish has often been confounded by voyagers with the true Flying-fish (*Exocœtus*), which belongs to an entirely different order.

The Bull-head or Miller's-thumb.—This odd little fish is very common, and may be found in most streams of England. It is called the

Bull-head on account of the great size of its head. For the same reason, it is often called the Miller's-thumb, because the flatness and width of its head is supposed to resemble the state of the thumb of a miller caused by constantly trying the meal between his finger and thumb.

It has a singular habit of diving under stones, and is quite contented if its head is under shelter, utterly regardless of the fact that its tail is waving about in the most open and undisguised manner imaginable. This peculiarity often renders it the victim of keen-eyed boys, who catch a sight of the tail of the Bull-head wriggling about just outside a stone, and then, with a steady and quiet grasp, seize the little fish before it can take the alarm.

Family II. Cottĭdæ.—(Gr. Κόττος, properly, a head; the Bull-head.)

COTTUS.

Gobĭo (Lat.) *the Bull-head or Miller's-thumb.*

The Common Perch is well known to anglers both as a "bold biting fish," and as a fish that does not yield up its life without endangering the person of its captor; for the formidable row of spinous rays belonging to the first dorsal fin have wounded the hands of many an incautious angler.

It is extremely voracious, so much so that after all the legitimate baĭt has been exhausted, it is a common practice for the fisherman to place on his hook the eyes of the perch already taken, which are

Sub-order II. Holodactўli.—(Gr. Ὅλος, entire; δάκτυλος, a finger.)

Family IV. . Percĭdæ.—(Gr. Πέρκη, *a* Perch.)

PERCA.

Fluviatĭlis (Lat. *of the river*), *the Perch.*

as eagerly bitten at, as the worms were formerly. An anecdote is related of a gentleman who struck at a Perch, but unfortunately missed it, the hook tearing out the eye of the poor creature. He adjusted the eye on the hook, and replaced the line in the water, where it had hardly been a few minutes before the float was violently jerked under the surface. The angler of course struck, and found he had captured a fine Perch. This when landed was discovered to be the very fish which had just been mutilated, and which had actually lost its life by devouring its own eye. It is quaintly observed by Izaak Walton, that "if there be twenty or forty in a hole, they may be at one standing all caught one after another, they being like the wicked of the world, not afraid though their fellows and companions perish in their sight."

The Perch seldom exceeds two pounds and a half in weight, and a Perch weighing a pound and a half is considered a very fine fish.

The MACKAREL.—The elegant shape and resplendent colours of the Mackarel point it out as one of the most beautiful fishes known. Nor is it only valuable for its beauty, as it is highly prized as an article of food in most parts of the world.

Family XIII. Scomberidæ.—(Gr. Σκόμβρος, a generic name for the Tunny.)

SCOMBER.

Scombrus (Latinized form of Σκόμβρος), *the Mackarel.*

Vast shoals of Mackarel visit our coasts, and myriads are taken by fishermen both by nets and with lines. The line of nets frequently exceeds a mile in extent, and of course the number of fish contained in this enormous net must be beyond all calculation. On several occasions, the meshes of the net were completely choked up by fish hanging by their gills, and the net acted like a dredge, sweeping up myriads more fish in a solid mass. In 1808, the whole net and its cargo sunk and were lost to the too successful fishermen.

The profits of the fishery vary exceedingly; sometimes the boats will hardly take a single mackarel, and at other times, or even in different spots, the draught of fish will nearly fill the boat. In 1834, one boat sold in one night nearly one hundred pounds' worth of mackarel.

The fish require to be used soon after they are taken out of the water, as the flesh is very tender, and easily injured by exposure to the air, or by carriage to any great distance.

When the fishermen employ the line for the capture of the mackarel, the hook is baited with a strip cut from a dead mackarel, and is suffered to trail overboard. The fish bite eagerly at this cannibal kind of bait, and are frequently taken by baiting the hook with a strip of scarlet leather or cloth.

The TUNNY is a tolerably large fish, averaging four feet in length, and is very common in the Mediterranean. Large fisheries are established during May and June, at which season immense shoals of these fish rove along the coast. The most approved method of fishing is by the "*madrague*" or "*tonnaro.*" A large number of long and deep nets are

THYNNUS.—(Gr. Θύννος, a Tunny.)

Thynnus, *the Tunny.*

placed along the shore, one edge being fixed to the bottom of the sea by anchors and weights, and the other edge kept at the surface of the water by corks. A wall is thus formed, stretching along the coast for nearly a mile in length. The tunnies swimming along the coast pass into this net, and continue their course until they are stopped by other nets placed across the principal net, and dividing it into chambers. From chamber to chamber the unfortunate fishes are driven through openings permitting their entrance, but preventing egress, until they arrive at the last chamber, called significantly the "chamber of death." A strong net, placed horizontally, enables the fishermen to draw the tunnies to the surface, when a shower of blows from poles and similar weapons soon destroys the entire shoal.

This fish is not unfrequently found on the English coast.

The SWORD-FISH.—The well-known Sword-fish inhabits every part of the Mediterranean Sea, and has several times been seen near the shores of England and Scotland.

The "sword" for which this fish is so famous, is an elongation of the upper jaw, of great strength, and capable of doing considerable injury to any object against which it directs its attacks. In the British Museum is a portion of the bottom of a ship, pierced completely through by the "sword" of one of these fish. Its unfortunate owner must have instantly perished by the shock, for the sword was imbedded almost to its base, and broken short off. In one instance, a Sword-fish attacked a whaling-ship, and drove its weapon "through the copper sheathing, an inch-board sheathing, a three-inch plank of hard wood, the solid white oak-timber of the ship twelve inches thick, through another two-and-a-half inch hard oak ceiling plank, and lastly, perforated the head of an oil-cask, where it still remained immovably fixed, so that not a single drop of oil escaped."

In the Mediterranean, the fishermen eagerly chase the Sword-fish. The harpoon and line are used, much in the same manner as in the whale

XIPHĬAS.—(Gr. Ξιφίας, shaped like a sword; the Sword-fish.)

Gladĭus (Lat. *a Sword*), *the Sword-fish.*

fishery. The Sicilian fishermen have a strange superstition that if the Sword-fish were to hear a word of Italian, it would instantly dive and escape them. They therefore restrict their vocal sounds to an unintelligible chant. It it said that the whale is an object of particular enmity to the Sword-fish, and that ships are struck by it, being mistaken for whales.

The length of this fish is usually from twelve to fifteen feet. It is said to feed principally on tunnies, pursuing the shoals, and transfixing the fish with its sword.

The STICKLEBACK.—There are six species of Sticklebacks known to inhabit England, the habits of all being very similar. They are most pugnacious little creatures, and will fight on the smallest provocation, dashing at each other and endeavouring to tear open their adversary's side with the sharp spikes that adorn their sides. The brilliant colours with which they are decorated, only belong to the males, and not to them if they have been vanquished. In such a case, the conqueror looks more brilliant than before, and sails about with as much dignity as can be assumed by an animal an inch and a half in length. The unfortunate individual who has been defeated sneaks off into some corner, and soon loses his beautiful colouring, his crimson, green, and gold panoply changing into a very dull matter-of-fact grey.

Gasterostĕus.—(Gr. Γαστήρ, the belly; ὀστέον, a bone.)

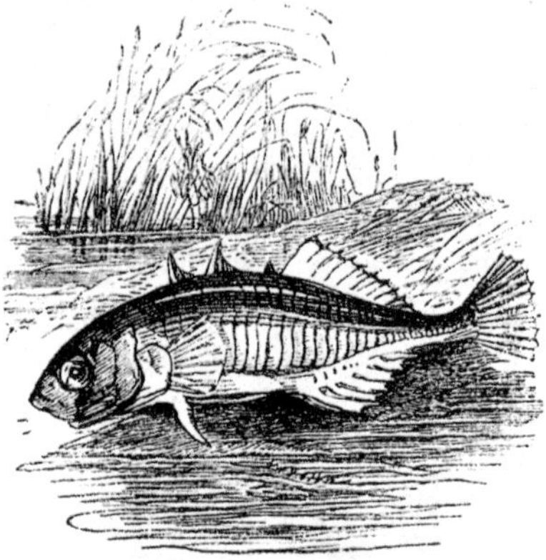

Aculeātus (Lat. *thorny,*) *the Stickleback.*

The courage of the Sticklebacks is not only exerted against their own species. These little fish are quite aquatic lemmings, and will boldly attack a stick if it is placed near their haunts. I have often known them dash with such violence at a stick which I have presented to them, that the blow of their head against the wood produced a sensible jar.

Family XIV. Zeïdæ.—(Gr. Ζεύς, Jupiter.)

ZEUS.

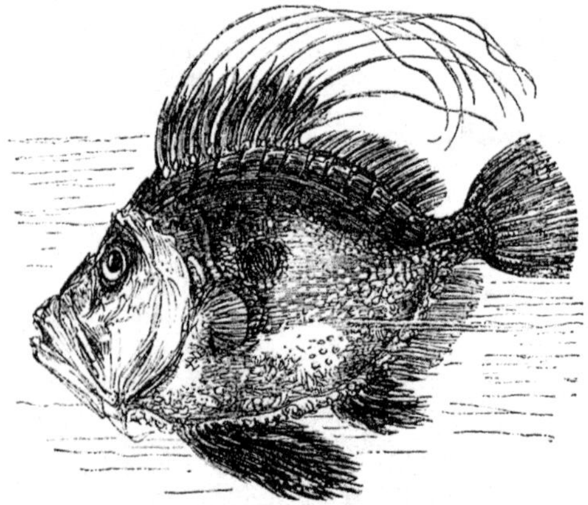

Faber (Lat. *a Workman ; sometimes used for the fish*), *the John Dory.*

The John Dory, rendered illustrious by Quin the comedian, who was not less known for his comic powers than for his love of good living, is found plentifully off the coasts of Cornwall and Devonshire. The derivation of its name is not quite certain, but in all probability it is derived from the French, *dorée,* or golden, in allusion to its peculiar golden yellow colour.

Traditions vary as to the spots so conspicuous on its side. Some strenuously assert that this was the fish caught by St. Peter when he took the tribute-money out of its mouth, and upon whose sides the marks of his finger and thumb were left. The Haddock, however, vies with the Dory for this honour. Other traditions are quite as vigorous in their assertion, that St. Christopher produced these marks while crossing an arm of the sea, bearing the Saviour in his arms.

Family XVII. Syngnathidæ.—(Gr. Σύν, together; γνάθος, the jaw.)

Hippocampus.—(Gr. Ἱππόκαμπος, derived from ἵππος, a Horse; κάμπη, a joint or a caterpillar, a Sea-horse.)

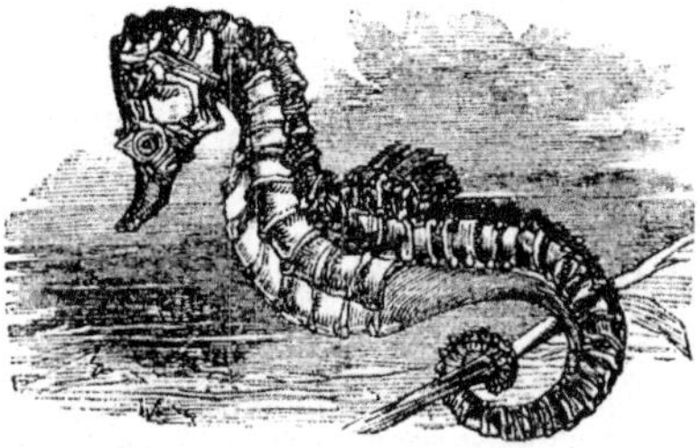

Brevirostris (Lat. *short-beaked*), *the Sea-horse.*

The singular fish called the Sea-horse has often been found off the southern coasts of England. The habits of this fish are very singular and interesting. A pair were kept alive for some time in a glass vessel, and exhibited considerable activity and intelligence. They swam about with an undulating kind of movement, and frequently twined their tails round the weeds placed in their prison. Their eyes moved independently of each other, like those of the chameleon, and the changeable tints of the head closely resemble that animal.

More than once, these curious fish have been seen curled up in oyster shells.

The singular creatures called Pipe-fish also belong to the Syngnathidæ.

The Remora, or Sucking-fish, is remarkable for the peculiar apparatus situated on the upper part of its head. By this it can adhere to any object so firmly that it is a difficult matter to make it loose its hold. It is often found adhering to large fish or to the bottoms of ships, probably in both instances for the sake of the fragments of food rejected by the one, or thrown overboard from the other.

The older writers on Natural History fully believed that one Remora had the power of arresting the swiftest ship in its course, and fixing it firmly in the same spot in spite of spread canvass and swift gales. As the Remora is about the same size as a herring, our ancestors naturally considered this a very curious circumstance, and wrote no few poems on the subject. The following *true* account of this fish is extracted from Macgillivray's Voyage of the Rattle-snake:—

"Small fish appeared to abound at this anchorage (the Calvados group of islands). I had never before seen the Sucking-fish (*Echeneis remora*) so plentiful as at that place; they caused much annoyance to our fishermen by carrying off baits and hooks, and appeared always on the alert, darting out in a body of twenty or more from under the ship's bottom when any offal was thrown overboard. Being quite a nuisance, and useless as food, Jack often treated them as he would a shark, by sprit-sail yarding,* or some less refined mode of torture. One day, some of us while walking the poop had our attention directed to a sucking-fish about two and a half feet in length, which had been made fast by the tail to a billet of wood by a fathom or so of spun yarn, and so turned adrift. An immense striped shark, apparently about fourteen feet in length, which

Family XXII. Echeneïdæ.—(Gr. Ἐχενηΐς; from ἔχω, I hold; ναῦς, a ship.)

ECHENĒIS.

Remŏra (Lat. *properly, a delay*), *the Sucking-fish.*

had been cruising about the ship all the morning, sailed slowly up, and turning slightly on one side, attempted to seize the apparently helpless fish, but the sucker with great dexterity made himself fast in a moment to the shark's back. Off darted the monster at full speed, the sucker holding fast as a limpet to a rock, and the billet towing astern. He then rolled over and over, tumbling about; when, wearied with his efforts, he lay quiet for a little. Seeing the float, the shark got it into his mouth, and disengaging the sucker by a tug on the line, made a bolt at the fish; but his puny antagonist was again too quick, and fixing himself close behind the dorsal fin, defied the efforts of the shark to disengage him, although he rolled over and over, lashing the water with his tail until it foamed all around. What the final result was, we could not clearly make out."

The Angler, or Fishing Frog, as it is more generally called, is not uncommon in all the European seas. The peculiar formation of its pectoral fins enables it to crawl for some distance on land.

* "Sprit-sail yarding" is an ingenious torture invented by sailors for the especial benefit of the shark. It is accomplished by thrusting a small spar transversely through the poor creature's mouth, so that, being unable to procure its food on account of the impediment, it dies by degrees from the combined effects of hunger and pain.

On its head are two elongated bony appendages, curiously articulated to the skull by a joint formed something like the links of a chain, and capable of movement in any direction. The Angler couches close to the bottom of the sea, and by the movement of its pectoral fins stirs up the sand and mud, and agitates the bony appendages amid the turbid cloud

Family XXIII. Lophiïdæ.—(Gr. Λόφος, a crest.)

LOPHIŬS.

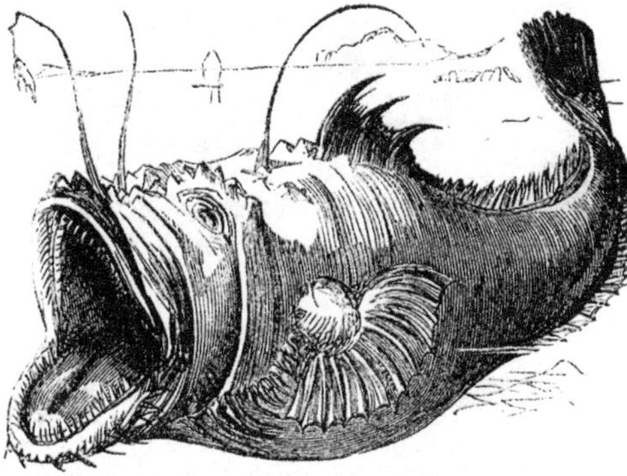

Piscatorĭus (Lat. *fishing*), *the Angler.*

produced. The small fishes, observing the muddy water, and taking the filaments for worms, approach to seize them, and are instantly engulphed in the capacious jaws of the crafty Angler.

The voracity of the Angler is so great, that when caught in a net together with other fish, it generally devours some of its fellow-prisoners —a useless act, for the fishermen mostly open its stomach, and recapture the flounders and other fish found in its interior.

The Malacopterygian fishes have their fin membranes supported by flexible rays. The Abdominal Malacopterygii have their ventral fins situated on the belly, without any connexion with the bones of the shoulder.

The COMMON CARP is a well-known inhabitant of our ponds, lakes, and sluggish rivers. It is a very shy and wary fish, rejecting one day a bait which had been freely taken the day previous. In 1847, while fishing in a small pond near Oxford, I took in one hour six or seven carp, weighing from half a pound to nearly three pounds each. A few days afterwards, although the weather was equally propitious, the carp were not, and the whole day was spent without even a bite.

It lives to a great age, and when very old its scales turn grey just as

Order II. . . *MALACOPTERYGĬI.*—(Gr. Μαλακός, soft; πτερύγιον, a fin.)
Sub-order I. ABDOMINALĬA.—(Lat. *belonging to the abdomen.*)
Family I. . . Cyprinĭdæ.—(Gr. Κυπρῖνος, a Carp.)

CYPRĪNUS.

Carpĭo (Lat.) *the Carp.*

human hairs do. In several places in France numbers of Carp were kept until they attained an enormous size. These great sluggish fish were accustomed to come to the water's edge in order to be fed at the call of their keeper. Feeding the Carp was almost a hereditary amusement of the later kings of France.

Very few fish are so tenacious of life as the Carp. It is the custom in Holland to keep these fish in nets filled with wet moss. They are fed with bread and milk, and are preserved in health by frequent immersion in water, in order to keep the moss thoroughly wet.

Two or three pounds is the average weight of a good Carp, but individuals have been known weighing upwards of eighteen pounds. It is enormously prolific, as the roe of one female weighing nine pounds was found to contain six hundred thousand eggs. Of course comparatively few of these eggs arrive at maturity, by far the greater number being eaten by other fish.

The GOLD-FISH or GOLDEN CARP, is another species of the genus Cyprinus. It was originally brought from China, about two hundred years since, when it was considered a great curiosity; now, however, it is quite common, and is found to live in ponds even when the surface of the water is thickly covered with ice. The ponds in Christ Church College, and the Botanic Gardens, Oxford, are thickly populated with these beautiful fish, which increase with the most marvellous rapidity. The pond in the centre of the Clarendon Printing Office was stocked with these fish, and as the spare water from the steam-engine used in the works passed into the pond, they throve amazingly. One unfortunate

CYPRĪNUS.

Barbus (Lat. *Barba*, a beard), *the Barbel.*
Aurātus (Lat. *gilded*), *the Gold-fish.*

morning, the surface of the pond was covered with Golden Carp, all floating dead. Some verdigris had formed in some part of the engine, had been washed into the pond, and had poisoned all its finny inhabitants.

The Barbel is found in most of the European rivers. Its flesh is coarse and unsavoury, but it is eagerly sought after by anglers, as the spirit and vigour displayed by it when hooked afford fine sport. It is peculiarly clever at breaking the line, a feat sometimes accomplished by a violent blow of the tail, and sometimes by contriving to twist the line round a root or post, and giving a sudden jerk.

It feeds principally on larvæ and molluscs, inhabiting the banks, and obtains them by rooting in the sand with its snout. The barbels, or beards, hanging from the upper jaw, doubtless assist in these investigations. It frequently grows to a very great size, weighing from fifteen to eighteen pounds, and measuring upwards of three feet in length. Many are captured by nets during the summer, at which season they frequent the weedy parts of the river in shoals; but in winter they retire to the shelter afforded by banks and old woodwork. Several good swimmers have been known to dive after the Barbel, as they lay pressed against the banks, and to bring up one each time, not unfrequently appearing with two, one in each hand.

The Gudgeon.—The ease with which the Gudgeon is taken has passed into a proverb. This pretty little fish is usually found in shallow parts of rivers, where the bottom is gravelly. If the gravel is stirred up, the Gudgeons immediately flock to the place, and a worm suspended amid the turbid water is eagerly snapped at by them. The fishermen usually take them in nets, and keep them alive in well-boats. They are largely purchased as baits for trolling.

The flesh of the Gudgeon is particularly delicate, and although its length rarely exceeds seven inches, yet from the ease with which numbers can be obtained, it forms a dish by no means to be despised.

GOBĬO.—(Lat. *a Gudgeon.*)

ABRĂMIS.—(Gr. Ἀβραμίς, a Bream.)

Fluviatĭlis (Lat. *of the river*), *the Gudgeon.*

Brama (Lat.) *the Bream.*

The BREAM is very common on the Continent, but in England is only found in certain rivers and lakes, such as the Medway and Trent, and the lakes of Cumberland and Westmoreland. It is also found in the lakes of Ireland.

The breadth of the Bream is greater in proportion to its length than that of most fishes. It affords excellent sport to the angler, biting readily and resisting vigorously when hooked. The most approved method of catching this fish is by preparing the spot with ground bait for a day or two previous; the Bream then assemble in numbers and bite freely at a bait. In Ireland the Bream taken were accustomed to be given to the poor, who split and salted them for winter provision.

Its length rarely exceeds ten or twelve inches, nor is it of any value for the table.

The TENCH.—The habits of the Tench are not unlike those of the carp, excepting that it seems even more sluggish than that fish. It especially delights in muddy banks of ponds, where the weeds grow

TINCA.—(Lat. *a Tench.*)

Vulgāris (Lat. *common*), *the Tench.*

thickly. Roget gives an account of a Tench that had been taken out of a pond almost filled up with stones and rubbish, and which had actually grown into the shape of the hole where it had been confined, evidently for many years. The weight of that fish was eleven pounds nine ounces. Four hundred tench and as many perch were also taken out of the same pond. This fish is even more tenacious of life than the carp.

The ROACH is very common in most rivers of this country, and is generally spread over the temperate parts of Europe. It is by no means a large fish, rarely exceeding two pounds in weight, and but seldom attaining that size. These fish usually live in small shoals, and pass from one part of the river to another.

LEUCISCUS.—(Gr. Λευκίσκος; from λευκός, white.)

Rutĭlus (Lat. *shining red*), *the Roach.*
Leuciscus, *the Dace.*

The Roach is not unlike the Dace, but may be easily distinguished by its bright red ventral fins, those of the dace being silvery white. It is rather a favourite with anglers, as it bites or rather nibbles at the bait in such a dainty and delicate manner, that the disappointed fisherman not unfrequently finds the bait gone without the movement of his float betraying the theft. A quick eye and a dexterous hand are required for this sport. The float is so balanced as barely to appear above the surface of the water, for, unlike the perch, that dashes at the bait and boldly jerks the float at once under water, the Roach does little more than swim under the bait as far as it can, and then just gives a gentle nibble, repeating the process until the bait has entirely left the hook.

The DACE.—The habits of the Dace are so similar to those of the Roach as to need but little description. It is usually found wherever the roach resides, and, like that fish, swims in shoals. It makes an

THE DACE.*

excellent bait for trolling, as the silvery whiteness of its scales renders it a conspicuous object, and serves to attract the pike. It seldom exceeds nine or ten inches in length.

The BLEAK and the MINNOW both belong to the genus Leuciscus. The former fish is remarkable for the use made of its scales, which when washed in water deposit a powder much used in the manufacture of artificial pearls.

The CHUB is also common in most of our rivers. It affords good sport to the angler, both with a fly and with a bait.

The usual bait employed is a cockchaffer, which, when fastened to the hook and artistically made to dance on the surface of the water, is a temptation that few Chub can resist. This method of fishing is termed "dibbing," and the peculiar movement is communicated to the bait by tapping the butt end of the rod, while the cockchaffer or moth just rests on the surface of the water.

Its flesh is very coarse, and requires some skill on the part of the cook to make it fit for the table. Its weight rarely exceeds five pounds, but it is very powerful, and requires a strong line and skilful management on the part of the angler. A well-known piscator at Oxford, while fishing with the fly from a small skiff, succeeded in hooking a Chub, apparently weighing about four pounds, which actually towed him up and down

* Two drawings of this fish having been made, both have been inserted.

LEUCISCUS.

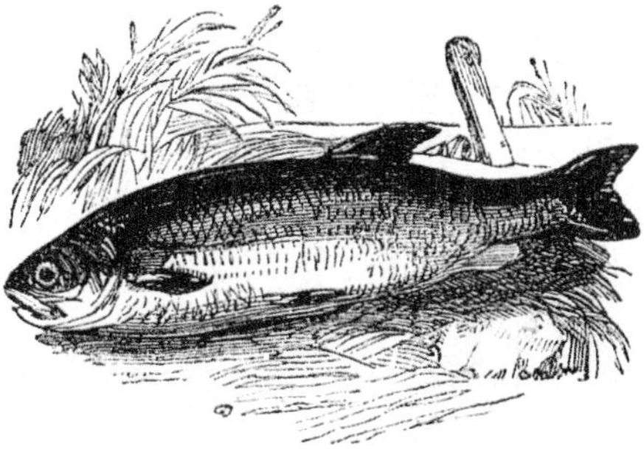

Cephălus (Gr. Κέφαλος, a large-headed fish), *the Chub.*

stream for some time, until the line, not calculated for Chub, snapped, and the fish of course escaped.

In some counties the LOACH goes by the name of "Beardie," in allusion to the little fleshy particles that hang from its lips. It has also the name of Groundling, on account of its habit of living close to the bottom of the water.

It is a common fish, and may be taken in most streams, especially if the bait is drawn over the bed of the stream. The principal peculiarity about the fish, is the comparatively great breadth of the tail where it joins the spine. This formation, together with the generally pellucid appearance of its body, at once distinguish it from any other fish.

COBĪTIS—(Gr. *κωβίτης, dim. of κωβιός, a Gudgeon.*)

Barbatŭla (Lat. *bearded*), *the Loach.*

The PIKE.—This fierce and voracious fish is now common in most rivers and lakes in England, although it was formerly so rare as to be rated at ten times the value of turbot.

It affords much sport to anglers, who generally employ a method of fishing called "trolling." A gudgeon, roach, or large minnow is so fixed to a number of formidable hooks, that when drawn through the water, it spins rapidly round, and attracts the notice of the watchful Pike, who dashes at the glittering bait with a violence that jars the rod down to the

Family II. Esocidæ.—(Lat. *Esox*, a Pike.)

ESOX.

Lucius (Lat. *a Pike*), *the Pike.*

very butt. Off swims the pike to his place of concealment, leisurely turns the head of the bait downwards, and swallows it. Now, to swallow the fish is easy enough, but the array of barbed hooks proves an effectual obstacle to the endeavours of the Pike to get rid of the unwelcome morsel as soon as the angler jerks the line, and gives the Pike to understand that hooks have points. The deluded Pike now endeavours to break the line, but a good fisherman foils all his efforts, and at last lands him, wearied and bleeding, but ferocious to the last.

The method of fishing for Pike called "trimming" is hardly worth mention. A line baited with living fish is fastened to a float, and suffered to lie on the surface of the water. The Pike, seeing the bait swimming about, dashes at it and hooks itself in the effort.

This fish varies in size from two or three pounds' weight, to twenty or thirty; but a Pike weighing fifteen pounds is considered a very fine fish. Above that weight they are almost useless for the table. A Pike weighing less than two pounds is called a jack. In the Ashmolean Museum at Oxford is a Pike weighing thirty pounds, that was taken in the lake at Blenheim Park. Another Pike, weighing twenty-five pounds, was caught near Oxford a few months ago, and I have seen the skin of one that weighed thirty-five pounds when first caught.

The appetite of this fish is almost insatiable. Mr. Jesse threw to one Pike of five pounds' weight, four roach, each about four inches in length, which it devoured instantly, and swallowed a fifth within a quarter of an hour. Moor-hens, ducks, and even swans have been known to fall a prey to this voracious fish, its long teeth effectually keeping them prisoners under water until drowned.

The FLYING-FISH.—This fish, so celebrated in most books of voyages, is found in the warmer latitudes, but has several times been seen off our coasts. The so-called "flight" is very similar to that of the flying squirrels and dragons, the fish merely springing out of the water with a violent impetus, and sustaining itself in the air by means of its enormous pectoral fins. It is not able to alter its course while in the air, nor to

EXOCŒTUS.—(Gr. 'Εξώκοιτος, sleeping out [of the sea].)*

Volitans (Lat. *flying*), *the Flying-fish.*

rise a second time without repeating its course through the water. The reader will notice the remarkable fact, that individuals of three wingless classes, the Mammalia, the Reptiles, and the Fishes, have each the power of sustaining themselves in the air.

The "flight" of this fish seldom exceeds two hundred yards. The unfortunate creatures are pursued in the water by "Dorados," erroneously called dolphins, and other fishes of prey. To escape their finny tyrants, they spring into the air, and for a while escape. But the gulls and albatroses are on the watch, and pounce on the Flying-fish from above, so that the persecuted creatures are tolerably sure to fall a prey to one or the other of their foes.

* The ancients believed that this and some other fishes slept on the beach.

The usual height of flight is about two or three feet above the surface of the water, but it has frequently been known to exceed fourteen feet, and in one instance a Flying-fish came skimming into the ports of a large man-of-war, nearly twenty feet above the water.

The size of the fish is about the same as that of a herring. Sailors are always glad to capture it, as its flesh proves an agreeable change from the eternal salt junk, by which the power of the sailor's teeth is woefully tried.

The food of this fish is molluscs and small fishes.

Family IV. Salmonĭdæ.

SALMO. — (Lat. *a Salmon.*)

Salar (Lat. *a Salmon*), *the Salmon.*

The SALMON is a migratory fish, annually leaving the sea, its proper residence, and proceeding for many miles up rivers for the purpose of depositing its spawn. This duty having been accomplished, it returns to the sea in the spring. The perseverance of this fish in working its way up the stream is perfectly wonderful. No stream is rapid enough to daunt it, nor is it even checked by falls. These it surmounts by springing out of the water, fairly passing over the fall. Heights of fourteen or fifteen feet are constantly leaped by this powerful fish, and when it has arrived at the higher and shallower parts of the river, it scoops furrows in the gravelly bottom, and there deposits its spawn. The young, called "fry," are hatched about March, and immediately commence their retreat to the sea. By the end of May the young salmon, now called "smolts," have almost entirely deserted the rivers, and in June not one is to be found in fresh water. Small Salmon weighing less than two pounds are termed "salmon peel," all above that weight are called "grilse."

The havoc wrought among Salmon by foes of every description is so enormous, that notwithstanding the great fecundity of the fish, it is a matter of surprise that so many escape destruction; for although the fish are preserved from their human foes by many stringent regulations, yet other foes, such as otters, who devour the large fish, and other fish who devour the spawn, have but little respect for laws and regulations.

While in the rivers, multitudes of Salmon are annually caught, usually by stake nets, which are capable of confining an immense number of fish at one time Salmon spearing is a favourite amusement. This animated and exciting sport is usually carried on by torch-light. The torches, when held close to the surface of the water, illumine the depths of the river, and render every fish within its influence perfectly visible. The watchful spearman, guided by slight indications bearing no meaning to an unpractised eye, darts his unerring spear, and brings up in triumph the glittering captive, writhing in vain among the barbed points. In the northern rivers this destructive pursuit is carried on to a great extent, more than a hundred salmon being frequently taken in an evening. Anglers also find considerable sport in using the fly for this beautiful and active fish, whose strength makes it no mean antagonist.

SALMO.

Farĭo (Lat. *a Trout*), *the Trout.*

The COMMON TROUT is found in many rivers in this country, always preferring rapid, shallow, and sparkling streams, especially if there should be little falls at intervals. The Derwent and the Dove are particularly famous for their trout. The latter river is quite the *beau ideal* of a trout stream. It never seems to know its own mind for half a mile together.

Sometimes it is rapid, frisking over stones and round trees, and throwing up the sparkling foam in all directions. Presently it has changed into a silent, slow, melancholy river, with dark pools of unknown depth, shaded by overhanging trees, and suggestive of murders successfully concealed. Everywhere are the trout. Lying quietly under the shelter of some large stone, while the water is leaping round them, are the moderate sized trout, darting off like meteors to snatch at a passing fly, and as quickly returning to their concealment. In the deeper pools are the larger fish, who are too sagacious to be deceived by the artfully made, fly of the professed angler, yet often fall victims to the less scientific but more successful plough-boy.

Several of my schoolboy years were spent near the banks of the Dove, which river, of course, formed one of our favourite haunts. We were accustomed to take the large trout by the rather unsportsmanlike, but very amusing method of "tickling." It was excessively amusing to watch the angry countenances of London anglers, who came to the Dove bedizened with all the appurtenances of rods, lines, baskets, &c., and who, after whipping the water most perseveringly for the whole morning without a single bite, while resting their tired arms, saw the country boys seated on the bank, armed with a long stick and a line barely two feet long, adding every minute to the heap of glittering fishes at their side.

The usual method of fishing for trout is with a fly, but trolling with a minnow is often successfully used, nor does the trout reject a well-selected and properly arranged worm.

The brilliant speckled tints of this beautiful fish vary much according to the locality and the time of year. In May the fish assume their brightest colours and their most delicate flavour. The size of the fish also varies exceedingly, being from half a pound in weight and about eight inches in length, to ten or fifteen pounds' weight.

The Smelt belongs to this family, and in its progress to the sea is destroyed in great quantities in mill-ponds, &c.

The value of the HERRING family to man is almost incalculable. The PILCHARD and the Herring are very similar in appearance, but may be easily known by the position of the dorsal fin, which in the Pilchard is so exactly in the centre of the body, that if the fish is held by it, the body exactly balances; while in the herring, the dorsal fin is placed rather backwards, so that when suspended, the fish hangs with its head downwards.

Unlike the herring, which visits every part of our coasts, the Pilchard is only found on the shores of Devonshire and Cornwall. Here, however the enormous shoals that annually make their appearance fully compensate for the limited space occupied by them. Occasionally a few shoals are seen on the southern coast of Ireland. The coasts of France and Spain are tolerably frequent resorts of this fish.

The fish are usually taken in an enormous building of nets, called "sean nets." The nets used in the sea fishery are two, a large net called the "stop sean," about a quarter of a mile in length, and a hundred feet in depth; and a smaller net, called the "tuck sean," about a furlong in length, and a hundred and twenty feet in depth, the average value of the two nets being 500*l.*

When the fishermen see a shoal of pilchards approaching, they immediately set out in two fishing boats, one of which carries the tuck sean

Family V. Clupeïdæ.—(Lat. *Clŭpĕa*, a Herring.)

CLUPĔA.

Pilchardus (Lat. *the Pilchard.*)

and the other the stop sean. Guided by signs from the master-seamen, they silently surround the shoal with the nets, the larger of which is used to enclose a large number of fish, and the smaller to pass within the other net, to bring the mass of fish into a small compass, and finally to prevent them from escaping until the fishermen have leisure to remove them to the boats.

When landed, the pilchards are taken to the store nouses, salted, and after remaining in heaps for five or six days, are pressed into casks by powerful levers. During the pressure, which lasts about a fortnight, fresh layers of fish being added as the former are pressed close, an abundance of excellent oil escapes from holes made in the cask for the purpose. The entire refuse of the fish, consisting of the superabundant salt, the scales and other rejected portions, is sold to the farmers as a valuable manure. The refuse of each pilchard is calculated to manure one square foot of land.

The HERRING makes its annual appearance in the northern parts of Scotland about June. This most valuable fish arrives in enormous shoals, five or six miles in length and three or four in breadth. Their advent is heralded by various sea birds, such as the gannets and gulls, which constantly hover over the shoals and commit unceasing devastations among them. Yet in spite of the myriads destroyed by birds and fishes, in spite of the shoals captured by man, in spite of the vast quantity of spawn devoured by other fishes, their numbers seem quite undiminished, and each year they are led by the instinct inculcated in them by Providence, to visit the shore in incalculable numbers, not only to yield to man an unfailing supply, but to make the necessary provision for the increase of their number.

The fishery is conducted by boats and nets, the whole fitting up of each boat costing little less than 1,000*l*. To add to the expense, the whole apparatus must be renewed every four or five years, as, independently of the injuries inflicted by the sea and the weight of fish, the dog-fish, which unremittingly follows the shoals of herrings, is often entangled in the nets together with its intended victims, and by its sharp teeth and vigorous struggles makes sad ravages among the nets.

CLUPĔA.

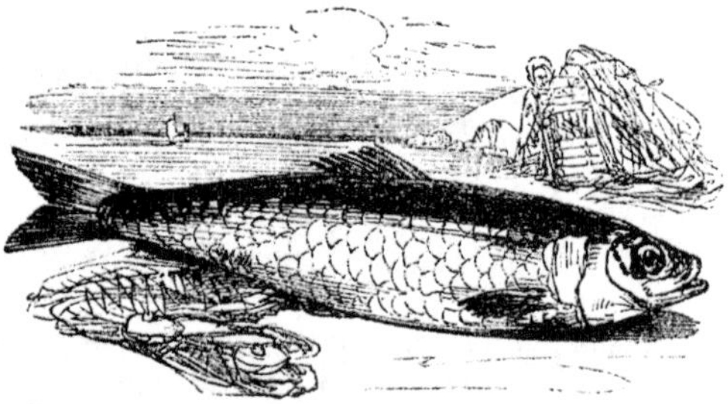

Harengus (Lat. *the Herring*).

When taken out of the water, the herring dies almost immediately, as do all fish that live near the surface of the water. Those, on the contrary, as the carp, tench, eels, and the flat fish, who reside at the bottom, are able to sustain life for a much longer period when taken out of their native element. It is therefore necessary that the herrings should be cured as soon as possible. The "White Herrings" are cured in the boats, but the "Red Herrings" are taken on shore, and suspended in the smoke of a wood fire for twenty-four hours, in addition to the salting that both they and the White Herring undergo.

The well-known SPRAT (*Clupea Sprattus*) also belongs to the genus Clupea, and, like the herring, visits our shores in large shoals. The Sprat fishery commences in the beginning of November. Not only are enormous quantities of this small but useful fish used as food, and sent into all parts of this country, but they are very largely used as manure; fish, according to the researches of Sir H. Davy, being a most powerful manure, retaining its fertilising influence for a long time. Many thousand tons' weight of sprats are annually used for this purpose.

The White-bait belongs to the same family.

ENGRAULIS.—(Gr. Ἔγγραυλις).

Encrasichŏlus (Gr. Ἐγκρασίχολος, mixed with bitter; *the Anchovy*, from its taste), *the Anchovy*.

The little ANCHOVY is a fish of no small importance, being very largely used in various sauces, besides the numbers that are preserved in pickle. It is common in the Mediterranean, and is also found on our coasts. The upper jaw of this fish is longer than the lower one; the entire length of the fish is usually from four to five inches, but it has been seen measuring upwards of seven inches.

Sub-order II. SUB-BRACHIĀTA.—(Lat. *sub*, under; *brachium*, an arm.)

Family VI. . Gadĭdæ.—(Gr. Γάδος.)

MORRHŬA.

Callarĭas (Gr. Καλλαρίας), *the Cod.*

The COD.—In this Sub-order the bones of the ventral fins are placed *under*, and support the bones of the shoulder.

The well-known Cod-fish is principally found on the coasts of Newfoundland, but is taken in great numbers on our own shores. The hook is generally employed for the capture of this fine fish. An immense number of hooks, each baited with a whelk or limpet and attached to short lines, are fastened at intervals along a rope, which is stretched, or shot, as it is termed, across the tide, in order to prevent the hooks from getting entangled. Such is the voracity of the fish, that nearly five hundred fish have been taken by one man in the course of ten hours. The intense cold renders the Cod fishery a service of great hardship.

When taken, the fish are placed in a well boat, through which the salt water has a free passage, so that the cod-fish are brought to Billingsgate still living. Several successful experiments have been made to preserve this fish in salt water ponds, in which it appears to thrive well. The fecundity of this fish is almost incredible, the roe of one fish having been ascertained to contain *nine million eggs*. The Whiting belongs to this family.

In the FLAT-FISH we see a most extraordinary instance of adaptation of structure to peculiar circumstances. We have all seen Flat-fish, and all know that the upper side is dark, and the under side nearly white. The word 'side' is used advisedly, as these curious fish actually lie *on their sides* at the bottom of the water while undisturbed, or merely feeding. When, however, they are alarmed, they rapidly assume the vertical position, and dart off with great speed. The dark upper surface serves to protect them from becoming too visible to enemies above. The two eyes are also placed on the upper side of the head for obvious reasons. In fact, the whole fish appears as if it had been laid on its side, and rolled flat, the head also being twisted round, and the lower eye removed to the upper surface.

The TURBOT is found on the coasts of most parts of England, but the fisheries are nearly exclusively confined to the southern coasts of Ireland.

The fishery is conducted both by nets and lines. The net, called the haul-net, drags from the bottom not only turbots but other flat-fish, such as soles and plaice. The line, used when the bottom of the sea is too deep or rocky for the net, is armed with many hooks, baited with smelts and other small fish. The lampern, or river lamprey, was formerly in very great use as a bait, as its brilliant silvery appearance, and its great tenacity of life, rendered it peculiarly fit for the capture of the voracious but dainty turbot, who, rejecting all stale or discoloured baits, eagerly devours them if bright coloured and moving. The fishermen state that the turbot will not touch a bait that has been bitten by any other fish. On the English coasts one turbot-line frequently extends for three miles in length, and is furnished with 2,500 hooks, which are attached to the main line by small horse-hair lines, each twenty-seven inches in length. This enormous line is "shot" across the current at the turn of the tide.

Family VII. Pleuronectĭdæ.—(Gr. Πλευρόν, a rib or side; νηκτόν, power of swimming.)

Psetta.—(Gr. Ψῆττα, a Turbot.)

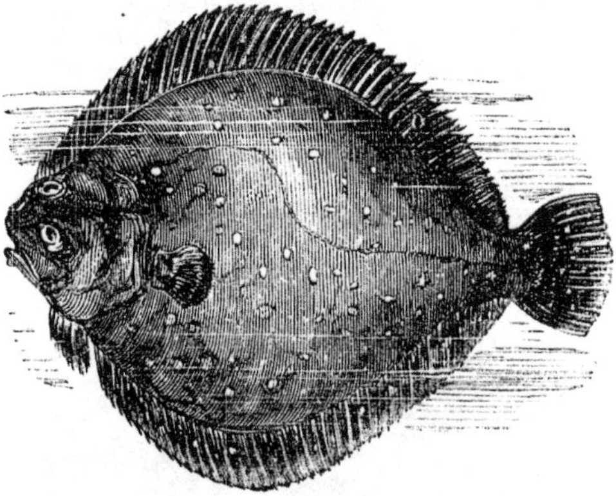

Maxĭma (Lat. *greatest*), *the Turbot.*

Each boat possesses a double set of lines, so that one line is "shot" and another "hauled" every turn of the tide.

The little star-like bones imbedded in the upper part of the skin of this fish are very curious. The dark side of the turbot is the left, on which the eyes are also placed. Reversed turbots, and even turbots dark on both sides, are not at all uncommon.

Solĕa.—(Lat. *the Sole of a shoe.*)

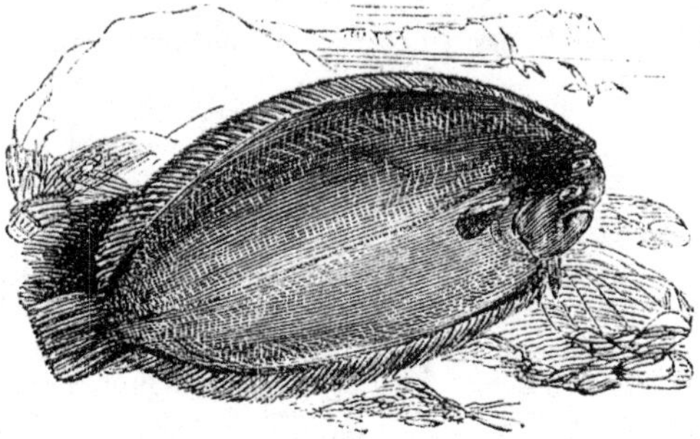

Vulgāris (Lat. *common*), *the Sole.*

The Common Sole is too well known to need much description. This fish is the reverse of the turbot, having the eyes and colour on the right side; although, as in the turbot, varieties are not rare. It is in season during most parts of the year, except a few weeks in March or April. Although it is a marine fish, it seems to thrive well in river-water, or even in a pond. Mr. Arnold kept several in a pond in Guernsey, where the soles became twice as thick in proportion to their length as those living in the sea.

Sub-order III. APŎDA.—(Gr. 'A, privation; πούς, a foot.)
Family IX. . . Murænĭdæ.—(Gr. Μύραινα, a Sea-eel.)

ANGUILLA.—(Lat. *a little Eel.*)

Acutirostris (Lat. *sharp-beaked*), *the Sharp-nosed Eel.*

The EELS form the sub-order of the APODA, or footless fish, so called from the absence of ventral fins.

These fish assume a form very similar to the serpents. Although on a hasty examination they seem to be devoid of scales, yet when the skin is dried, very minute scales may be seen through the semi-transparent outer skin, and may be easily detached by carefully separating the two skins.

Eels inhabit muddy ponds and rivers, and are common in many canals. They are susceptible of cold, and constantly descend the rivers to deposit their spawn in the sea, after which, the young when hatched work their way up the rivers, thereby precisely reversing the habits of the salmon. They are capable of living out of water for a long time, and often make voluntary land excursions, either for the purpose of avoiding an insurmountable fall, or in search of frogs or worms, on which they feed. In the winter, while they are lying torpid in the mud, multitudes are taken by eel-spears—many-pronged instruments, whose prongs are feathered with recurved barbs, which, when pushed into the mud, entangle the eels, and effectually prevent their escape.

There are supposed to be four species of English eel; namely, the Sharp-nosed, the Broad-nosed, the Snig, and the Grig.

The CONGER EEL is found in all the rocky parts of the British coasts, and is exceedingly common on the coasts of Cornwall.

CONGER.—(Lat. *from* Gr. Γόγγρος, a Conger Eel.)

Vulgāris (Lat. *common, the Conger*).

It is usually caught with a hook, the best bait of which is a sand-launce, a little fish belonging to the same family as the eels, and which buries itself five or six inches deep in the sand when the tide ebbs, and releases itself on the next flood tide. The fishermen rake it out of the sand with iron hooks. A pilchard is a common bait for the Conger.

The size of this fish is sometimes very great. Yarrell mentions, in his "British Fishes," that "specimens weighing eighty-six pounds, one hundred and four pounds, and even one hundred and thirty pounds, have been recorded, some of them measuring more than ten feet long and eighteen inches in circumference. They possess great strength, and often form very formidable antagonists if assailed among rocks, or when drawn into a boat with a line."

Family X. Gymnotĭdæ.—(Gr. Γυμνός, naked; νῶτος, the back.)

GYMNŌTUS.

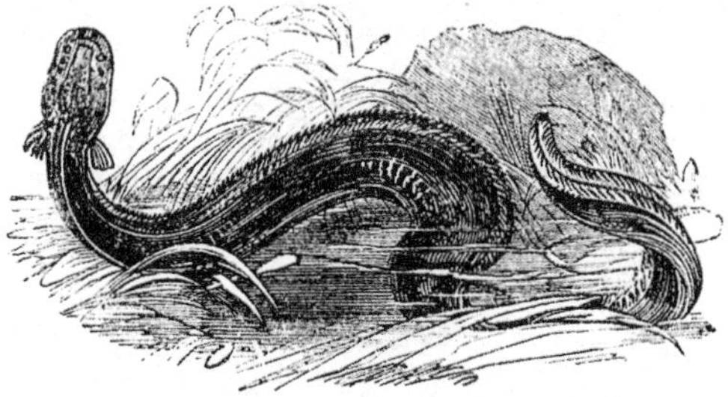

Electrĭcus (Lat. *electric*), *the Electric Eel.*

The ELECTRIC EEL.—This curious fish, which exhibits the singular phenomenon of voluntary electric power residing in a living animal, is an inhabitant of the fresh-water rivers and ponds of Surinam, and other parts of South America, where it was first discovered in the year 1677.

This power of emitting an electric shock, is apparently given it in order to enable the creature to kill its prey. Those who have seen the Electric Eel in the Polytechnic while being fed, will have little doubt of this. The fish given to it are, directly it becomes aware of their presence, instantly struck dead, and then devoured. This specimen is unfortunately blind, but it has learned to turn in the direction of a paddling in the water, made by the individual who feeds it. The fish is scarcely in the water before a shock from the Gymnotus kills it. The usual length of the Gymnotus is about three feet.

Captain Stedman, in his account of Surinam, gives an account of the electric eel, which he, of course, had many opportunities of seeing. He attempted, for a trifling wager, to lift up a gymnotus in his hands, but according to his own words:—

"I tried about twenty different times to grasp it with my hand, but all without effect, receiving just as many electrical shocks, which I felt even to the top of my shoulder. It has been said that this animal must be touched with both hands before it gives the shock, but this I must take the liberty of contradicting, having experienced the contrary effect." The eel mentioned was a small one, only two feet long; but one that had arrived at its full growth would have given a very much stronger shock. An English sailor was fairly knocked down by a shock from one of these eels, nor did he recover his senses for some time. It is said that the shock can pass up a stick, and strike the person holding it. Mr. Bryant and a companion were both struck while pouring off the water from a tub in which an electric eel had been placed.

Humboldt, in his "Views of Nature," gives a very animated description of the method employed by the Indians to take these formidable creatures—a method equally ingenious and cruel. Knowing from experience that the powers of the gymnotus are not adequate to a constant volley of shocks, they contrive that the shocks shall be expended on the horses instead of themselves.

Having found a pool containing electric eels, they force a troop of wild horses to enter the pool. The disturbed eels immediately attack the intruders, and destroy many of them by repeated shocks; but by constantly forcing fresh supplies of horses to invade the pool, the powers of the gymnoti become exhausted, and they are then dragged out with impunity.

The Short Sun-fish.—This order derives its name from the curious structure of the jaws, which are fixed together in a very peculiar manner.

The Short Sun-fish has been frequently taken on almost all parts of our coasts. It is of a most singular shape, looking as if three-fourths of a very large fish had been cut off, leaving only the head and shoulders, something like a marine Baron Munchausen's horse.

Order III. *PLECTOGNĂTHI.*—(Gr. πλεκτός, plaited; γνάθος, the jaw.)
Family I. Diodontĭdæ.—(Gr. Δίς, double; ὀδούς, a tooth.)

ORTHAGORISCUS.—(Gr. Ὀρθαγορίσκος, a Sucking pig.)

Mola (Lat. *a Mill-stone*), *the Short Sun-fish.*

It attains to a very large size, and has been known to weigh three hundred pounds, its length being only four feet five inches.

It lives mostly at the bottom of the sea, but frequently rises to the surface, and lies, perhaps, asleep, floating with the tide. Sailors in this case are fond of trying their skill with a harpoon. When struck, it uses very powerful but exceedingly awkward efforts to escape. The sailors, of course, eat it, as they do almost anything.

The STURGEON.—The remaining fishes belong to the Cartilaginous sub-class; that is, their skeletons are composed of cartilage, and not of true bone.

The first sub-order possess free gill-covers, like those of all the preceding fish; but the remainder breathe by means either of slits, as in the sharks, or holes, as in the lampreys.

The Sturgeon is remarkable for the rows of bony plates extending along the body. It is exceedingly common in the northern parts of

Sub-class II. PISCES CHONDROPTERYGĬI.—(Gr. Χόνδρος, cartilage; πτερύγιον, a fin.)

Sub-order I. ELEUTHEROPŌMI.—(Gr. Ἐλεύθερος, free; πῶμα, a lid or cover.)

Family I. . Acipenserĭdæ.—(Lat. *Acipenser*, a Sturgeon.)

ACIPENSER.

Sturĭo, *the Sturgeon.*

Europe, where regular fisheries are organized for its capture. Almost every part of it is used. Isinglass is obtained by drying and shredding the air-bladder; caviare is made of the roe of the female, and the flesh is extensively preserved both by pickling and salting, besides the large quantities that are consumed fresh. The flavour of its flesh is said not to be unlike veal.

It has occasionally been taken on our coasts, usually by entangling itself in the nets, and although it then does some injury to the nets by its violent struggles to release itself, it is otherwise perfectly harmless. Yarrell mentions that a sturgeon measuring eight feet six inches in length, and weighing two hundred and three pounds, was taken in a stake net near Findhon in 1833. A specimen was once caught in the Esk, weighing four hundred and sixty pounds. The female always deposits her eggs in fresh water, and the young, when hatched, descend to the sea, and are supposed not to return again until, in their turn, they seek the fresh water in order to deposit their spawn.

The SHARKS and RAYS have no gill-covers, but the water passes through five elongated apertures on each side of the head.

The Sharks are proverbially ferocious and dangerous creatures, and are the pest of those seas which they infest. Their mouths are furnished with several rows of sharp jagged teeth, which can be raised or depressed at pleasure, and which can cut through a limb or even the body of a man with the greatest ease. The mouth of these fishes is placed beneath the head, so that a shark cannot seize a prey at the surface of the water without turning on its side, which evolution often gives time for its expected prey to escape.

Sub-order II. TREMATOPNĔI.—(Gr. Τρῆμα, a hole pierced through anything; πνέω, I breathe.)

Family I. . . Scyllĭdæ.—(Gr. Σκύλια, a Dog-fish.)

SCYLLĬUM.

Canicŭla (Lat. *a little Dog*), *the Little Spotted Dog-fish.*

The LITTLE SPOTTED DOG-FISH is the most common of the Sharks that visit our shores. It is principally known on account of the havoc it makes among the fish during the seasons of the various fisheries, for which reason it is most especially detested by the unfortunate fishermen, who not unfrequently, together with their expected spoil, draw up a few dog-fish in their nets. The dog-fish, on finding themselves entangled, immediately commence tearing the nets to pieces with their sharp and powerful teeth.

The empty eggs of this fish are often found washed up on the sea-shore, and called by the name of "mermaids' purses." They are oblong, and furnished at each corner with a long semitransparent convoluted tendril, the use of which is apparently to entangle and fix the egg among the sea-weed, and thus prevent it from being washed on shore until the young is hatched.

A considerable quantity of oil can be obtained from the brain of the dog-fish, and the skin, in common with that of other cartilaginous fishes, is made into shagreen.

The WHITE SHARK is a well-known scourge of the Mediterranean Sea and the Atlantic Ocean. This is the creature so detested by sailors, who, when they have caught a "shirk," subject it to every possible indignity.

This voracious creature has been known to swallow an entire man, and as it is in the habit of lurking about ships for the sake of the scraps thrown overboard, and almost invariably swallows whatever is cast over the side, the contents of its stomach are often of a most heterogeneous

Family II. Squalidæ.

SQUALUS. — (Lat. *a Shark.*)

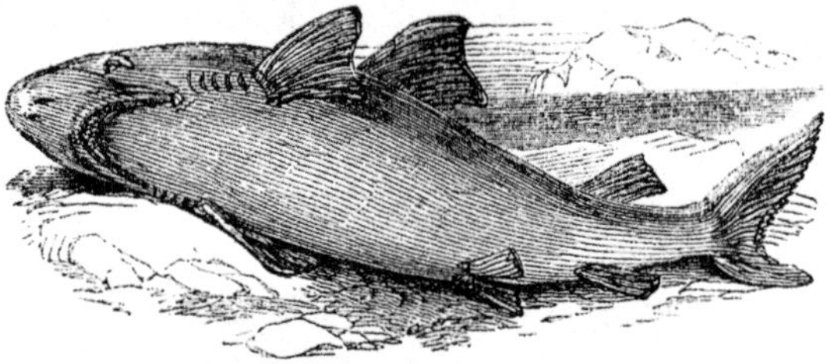

Carcharias (Gr. a Shark; from Καρχαρος, jagged; in allusion to its teeth), *the White Shark.*

description. The sailors always amuse themselves by seeing what the shark had "stowed away," and the substances thus brought to light have been most curious. The entire contents of a lady's work-basket, down to the scissors, were found in the interior of one shark, and another had actually swallowed an entire bull's hide—a circumstance which led the operating sailor to remark that the shark had swallowed a bull, but could not "disgest" the hide.

The amphibious South Sea Islanders stand in great dread of the Shark, and with good reason, for not a year elapses without several victims being offered to the rapacity of this terrific animal. Nearly thirty of the natives of the Society Islands were destroyed at one time by the sharks. A storm had so injured the canoe in which they were passing from one island to another, that they were forced to take refuge on a raft hastily formed of the fragments of their canoe. Their weight sunk the raft a foot or two below the surface of the water, and, dreadful to say, the sharks surrounded them and dragged them off the raft one by one, until the lightened raft rose above the water and preserved the few survivors.

The HAMMER-HEADED SHARK* inhabits the same latitudes. This curiously constructed fish closely resembles the white shark in all respects but the head, which is widened out at each side, exactly like a double-headed hammer or mallet. The eyes, being placed at each extremity of the head, must of course possess a very extended power of vision.

The THRESHER, a fish which has a curious habit of springing out of

* See page 383.

SPHYRNĬAS.—(Gr. from Σφῦρα, a Hammer.)

Zygæna (Gr. Ζύγαινα), *the Hammer-headed Shark.*

the water, and inflicting a violent blow with its tail on any object that annoys it, belongs to the Shark tribe.

Family IV. Pristĭdæ.—(Gr. Πρίστις, the Sawfish.)

PRISTIS.

Antiquōrum (Lat. *of the Ancients*), *the Sawfish.*

The SAWFISH is found in the greatest perfection in the tropical seas, although it also inhabits the Mediterranean. The weapon from which the fish derives its name, is a flat, long prolongation of the head, on each edge of which are set hard tooth-like projections, curiously inserted into the bone.

This fish has been known to employ its saw in the attack of the whale, burying the apparently inappropriate weapon to the very root in the body of the whale; nor are instances wanting where the saw has been found firmly imbedded in the hull of a ship.

The strength of the Sawfish is very great. Captain Wilson gives an account of the capture of a Sawfish, measuring twenty-two feet in length, and weighing nearly five tons. After the fish had been entangled in a net for several hours, making violent efforts to escape, Captain Wilson got a rope firmly fixed round its saw, and set thirty men to haul at the rope. The whole thirty could not move it one inch, nor was it until one hundred men had been pulling at the rope for nearly the whole of the day, that they succeeded in dragging it on shore. Even then it made such violent strokes with its saw, that they were forced to fasten strong guy ropes to prevent it from cutting them to pieces. It was finally disabled by a Spaniard, who cut through the joint of the tail.

Family V. Raïdæ.—(Lat. *Raia*, a Skate or Ray.)

TORPĒDO.—(Lat. *Cramp or numbness.*)

Scutāta (Lat. *shielded*), *the Torpedo.*

The TORPEDO may fairly be considered a British fish. It affords a second instance of the electric power residing in a fish. The organs that produce the electric shock are shown externally by two elevations extending from the eyes about half down the body.*

Although it has once or twice been caught on our coasts, it is usually found in the Mediterranean, where its powers are well known, and held in some awe. The shock that the Torpedo gives, of course, varies according to the size of the fish and its state of health, but a tolerably large fish in good health can, for the time, disable a strong man. From the effects of its shock, it is in some parts called the Cramp-fish.

* Those who would wish to examine the structure of this most singular organ are referred to the Museum of the College of Surgeons, where is a series of beautiful wax models, admirably illustrating the entire structures.

Colonel Montagu notices a Torpedo caught on a turbot line, at Tucky. It weighed about one hundred pounds, and completely puzzled the fisherman, who found it hanging dead on the hooks, and had never seen such a creature before. Colonel Montagu quaintly remarks, that had it not been dead, the fisherman would certainly have had a shock that would have made him remember the species again.

The Rays are at first sight not unlike the turbot and sole, but a closer examination will show that the Rays really swim with their backs upwards, whereas the turbot swims on its side. The movement of the Ray is very curious, and is admirably expressed by the word "sluddering"—used by an old fisherman.

The Skate is caught in abundance on our shores, and in England is in much request as an article of food, although in Scotland it is used principally for bait.

RAIA.

Clavāta (Lat. from *Clavus*, a nail), *the Thornback Skate.*

The Thornback Skate derives its name from the spiny armature of the tail, with which the fish defends itself most vigorously by bending itself almost into a semicircle and lashing about with its tail. The female of the Thornback Skate is termed a Maid. It often attains to a large size, the largest known being twelve feet in length, and nearly ten in width.

The jaws of the Rays are exceedingly powerful, and enable them to crush with perfect ease the various shell-fish on which they feed.

The STING RAY is another species, which is armed with a serrated bone in its tail, with which it can inflict painful and even dangerous wounds.

Sub-order III. CYCLOSTŎMI.—(Gr. Κύκλος, a Circle; στόμα, the mouth.)
Family I. . . . Petromyzonĭdæ.—(Gr. Πέτρος, a Stone; μύζω, I suck.)

PETROMȲZON.

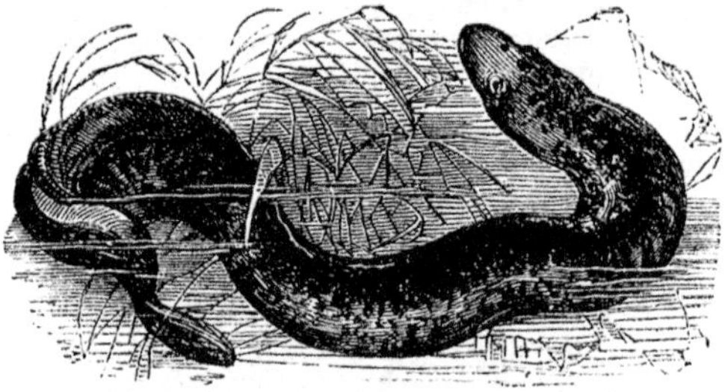

Marīnus (Lat. *Marine*), *the Lamprey.*

The LAMPREY.—These curious fishes, in many respects the lowest in organization of the vertebrate animals, are chiefly remarkable for the singular construction of the mouth, which, formed like that of the leech, enables the Lampreys to hold firmly to any object by suction. The breathing apparatus appears externally to consist of fourteen small apertures, seven on each side of the neck. Their progress through the water is accomplished by a rapid undulating movement.

The MARINE LAMPREY is found in the Mediterranean, and in most of the northern European rivers. It has also been discovered in America. A few are caught in the Thames almost every year, but the Severn is its usual haunt. Like many other fishes, it travels for many miles up rivers for the purpose of depositing its spawn, at which time it is considered to be in the highest perfection.

The spawn is deposited in furrows, some excavated by the parent Lampreys, who, by the help of their sucker-like mouths, rapidly remove even large stones.

The LAMPERN, or RIVER LAMPREY, is plentifully found in many rivers of England. It is extremely common at Ashbourne in Derbyshire, inhabiting the Dove and its tributary brooks. Strange to say, the inhabitants of Ashbourne held it in some abhorrence, and there was only one individual possessing sufficient strength of mind to eat them. He found them a most agreeable addition to his ordinary diet. The Lamperns, or Lampreens as they were called, used to lie in masses of eighteen or twenty together in a hole, and if disturbed, set off down the stream with some speed.

LAMPETRA.—(Gr. Λαμβάνω, I hold; πέτρος, a Stone.)

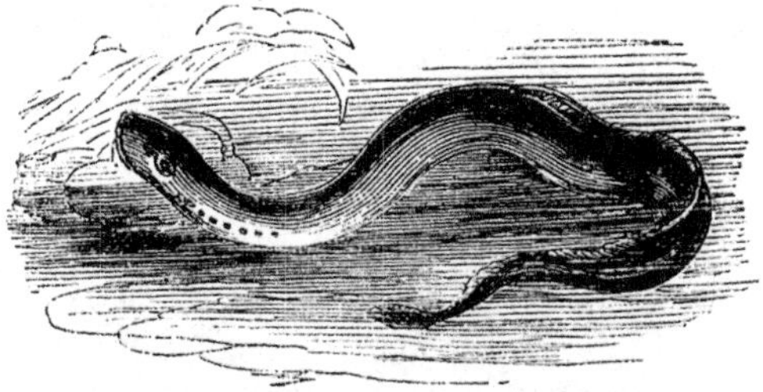

Fluviatĭlis (Lat. *of the river*), *the Lampern.*

It was formerly held in great repute as bait for turbot, cod, and other fish, but in consequence of the diminished supply other substances have been employed. Its length is usually from twelve to fifteen inches. In some counties it is called Seven-eyes, in allusion to the breathing apertures in the neck.

The MYXINE, which, although a decided fish, was classed by Linnæus among the worms, occurs frequently on the eastern coasts of this country. The fishermen find it within the bodies of fish attached to the lines. The Scarborough fishermen call such fish "robbed," as the Myxine, in the course of a single tide, will devour the whole fish, except the skin and bones. It is usually found in the body of the cod.

Family II. Myxinĭdæ.

MYXĪNE.—(Gr. Μυξῖνος, from μύξα, slime.)

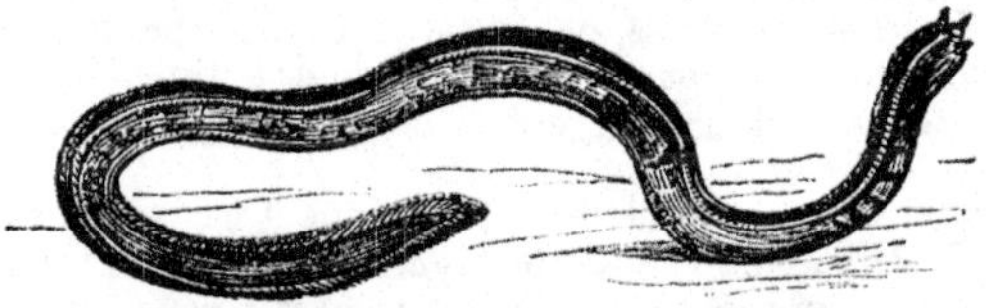

Glutinōsa (Lat. *glutinous*), *the Myxine or Glutinous Hag-fish.*

It is quite blind, but is supposed to derive considerable aid from the eight barbules ranged round its mouth. Six individuals have been found in the body of a single haddock.

Along the under surface of the body are two rows of pores, from which the Myxine is enabled to throw out a most copious gelatinous secretion, apparently for the purpose of escape from its enemies. The length of the Myxine is from twelve to fifteen inches.

Division II. INVERTEBRĀTA.—(Lat. without vertebræ.)
Class VI. . MOLLUSCA.—(Lat. from *mollis*, soft; properly, a soft nut.)
Order . . . *CEPHALOPŎDA*.—(Gr. Κεφαλή, the head; πούς, a foot.)
Family . . Octopĭdæ.—(Gr. 'Οκτώ, eight; πούς, a foot.)

OCTŌPUS.*

Vulgāris (Lat. *common*), *the Cuttle-fish.*

THE MOLLUSCA have neither spine nor bones, the nervous system consisting of a number of nervous knobs called "ganglia," which give off filamentous nerves in different directions.

Few Molluscs possess eyes, but in one or two, as the snails and slugs, those organs are to be found, and in the higher Molluscs, such as the Cuttle-fish, we see not only large and brilliant eyes, but also organs of hearing.

The CEPHALOPODA, so called from the organs of movement surrounding the head, are divided into *naked* and *testaceous*,† or covered with a shell.

The COMMON CUTTLE-FISH is an example of a naked cephalopodous mollusc. This repulsive looking creature is common on our shores, and is, in spite of its unpleasant appearance, often used for food. Its eight long and flexible arms are covered with suckers of various sizes, enabling their owner not only to fix itself firmly to the rocks on which it dwells, but to seize and retain with the greatest tenacity any unfortunate fish or shell that may happen to come within its reach. Its powerful parrot-like beak enables it not only to devour fishes, but even to crush the shells and crustacea that are entangled in its deadly embraces. In this country,

* Or, OCTĂPUS.

† Derived from Lat. *testa*, a shell.

the Cuttle does not grow to any great size, but in the Indian Seas it is absolutely dangerous, and the crews of boats are forced to be armed with a hatchet, to cut off the arms of the cuttle-fish.

There are few who have not heard of the colour called "sepia." This is, or ought to be, prepared from a black pigment, secreted by the Cuttle-fish, and used in order to escape its foes, by blackening the water with the ink, and hurrying off under shelter of the dense cloud of its own creating. Dr. Buckland actually drew a portrait of a fossil Cuttle-fish with some of its own ink that still remained in its body.

The substance sold in the shops as cuttle-fish bone is a chalky substance secreted from the mouth of the fish, and composed of an infinite number of plates, joined by myriads of little pillars.

At a meeting of the Ashmolean Society at Oxford, Dr. Buckland, while exhibiting some relics of a huge fossil Saurian, said, "I know where that fellow lived, I know where he died, and moreover, I know what he had for dinner on the day that he died. He had a cuttle-fish for dinner, and here is its bony ring, which I found in the Saurian's stomach."

The entire body is soft, and encased in a coarse, leather-like skin, unprotected by any shell.

ARGONAUTA.—(Lat. *a sailor in the ship Argo.*)

Argo (Lat.), *the Argonaut or Paper Nautilus.*

The ARGONAUT, or NAUTILUS, is an example of the testaceous Molluscs. This curious creature, about which so many marvellous and poetical tales have been told, is very abundant in the Mediterranean.

It has been clearly proved that the Nautilus does *not* urge itself along the surface of the water by the expanded arms used as sails. These arms are in fact used to cover the shell, and it is from these that the beautful shell is secreted. The Argonaut propels itself through the sea by violently ejecting water from the tube with which it, as well as the cuttle-fish, is furnished for that purpose. The colours of the living animal of the Nautilus are exceedingly beautiful.

The arms of this creature

are furnished with suckers. Its shell, when the poulp (as the living Argonaut is called) is still existing, is flexible and semi-transparent; but when the animal is taken away the shell soon becomes rather opaque and is very fragile.

The fossil Ammonites belong to the testaceous Cephalopoda.

Shells are secreted from a part of the inhabitant called the "mantle," and of course, as the shell is always added round the rim, as may be seen by taking a small snail in the spring, it naturally follows, that as the animal becomes larger, so the mantle becomes larger, and secretes a larger ring of shell.

Many shells, as that of the oyster, are deposited in layers, a fine membrane interposing between each layer: they are therefore called membranous shells. Most membranous shells are lined with a brilliant enamelled substance, called "nacre:" "mother of pearl" is the nacre of the pearl oyster. That of the fresh-water mussel is a beautiful azure.

The other structure of shells is called "porcellaneous," because they look like porcelain or china. The common cowrie is a well-known instance of a porcellaneous shell. Some shells are so transparent as to resemble glass, and are therefore called "vitreous." *

Shells are divided into Univalve, or one-valved shells, such as the snail; and Bivalve, or two-valved shells, such as the oyster. Those of the Univalve Molluscs are capable of protecting themselves when withdrawn inside the shell by a horny plate called the "operculum," † which completely closes up the aperture, and which may be seen in the periwinkle. The closing membrane found in the common snail, if taken in the winter, is called the epiphragma, ‡ and is supposed to be hardened mucus.

The Gasteropoda move by means of a fleshy disc or foot on the under surface of the body, and by the alternate expansive and contractive movements of this foot, the creature is enabled to crawl. The Gasteropoda inhabit both land and water, unlike the bivalves, which are exclusively inhabitants of the latter element.

If the shell of a Gasteropodous mollusc be broken, it has the power of repairing the injury by secreting fresh layers of shell from the mantle within.

The Slugs are well-known invaders of our gardens, and, together with the snail, the caterpillar, and the mysterious "blight," are objects of the gardener's most intense hatred. The Black Slug is usually found by hedge-banks, and in grassy meadows. It seldom ventures out by day, especially if the day be bright; but at night, when the dew is on the ground, it may be seen trailing its dark length through the herbage, or

* Derived from Lat. *vitrum*, glass. † Lat. a cover or lid.
‡ Gr. Ἐπίφραγμα, a cover.

Order . . *GASTEROPŎDA.*—(Gr. Γαστήρ, the belly; πούς, a foot.)
Sub-order, PULMOBRANCHIĀTA.—(*Pulmo*, the lungs; *branchiæ*, gills.)
Family . Limacĭdæ.—(Gr. Λείμαξ, a Slug.)

LIMAX.

Ater (Lat. *black*), *the Black Slug.*

eagerly devouring the leaves. While employed at night in decoying moths, by means of a fragrant compound of sugar, beer, and rum, spread on the trunks of trees, I used constantly to find my bait attacked by huge slugs of all kinds, descending and ascending towards the sweet but dangerous banquet. The small Grey Slug (*Limax cinerea*) is more common in gardens than the black slug.

The COMMON SNAIL.—Several species of Snails inhabit this country, among which the Edible Snail (*Helix pomatia*), the Belted Snail (*Helix nemoralis*), and the common Garden Snail (*Helix aspersa*), are the most conspicuous. The Edible Snail was imported into England by the Romans, who prized them highly, and fattened them in a building erected for that express purpose, as indeed is now done in some parts of the Continent. This snail grows to a large size, nearly attaining the magnitude of an ordinary closed fist.

Family, Helicĭdæ.—(Gr. Ἕλιξ, twisted.)

HELIX.

Aspersa (Lat *sprinkled*), *the Common Snail.*

The eyes of the snail are placed at the extremity of the tentacula, or "horns" as they are usually called.

The common garden snail is so well known that no description of it is needed. It lays eggs very large in comparison with the size of the parent; they are about the size of small peas, round, soft, and semi-transparent. They are deposited about two inches below the surface of the earth.

This creature is very tenacious of life. A living snail was exhibited at the Ashmolean Society at Oxford, which had made a long sea voyage, packed up in cotton wool. An immersion in water soon brought the inhabitant to view, and when it was exhibited it was crawling about a box in perfect health.

Family, Turbinĭdæ.—(Lat. *Turbo*, a Whorl.)

SCALARĬA.—(Lat. *Scala*, a Ladder, or stairs.)

Family, Coniĭdæ.—(Gr. Κῶνος, a Cone.)

CONUS.

Pretiōsa (Lat. *valuable*), *the Royal Staircase Wentletrap.*

Generālis (Lat. *general*), *the Cone.*

The ROYAL STAIRCASE WENTLETRAP affords us an excellent and most beautiful example of the Turbinidæ. It is a native of the Chinese and Indian seas, and was formerly so scarce that a specimen two inches in length would sell for a hundred pounds. Even now, a very fine specimen cannot be obtained under six or seven pounds. For this reason, the specific name "pretiosa" was affixed to it by Lamarck.

As an example of the large family of CONES, we will take the common Cone, whose beautiful marbled colour and elegant shape render it a most attractive shell.

The MONEY COWRY and the WHELK.—The Cowries are not less celebrated for the elegance of their form, and the beauty of their markings, than for the curious circumstance that one species is used as current coin in Guinea and Bengal, thus being employed for the same purpose by two entirely distinct races of men, situated in different quarters of the globe. Their value is of course small in proportion to gold or silver. At the present time a rupee in Bengal is worth 3,200 Cowries, the value of the rupee being 2*s*. 3*d*. of our money.

The BUCCINIDÆ are so named from their fancied resemblance to a trumpet. The common Whelk is everywhere abundant on our coasts,

Family, Cypræïdæ.—(Gr. from Κύπρος, Cyprus.)

Family, Buccinïdæ. (Lat. *Buccina*, a Trumpet.)

ARICÏA.—(Lat.)

BUCCÏNUM.

Monēta (Lat. *the stamp on money*), *the Money Cowry.*

Undātum (Lat. *wavy*), *the Whelk.*

and is taken in such profusion that it is largely exported for food, and may be seen on the street stalls of the metropolis exposed for sale, like the oyster and periwinkle.

The proboscis of this creature is of a most singular structure, and by means of the numerous teeth with which it is armed, it is able rapidly to bore its way through shells, and then to feed upon the unfortunate inmate. The hermit crab often takes possession of the empty shells of the Whelk.

The famous Tyrian purple was obtained from one of the Buccinidæ, *Purpura imbricata.*

The beautiful THORNY WOODCOCK,* sometimes called by the name of Venus' comb, is an excellent example of the Muricidæ. This elegant shell is an inhabitant of the Indian Ocean.

The LIMPETS are spread over every latitude, except the Arctic regions. The common Limpet† is to be found on every rock and large stone at the sea-side. The mode of its attachment to the rocks is very curious, and well repays a careful examination. Every one who has seen a living

* See page 394. † Ibid.

Family, Muricĭdæ.—(Lat. *Murex*, the purple shell-fish.)

MUREX.

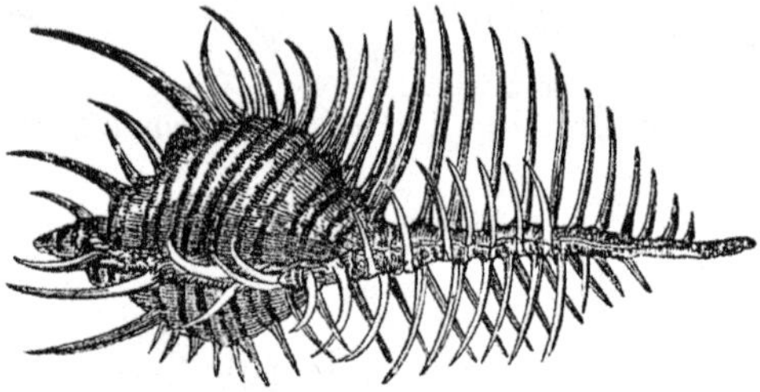

Tribŭlus (Lat. *a Thistle*), *the Thorny Woodcock.*

limpet knows how firmly it fixes itself to the rock. This is done by the inhabitant creating a vacuum on the under surface of its body, which causes the pressure of the atmosphere to keep it so tightly fixed to the

Order, *CYCLOBRANCHIĀTA.*—(Gr. Κύκλος, a circle; βράγχια, gills.)
Family, Patellĭdæ.—(Lat. *Patella*, a Porringer.)

PATELLA.

Vulgāta (Lat. *made common*), *the Limpet.*

rocks, that a blade of a strong knife is required to detach it. Frequently the margin of the shell adapts itself to the shape of the substance to which it adheres, proving that it must remain fixed in the same spot for a long time, and rendering it difficult to imagine from whence it can obtain sufficient nourishment to support life.

Sometimes a large shell may be picked up covered with limpets, that adhere firmly to it in spite of the rolling of the waves, and the tossings about to which it must necessarily be subjected.

Order, *CONCHIFĔRA.*—(Gr. Κόγχη, a Mussel-shell; φέρω, I bear.)
Family, Pectinĭdæ.—(Lat. *Pecten*, a Scallop.)

PECTEN.

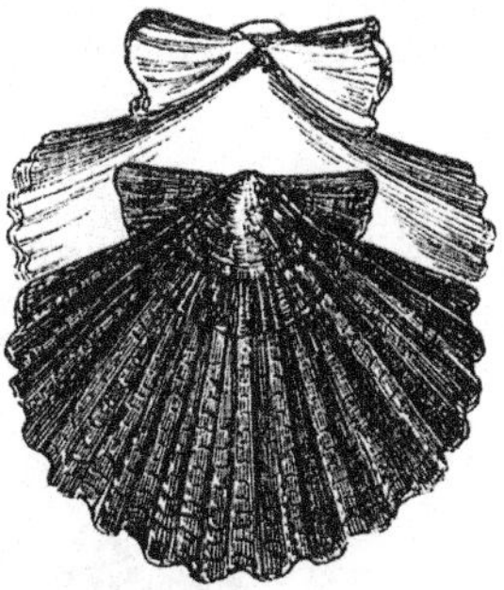

Jacobæus (Lat. *from a proper name*),* *the Scallop.*

We now arrive at the BIVALVE MOLLUSCS. It has been already stated that the Bivalves are all aquatic. These creatures are enabled to keep their shells firmly closed by means of a powerful muscle. Those who have attempted for the first time to open an oyster, must be convinced of the strength of this muscle. The two shells are united by a powerful and extremely elastic hinge, which after the death of the animal opens the shells widely.

The Bivalves do not enjoy such powers of locomotion as the Univalves, yet some, as the fresh-water mussel, can urge themselves along by means of a fleshy organ called the foot; and so powerful in some is this organ, that by means of it the animal can not only burrow in the sand, but actually leap out of a boat. The rapid opening and shutting of the valves is used by some, as the scallop, as a means of progression. It is believed that the Bivalves have no visual organs.

The common SCALLOP is found along our southern coasts, and in the seas of Europe. This shell was formerly used as the badge of a Pilgrim to the Holy Land.

> "———His pilgrim's staff he bore,
> And fix'd the Scallop in his hat before."

It is a singular fact, that in the stomach of the common Scallop is found an earthy deposit, which, when boiled in nitric acid in order to dissolve the animal and other portions, exhibits under a powerful microscope animalcules precisely similar to those which, in a fossil state, form the earth on which the town of Richmond in America is built.

The COMMON OYSTER has been for many ages considered as a delicacy for the table. In the times of the ancient Romans, we find that our "Native Oysters" were exported to Rome, and there placed in the Lucrine Lake, where they were fattened.

On our coasts the oysters breed in large beds, to which vast quantities of young oysters are conveyed by the fishermen, and suffered to increase without molestation. Newly-formed beds are untouched for two or three years. During the months of May, June, and July,† the oysters breed,

* Because the pilgrims to the shrine of St. James (Sancti Jacobi), called by the Spaniards Santiago, also bore the Scallop.

† Most people are acquainted with the proverb, that oysters are in season during the months in which is the letter R.

and are considered unfit for food. At this time the young, called "spat," are deposited in enormous numbers. They instantly adhere to the substance among which they fall; and this, whatever it be, is called "cultch," and is protected by severe penalties. About May the fishermen separate the spawn from the cultch, which is then thrown back into its former

OSTRĔA.—(Gr. Ὄστρεον, an Oyster.)

Edūlis (Lat. *edible*), *the Oyster.*

place. After May it is felony to disturb the cultch, as were it removed, mussels and cockles would rapidly take the place of the oysters.

The oysters are taken in the proper season by the "dredge," a kind of small net fastened round an iron frame-work, which scoops up the oysters and many other marine animals.

The part of the oyster called the "beard," is in reality the respiratory apparatus.

The PEARL OYSTER is the animal from which those highly valued ornaments, pearls, are extracted. The pearl is nothing more than "nacre," deposited in the shape of globular drops instead of being spread over the inner surface of the shell, in which case it is known as Mother-of-Pearl.

These valuable shells are found both in the Old and New World. Ceylon is very famous for its pearl fisheries. The fishermen are trained to remain a long time under water, and are assisted in their descent to the bottom of the sea by a heavy weight tied to their feet. They rapidly gather all the Pearl Oysters in their way into a basket, and when in want of air, give a signal to their friends above, who draw them to the surface by a rope. The Oysters are then left to putrefy for some weeks, when they are carefully washed, and the pearls extracted.

Family, Meleagrinĭdæ.—(Gr. Μελεαγρίς.)

MELEAGRĬNA.

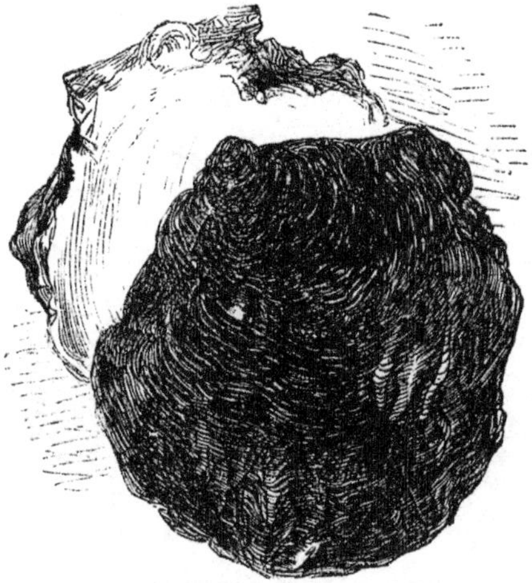

Margaritifĕra (Lat. *Margarita*, a pearl; *fero*, I bear), *the Pearl Oyster.*

The Chinese have a method of forcing oysters, or rather mussels, to form pearls, by artfully placing beads in their shells, round which a layer of nacre is deposited, and the beads then perfectly resemble real pearls.

Family, Mytilĭdæ.—(Gr. Μυτίλος, a Mussel.)

MYTĬLUS.

Edūlis (Lat. *edible*), *the Edible Mussel.*

The SEA MUSSELS are usually fixed where the tide leaves them alternately wet and dry, and it is worthy of notice that those "shell-fish" which are exposed to variations of this kind are enabled to close their shells so firmly as to prevent any evaporation. One species is extensively used as an article of food.

The river mussels occasionally produce pearls of some value. The nacre of these mussels is of a beautiful azure blue.

Order, *CIRRHOPŎDA*.—(Lat. *Cirrus*, a lock of hair; *pes*, a foot.)

PENTALASMIS.—(Gr. Πεντέ, five; ἔλασμα, a plate.)

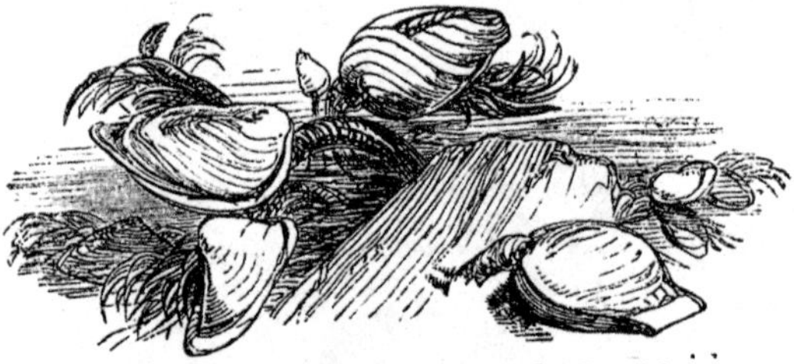

Anatifĕra (Lat. *Goose-bearing*), *the Bernicle.*

The BERNICLE.—At first sight, the Bernicle bears a close resemblance to a mussel-shell fixed to a long stem. On a closer examination, however, the difference is at once apparent. The shell is in fact composed of *five* pieces, and through the aperture of the shell are thrust two rows of arms, or "cirrhi," as they are more properly called. These cirrhi serve to entangle the small crustacea or molluscs which pass near their sphere of action, and which are then carried to the mouth and speedily devoured.

The Bernicle is always found adhering to some larger object, usually floating wood, and is very common on the hulls of ships. Although the perfect animal is permanently fixed, it has been discovered that the young are free and capable of locomotion; nor is it until a week or two has passed, that they finally settle themselves.

The name Anatifera, or Goose-bearing, has been given to this animal on account of the ancient story of the production of the Bernicle-goose. This fable has already been related under the article, Bernicle-goose.

Class VII. . CRUSTACĔA.—(Lat. *Crusta*, a crust or shell.)
Sub-class I. MALACOSTRĂCA.—(Gr. Μαλακός, soft; ὄστρακον, a shell.)*
Order I. . . *DECAPŎDA*.—(Gr. Δέκα, ten; πούς, a foot.)
Sub-order I. DECAPODA-BRACHYŪRA.—(Gr. Βραχύς, short; οὐρά, a tail.)
Family I. . Cancerĭdæ.—(Lat. *Cancer*, a Crab.)

CANCER.

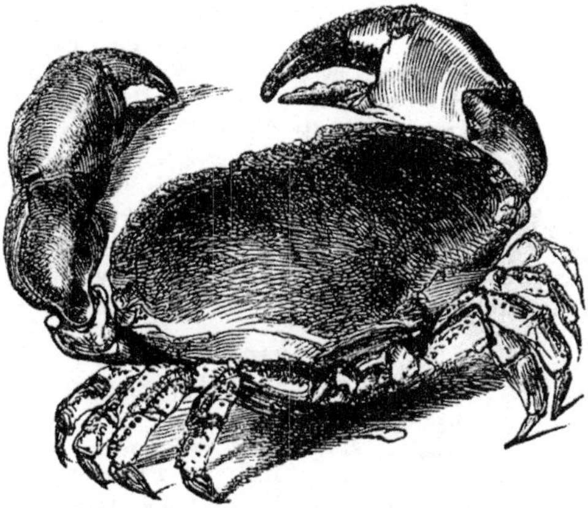

Pagūrus (Gr. πήγνυμι, I fix; οὐρά, a tail), *the Crab.*

THE CRUSTACEA are almost all aquatic animals. They have no internal skeleton, but their body is covered with a strong crust, which serves for protection as well as for strength. Their whole framework consists of a series of rings fitted to, and working in each other; some forming limbs, and others developing into the framework supporting the different organs. From this reason, they and the remaining animals, as far as the star-fishes, who have no limbs at all, are called "articulated" animals.

Their method of growth is very curious. Other animals, as they increase in size, experience no particular inconvenience. Not so the Crustacea. Their bodies are closely enveloped in a strong, unyielding mail, which cannot grow with them. Their armour is therefore cast off every year, and a fresh coat formed to suit their increased dimensions. Not only is the armour cast off, but even the covering of the eyes, the tendons of the claws, and *the lining membrane of the stomach, with its teeth.*

They all also possess the curious power of reproducing a lost or injured limb. In the former case, a fresh limb supplies the place of that lost; and in the latter case, the animal itself shakes off the injured joint, and

* So called because their shell is soft compared with that of the univalve or bivalve molluscs.

a new one soon takes its place. Lobsters, when alarmed, frequently throw off their claws.

The Decapods, as their name imports, are the fortunate possessors of ten legs, five at each side. They also possess three pairs of jaws, besides the teeth in the stomach. They breathe by means of branchiæ or gills, fixed at each side of the throat or chest, often erroneously called the head.*

The COMMON CRAB belongs to the short-tailed Decapods. It is abundantly taken on our coasts by fishermen, who employ for its capture a wicker basket called a "creel" or crab-pot. The crab-pots are made each with an aperture which permits the animal to enter, but forbids its egress—just like a common wire mouse-trap. A piece of a fish is fastened at the bottom of the creel, and the whole apparatus let down to the bottom of the sea, guarded by a line connected with a float, by means of which the fishermen draw it up and then remove its contents. Each float has a peculiar mark, by which the fisherman knows his own. When taken, the crabs are kept alive in well-boats, until wanted.

Sub-order II. DECAPŎDA-ANOMOŪRA.—(Gr. Ἄνομος, unaccustomed; *οὐρά*, a tail.)

Family III. Pagurĭdæ.—(Gr. Πήγνυμι, to fix; *οὐρά*, a tail.)

PAGŪRUS.

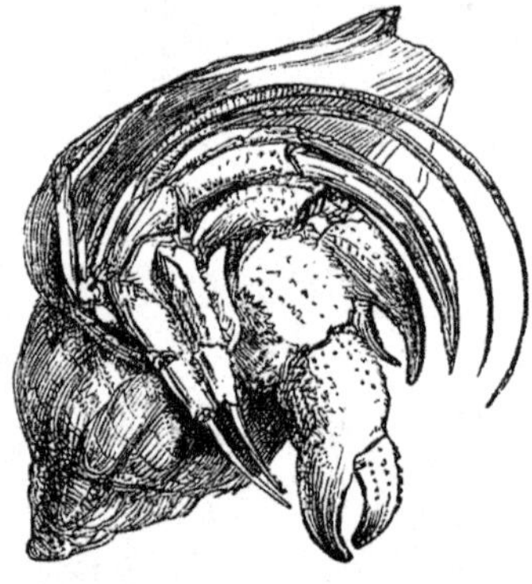

Bernhardus (Lat. *proper name*), *the Hermit Crab.*

The HERMIT CRAB is not so well protected as most of his relations, for his tail has no shelly armour. He is therefore forced to protect his undefended tail by putting it into an empty shell, usually that of a whelk, and then walks about, dragging his curious house after him. Sometimes, two hermit crabs wish to obtain possession of the same shell, and then there is a battle royal. When the crab grows larger, he only has to change his old shell for a new one, and it is very amusing to see these creatures slipping their tails first into one shell, and then into another, until they have pleased themselves with a good fit.

They are very common on our coasts, and may be found of all sizes, from the crab that fills a tolerably large whelk shell, to the little one whose habitation hardly exceeds the size of a pea.

The Land Crabs make annual excursions to the sea in large armies

* These animals have no distinct head; that and the thorax being merged into what naturalists call "cephalo-thorax," or head-thorax.

They go straight forward, and nothing except a house or such insurmountable barrier can stop them. Those of Jamaica are particularly celebrated.

Sub-order III. DECAPŎDA-MACROURA.—(Gr. Μακρός, long; οὐρά, a tail.)
Family V. . . Astacĭdæ.—(Gr. Ἀστακός, a Lobster.)

POTAMOBĬUS.—(Gr. Ποταμός, a river; βιόω, to live.)

Astăcus *the Cray-fish.*

The long-tailed crustaceans include the Lobster, Shrimp, &c.

The RIVER CRAY-FISH is common in most of our rivers and brooks. It resides in holes in the bank, sometimes excavated by itself, but more often the deserted habitations of water-rats. In rocky situations it lives under and among the stones. The excellence and delicacy of its flesh causes it to be much sought after. The usual method of catching these animals is by lowering a net to the bottom of the water, baited with a piece of meat. The cray-fish soon discover this and come in numbers to the bait, when the net is suddenly hauled up, and most of the cray-fish secured. Some, however, escape by darting off backwards, a movement produced by the violent bending of their tails. It is a favourite amusement with boys to search for them in their holes, and drag them from their concealment.

The common LOBSTER is found in great abundance on our coasts, usually in the clear rocky waters. The fishermen take great numbers of lobsters in baskets made on the same principle as those used for the capture of the crab. The powerful tail of the lobsters enables them to spring through a great distance if alarmed, and they have been seen to pass nearly thirty feet. They direct their course with wonderful accuracy, and can throw themselves through apertures hardly larger than the size of their bodies: of course they spring tail foremost.

The grasp of the lobster's claw is so tight that to break off the claws is often the only method of disengaging its hold.

ASTĂCUS.

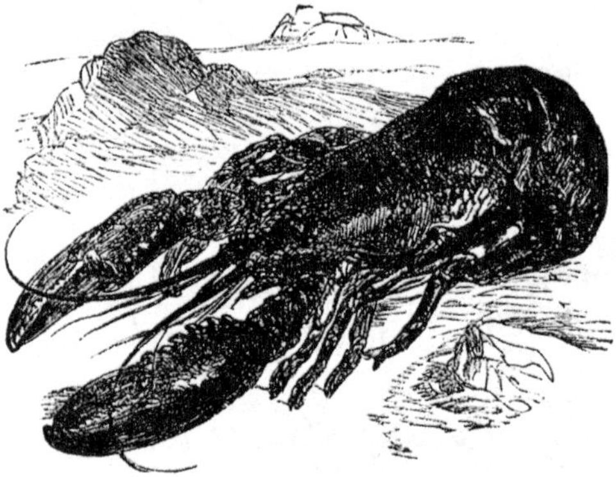

Gammărus (Lat.) *the Lobster.*

Although enormous quantities are destroyed every year, they are so prolific that the supply never fails.

The so-called lady's-fingers of the Lobster are its breathing apparatus.

The PRAWN and the SHRIMP are so familiar to every one as to need but little description. Both are taken in nets swept along the sandy bottom of the sea. The chief distinction in the appearance of these two creatures is the serrated or toothed ridge which runs along the back of the head, or rather carapace, of the Prawn. When in their natural state, they are of a brown colour, and only assume the pinkish hue when boiled. Spirits of wine has the same effect.

The Fresh-water Shrimp (*Gammarus Pulex*), and the Water-Flea (*Daphnia Pulex*), both so common in our rivers and ponds, are placed among the Crustaceæ.

Family VI. Crangonĭdæ.—(Gr. Κραγγών, a Shrimp.)

Family VIII. Palæmonĭdæ.—(Gr. Παλαίμων, a proper name.)

CRANGON. PALÆMON.

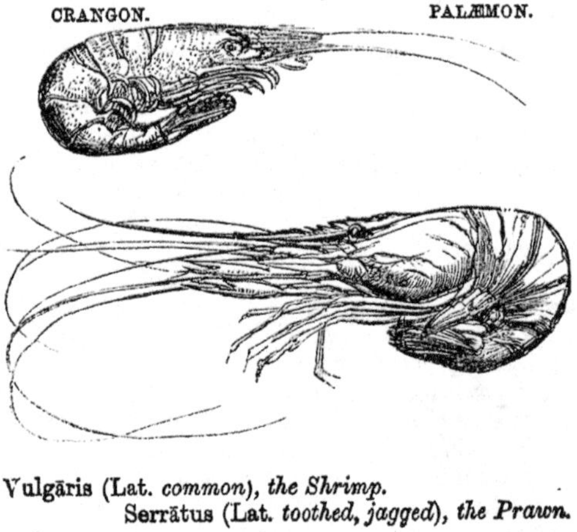

Vulgāris (Lat. *common*), *the Shrimp.*

Serrātus (Lat. *toothed, jagged*), *the Prawn.*

Class VIII. ARACHNĬDA.—(Gr. Ἀράχνης, a Spider.)
Order . . . *PULMONARĬA.**—(Lat. *Pulmo*, the Lungs.)
Family . . Araneĭdæ.—(Lat. *Aranĕa*, a Spider.)

MYGĂLE.—(Gr. Μυγάλη, the Shrew-mouse.)

Avicularĭa (Lat. Avicŭlus, *a little Bird*), *the Bird Spider.*

The Class ARACHNIDA, or the Spiders, are by many supposed to be insects. Such, however, is not the case. The Arachnida possess eight legs, while the true Insects only have six; they undergo no transformations, they possess no wings or antennæ (the place of the latter organs being supplied either by two jointed claws, as in the Scorpions, or by two fangs, as in the Spiders); and their eyes are simple instead of compound.

Could people divest themselves of the horror felt at the sight of these creatures, especially of the larger sort, they would be well repaid by the interesting instinct displayed by all the Spiders, who do not differ from each other more in form than in habits. Those of our own country afford an ample field, which has been as yet but imperfectly trodden. There are the Gossamer Spiders, who float high into the air, borne upon an almost invisible thread; the Water Spiders, who form an air-tight

* So called, because the animals belonging to this class breathe by means of air-sacs, called by Latreille, Pneumobranchiæ, or lung-gills. The Trachearia, on the contrary, breathe by means of tracheæ, or air-tubes branching through the whole system, like the insects.

dwelling under the wave; the Hunting Spiders, that creep stealthily upon their prey, and then spring on it like lightning; the beautiful Garden Spiders, who weave from their self-afforded stores their geometrical nets; the Pirate Spiders, who skim over the surface of the waters, and snatch up the drowning and helpless fly; together with many others, whose form and habits must be familiar to any observer of Nature.

On account of the limited space that can be appropriated to each Class, a short account of some of the principal species of this Class is all that can be given.

The enormous Spider represented above is a native of Surinam, and was brought into notice by that indefatigable naturalist, Madame Merian. Her account of it is very short. She relates that it carries about with it a habitation, resembling the cocoon of some of the moths, and that it is armed with sharp fangs and inflicts dangerous wounds, at the same time injecting into the wound a poisonous liquid. She also tells us that it feeds principally upon ants, but that in their absence it drags little birds out of their nests, and then, as she pathetically observes, "sucks all the blood out of their poor little bodies." * Here, however, it is generally supposed that Madame Merian has been imposed upon, as is evidently the case in another portion of her work, where she has drawn a curious insect, compounded of the head of a lantern-fly, and the body of a cicada. She seems to have had her doubts on the subject, for she takes care to say, "The Indians told me." †

THE GARDEN SPIDER.

* "Formicarum defectu, ipsas e nidis tollunt aviculas, omnemque corpusculis cruorem exsugunt."—Merian. Met. Insect. Sur. p. 18.

† "Persuasum mihi ab Indis est."—Id. p. 49.

The common Garden, or Geometrical Spider, as it is called from the mathematical regularity of its net, is an excellent example of the Spiders. The net is formed from a gummy substance secreted in an apparatus called the spinneret, through the holes of which the gummy secretion is drawn, and becomes hard when exposed to the air. Each thread is composed of many thousand lines. When the web is completed, the Spider generally hides itself under a leaf or other convenient lurking-place, and from thence pounces upon any unwary fly that has entangled itself in the slender meshes. Should the fly be a large one, the Spider rapidly encircles it with fresh threads until it has bound its wings and legs to the body, and then breaking off the few threads that held it to the net, bears it off triumphantly to its hiding-place. Frequently the Geometrical Spider sits in the centre of the web, apparently enjoying the air, and if disturbed shakes the net so violently that its shape is completely obscured by the rapidity of the vibrations.

The House Spider makes a thicker and irregular web, and hides itself at the bottom of a silken tunnel communicating with the web. An acquaintance of mine had so far tamed a huge house spider, that it would come and take a fly out of his hand. He states, that as it sat at the bottom of its den, its eyes gleamed like diamonds.

Several endeavours have been made to procure silk from spiders, but although a sufficient quantity has been obtained to weave gloves from, yet spiders are so pugnacious that they cannot be kept together. The eggs of the Spiders are enclosed in a silken bag, and when hatched, the young keep closely together, and when dispersed by an alarm, soon re-assemble.

The Tarantula, whose bite was fabled to produce convulsions which could only be appeased by music, is a spider of considerable size, inhabiting the south of Europe. It lives in holes about four inches deep in the ground.

The Scorpion.*—These formidable creatures inhabit most of the hotter parts of the globe. They are quite as pugnacious as the spiders, and if several are placed in one box, they will fight until few survive, who immediately devour their fallen foes.

The maxillæ of the Scorpion are developed into large claws, like those of the lobster. With these, the Scorpion seizes its prey, and while holding it pierces it with its sting, which is situated at the extremity of its tail. The tail is composed of six joints, rendering it very flexible.

The sting of this creature is exceedingly painful, and with some persons dangerous; indeed, the sting of the large black Scorpion of Ceylon is said to cause death.

* See page 406.

Family, Scorpionĭdæ.—(Lat. *Scorpio*, a Scorpion.)

SCORPĬO.

Europæus (Lat. *European*), *the Scorpion.*

Order, *TRACHEARĬA.*—(From *Trachēa*, a word used to represent the tubes through which insects aud other animals breathe;—the windpipe.)

Family, Acarĭdæ.—Gr. Ἄκαρι, a Mite.)

LEPTUS.—(Gr. Λεπτός, *small.*)

Autumnālis (Lat. *belonging to Autumn*), *the Harvest-bug.*

The HARVEST-BUG.—These creatures are mostly minute, requiring the aid of a microscope fully to develop their form; but some are considerably larger, and their organs can be distinguished with the naked eye. In this order are included the common cheese-mite, the harvest-bug, the water mites, &c.

Class IX. . INSECTA.—(Lat. *Insecare*, to cut into.)
Sub-class I. INSECTA MANDIBULĀTA.—(Lat. *mandare*, to chew.)
Order I. . . *COLEOPTĔRA*.—(Gr. Κολεός, a sheath; πτέρον, a wing.)
Family I. . Cicindelĭdæ. (Lat. *Cicindēla*, properly, *a Glowworm*.)

CICINDĒLA.

Campestris (Lat. *of the fields*), *the Tiger-beetle.*

The TIGER-BEETLE.—The body of an insect is divided or *cut into* three parts, called the head, the thorax, and the abdomen. The body is defended by a horny integument, divided into rings and connected by a softer membrane. The legs are six in number. Many insects possess wings, and in all the rudiments of those organs are perceptible. The eyes are compound, that is, a number of eyes are massed together at each side of the head; and so numerous are they, that in the compound eyes of the ant are 50 lenses, in the house-fly, 8,000, in the butterfly, 17,000, and in the hawk-moth, 20,000.

The insects pass through three transformations before they attain their perfect form. The first state is called the *larva*,* because the future insect is masked under that form; the second is called the *pupa*,† on account of the shape often assumed; and the third is called the *imago*,‡ as being the image of the perfect creature. Insects breathe by means of air-tubes, called tracheæ, which penetrate to every part of the body, even to the extremities of the limbs, antennæ and wings. The air gains access to the tubes by means of small apertures called spiracles. The tubes are prevented from collapsing by a delicate thread wound spirally between the two membranes of which the tubes are composed. This wonderful and beautiful arrangement not only prevents the tubes from collapsing, but keeps them flexible. There are, according to Stephens, whose arrangement is the one usually followed, fourteen orders of insects. Examples will be given of each, and their names explained. The most perfect insects are placed first.

* From Lat. *Larva*, a mask. † From Lat. *Pupa*, a doll.
‡ From Lat. *Imago*, an image or effigy.

There are two great divisions of insects, namely, those which bite and eat solid food with jaws, as the beetles, locusts, bees, &c., and those which suck liquid food through a proboscis, as the butterflies, flies, &c. The first order of insects derives its name from the sheath or covering with which the wings are defended.* This is a very extensive order, as, exclusive of exotic and other foreign beetles, it has been discovered that no less than three thousand five hundred inhabit this country. The first in order of the British insects, are the Tiger-beetles, so called from their activity and voracity. The most common of these is the ordinary Green Tiger-beetle, that may be seen any hot summer's day, glancing in the sun on sandy banks. The exceeding beauty of this insect is beyond all description. The upper surface of the body is a deep, dead green, changing under the microscope to a glossy gold, shot with red and green; the surface of the abdomen covered by the wings, and the entire under-surface of the body, are brilliant emerald green, and when the insect is on the wing it sparkles in the sun like a flying gem. Nor is this the last of its attractions, for when handled it gives forth a scent closely resembling that of the verbena. It is indeed as beautiful among insects as the tiger is among beasts, and is, perhaps, the more ferocious of the two. It runs and flies with great activity, and takes to the wing as easily as a bee or fly, and is in consequence rather difficult to capture without a net. Its jaws are long, sharp, curved like a sickle, and armed with several teeth. Its eyes are large and prominent, enabling it see on all sides. Its length is rather more than half an inch.

Family, Carabĭdæ.—(Gr. Κάραβος, a Beetle.)

CARĂBUS.

Cancellātus (Lat. *chequered*), *the Ground-beetle.*

The GROUND-BEETLE is one of our largest and most beautiful beetles. Its general colour is a coppery green, and its wing-cases are ornamented with several rows of oblong raised spots. Its length is about an inch.

* This, as well as the general covering of insects, is composed principally of a substance called by chemists, *chitine*.

The BURYING-BEETLE.—This curious beetle derives its name from it habit of burying any small dead animal left on the surface of the ground. With such rapidity does it work, that two beetles have been known to cover up a sparrow within a few hours; and so unwearied are they, that if several Burying-beetles are placed in a vessel filled with earth, and kept constantly supplied with dead frogs, mice, &c., they will continue to bury them as long as the supply is kept up. The object of this remarkable instinct, so beneficial in its effects, is to furnish food for the young who are hatched from eggs laid in the body of the animal during its burial. In this way innumerable carcases, which would pollute the atmosphere, are removed, and made beneficial to the soil.

Family, Silphĭdæ.—(Gr. Σίλφη, a Burying-beetle.)

NECROPHĂGUS.—(Gr. Νεκρός, a dead body; φαγεῖν, to eat.)

Vespillo (Lat. *a bearer of the dead*), *the Burying-beetle.*

The LAMELLICORN BEETLES are exceedingly useful to mankind. Many of them act as scavengers and farmers, for they not only remove putrefying substances from the surface of the ground, but bury them beneath.

The STAG-BEETLE is the largest of British insects. Although so formidably armed, it is quite harmless, and only uses its enormous jaws to break the tender bark of trees, in order that the sap on which it feeds may exude. The mouth of this beetle is very small, and is furnished with a brush, with which it licks up the food. Several of these beetles lived for some time on moist sugar. During the winter, it hides in the earth, making for itself a kind of cave, very smooth inside.* This beetle is common in the New Forest.

The DOR-BEETLE is a very common English insect. At the approach of evening, it may be seen whirling round in the air with a dull humming sound. The country children call it the Watchman, comparing it to a watchman going his rounds in the evening. It usually lays its eggs on a rounded mass of cow-dung, and then buries the whole mass in the ground. When caught, it pretends to be dead.

The Dor-beetle is very tenacious of life. I have now in my cabinet a specimen of this insect, which I took on the wing. It had lost several

* In the Ashmolean Museum, Oxford, is an excellent specimen of the winter habitation of this beetle, with the beetle itself enclosed.

legs, one wing-cover or elytron, the *whole* of the contents of the abdomen, and part of the thorax. I suppose that a bird must have been eating it, and have been disturbed, for when thrushes, blackbirds, jackdaws, &c.,

Family, Lucanĭdæ.—(Lat. *Lucanus,* the Stag-beetle.)

LUCĀNUS.

Family, Geotrupĭdæ.—(Gr. Γή, the Earth; τρυπάω, to bore.)

GEOTRUPES.

Family, Melolonthĭdæ.—(Gr. Μηλολόνθη, a Cockchaffer.)

MELOLONTHA.

Cervus (Lat. *a Stag*), *the Stag-beetle.*

Stercorarĭus (Lat. *Dung*), *the Dor-beetle.*

Vulgāris (Lat. *common*), *the Cockchaffer.*

eat large beetles, they begin by picking off the wings, limbs, &c. I also took, in May 1852, a cockchaffer walking along very unconcernedly, who had lost both his wings and elytra, and all the contents of the abdomen.

The COCKCHAFFER needs not much description. Its larva works great mischief during the spring, as it feeds on the roots of plants, and cuts them off with its sharp sickle-like jaws. Where many of these "grubs" have been, the grass curls up, and dries like hay. One farmer actually collected eighty bushels of the grubs of the Cockchaffer on his farm. Fortunately the thrushes, blackbirds, rooks, and many other birds, are inveterate destroyers of the grubs, and devour myriads of them. It is for this purpose that these birds pull up the grass, and not to spoil or devour the herbage, as is generally supposed.

The huge Hercules and Atlas Beetles, and larger still, the Goliath Beetle, belong to the Lamellicorns.

The GLOWWORM may be seen in the warm summer evenings, shedding its pale green light on grassy banks. The female insect gives out a much stronger light than the male, and there is some light visible even in the larva. The light of this insect proceeds from the abdomen. The light given out by the firefly, another kind of beetle inhabiting South America, proceeds from three yellow tubercles placed on the throat. The grub or larva of the Glowworm is of a singular form, and is furnished with a brush at the extremity of the tail, with which it cleanses its body from dust or the slime of the snails on which it frequently feeds.

Family, Lampyrĭdæ.—(Gr. Λάμπω, to shine; οὐρά, a tail.)
LAMPȲRIS.

Noctilūca (Lat. *night-shining*), *the Glowworm*

Of the two insects represented in the engraving, that on the left is the male, the female being that on the right.

The DEATH-WATCH.—The formerly terrible Death-Watch is now generally known to be merely a small beetle. Indeed it is nothing more than the creature that perforates the round holes in old "worm-eaten" furniture and wood-work. The "ticking" is produced by striking the head against the wood. If there is a Death-Watch in the room, it is easy to incite it to begin to tick, by striking with the head of a pin on the panelling. There are several insects that produce this sound, the Anobium striatum, tesselatum, and pertinax. The last-named is so called from the pertinacity with which it simulates death if alarmed, preferring to suffer the severest treatment rather than give signs of life.

Family, Ptinĭdæ.—(Gr. πτηνός, winged.)
ANOBIUM.—(Gr. ἀνά, through; βιόω, I live.)

Tesselatum (Lat. *tesselated*), *the Death-Watch.*

The MUSK-BEETLE.—The beautiful Beetles of which the common Musk-Beetle is an excellent example vary considerably in size; some being several inches in length, while some are hardly one-quarter of an inch long. The extreme length of their antennæ is the most conspicuous property, and from that peculiarity they are at once recognised.

A small moth, Adela de Geerella, possesses the same peculiarity. The length of the moth is about a quarter of an inch, and the length of the antennæ more than an inch and a half. The antennæ wave about with every breath of air, as if the insect had become entangled in a spider's web, and escaped with some of the loose threads floating about it.

Family, Cerambycĭdæ.—(Gr. Κεράμβυξ, the Musk-beetle.)

CERAMBYX.

Moschātus (Lat. *musky*), *Musk-beetle.*

The Musk-beetle is a large insect, common in most parts of England. It is extremely common at Oxford, and is found in old willow-trees, with which Oxford is surrounded. Its peculiar scent, something resembling that of roses, often betrays its presence, when its green colour would have kept it concealed. When touched, it emits a curious sound, not unlike that of the bat, but more resembling the faint scratching of a perpendicularly-held slate pencil. Its larva bores deep holes in the trees, which are often quite honeycombed by them.

The ROVE-BEETLES form an exceedingly extensive section. Some are so small as to require the assistance of the microscope to discover their shape, and others, as those represented on the next page, are more than an inch in length. The small species are usually on the wing, and it is very amusing to see them alight, and with their flexible tails tuck their long and beautifully shaped wings under the elytra, run about for a moment, and then again take to flight. These are the creatures that cause so much annoyance by flying into one's mouth or eye in the warm months.

Family, Staphylinĭdæ.—(Gr. Σταφυλῖνος, the Rove-beetle.)
Creophĭlus.—(Gr. Κρεάς, flesh; φιλεῖν, to love.)

Maxillōsus (Lat. *large-jawed*), *the Rove-beetle.*

The Great Rove-beetle is commonly found upon decaying animal substances. It is most formidably armed with two large, curved, sharp mandibles, the bite of which is tolerably severe; and more than once, when the creature has been recently feeding upon putrid substances, dangerous results have followed.

I much regret that want of space has withheld me from giving accounts of many most interesting beetles particularly some of the Carabidæ, the Silphidæ, Ptinidæ, and the Water-beetles. These last inhabit the water, and swim with remarkable activity. They occasionally come to the surface for a fresh supply of air, which they carry down between the elytra and the upper surface of the abdomen. They fly very well, but the construction of their limbs prevents them from walking. They cannot be kept in a limited space, as they are very fierce and voracious, and in one case when a male and female were placed in a jar filled with water, only one day elapsed before the male was found dead and half devoured by his disconsolate widow.

The Earwig is placed in an order by itself, called Dermaptera, from the soft elytra. The wings are large and exceedingly beautiful, and the method of folding by which they are packed under the very small elytra is very curious. The use of the forceps seems principally for the purpose of folding the wings and placing them in their proper position

Order II. DERMAPTERA.—(Gr. Δέρμα, skin; πτέρον, a wing.)
Family. . Forficulĭdæ.—(Lat. Dim. of Forfex, a pair of Shears.)

FORFICŬLA.

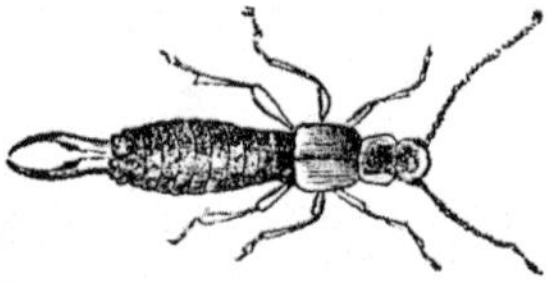

Forcipāta.—(Lat. *possessing forceps*), *the Earwig.*

under their cases. Ten species of earwigs inhabit England. The eggs of the earwigs are hatched, and the young protected by the parent.

The insect represented is chosen as being rather a rare species.

Order III. *ORTHOPTĔRA.*—(Gr. Ὄρθος, straight; πτέρον, a wing.)
Family . Locustĭdæ.—(Lat. *Locusta*, a Locust.)

LOCUSTA.

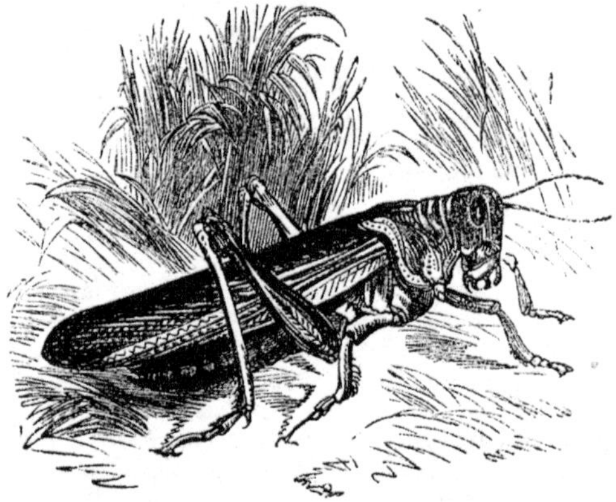

Tartarĭca (Lat. *of Tartary*), *the Locust.*

The Locust.—These pests of the warmer countries of the earth belong to the order called Orthoptera, because the wings are not folded transversely.

They fly in countless myriads, and where they descend, they devour every particle of green herbage—the trees are stripped of their leaves, the grass and corn is eaten to the very ground; for their jaws are so strong as to inflict a severe wound when the insect is incautiously handled. Nor does the mischief end with their life, for their dead bodies often accumulate in such numbers that the air is even dangerously infected. They infest Africa and Central Asia, but they annually make incursions to Europe, where the damage they occasion is much less reparable than in their native lands; for there the power of vegetation is so great that a few days repair the injuries caused by them, but in Europe a whole year is required for that purpose. The following account of these creatures is extracted from Mr. Cumming's South Africa:—

"On the following day I had the pleasure of beholding the first flight of locusts that I had seen since my arrival in the colony. We were standing in the middle of a plain of unlimited length, and about five miles across, when I observed them advancing. On they came like a snow-storm, flying slow and steady, about a hundred yards from the ground. I stood looking at them until the air was darkened with their masses, while the plain on which we stood became densely covered with them. Far as my eye could reach, east, west, north, and south, they stretched in one unbroken cloud; and more than an hour elapsed before their devastating legions had swept by. . . . .

"Locusts afford fattening and wholesome food to man, birds, and all sorts of beasts; cows and horses, lions, jackals, hyænas, antelopes, elephants, &c. devour them. We met a party of Batlapis carrying heavy burdens of them on their backs. Our hungry dogs made a fine feast on them. The cold frosty night had rendered them unable to take wing until the sun should restore their powers. As it was difficult to obtain sufficient food for my dogs, I and Isaac took a large blanket which we spread under a bush, whose branches were bent to the ground with the mass of locusts which covered it, and having shaken the branches, in an instant I had more locusts than I could carry on my back; these we roasted for ourselves and our dogs."

Our common grasshoppers belong to this order, bnt require no description.

Although the FIELD-CRICKET is not a very rare insect, it is but seldom caught, as it is very shy, and its holes, to which it instantly resorts when alarmed, are very deep and crooked. The female of this insect is armed with a very large ovipositor, nearly as long as its abdomen.

Although the Field-Cricket is endowed with large and strong jaws, it does not make use of them when handled, but only employs them in fighting with its own species. The hinder legs are long and powerful, appearing to be excellent limbs for leaping. The Field-Cricket, however, never seems to use them for leaping, but contents itself with crawling ignominiously. These qualities render the Field-Crickets an easy prey when they are once driven from their holes; but when they are only at the mouth of their caverns, as is their usual custom, they dive back so fast that they cannot be captured. Force is useless to get these creatures out of their holes, as the stroke of a spade would probably crush them to pieces, and their holes are too deep for them to be reached by any instrument.

Family, Achetĭdæ.—(Gr. Ἀχέτας, a Chirper, *i. e.* the Grasshopper.)

ACHĔTA.

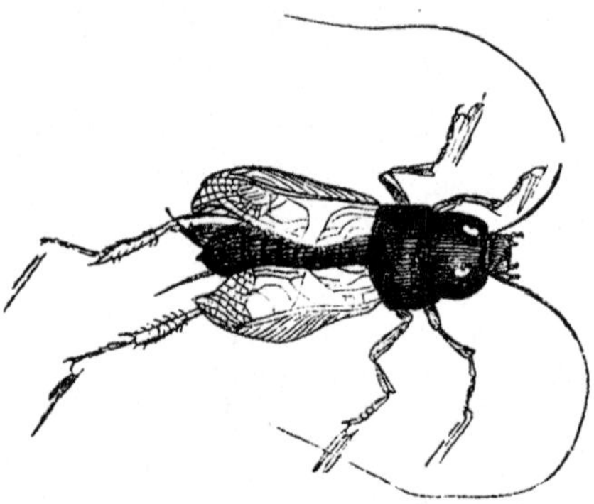

Campestris (Lat. *belonging to the field*), *the Field Cricket.*

The best method for inducing the Field-Cricket to leave its dwelling-place, is by thrusting a pliant stalk of grass into the hole, which accommodates itself to the windings of the passage, and brings out the inhabitant in great wrath to see what intruder has been trespassing.

It is a noisy insect, and utters its shrill cry while seated at the mouth of its hole, but instantly ceases if alarmed, and disappears into its hiding place. From the middle of May to the middle of July, its cry may be heard by night as well as by day.

The House-Cricket.—This well-known insect delights to live in places that are always warm, and consequently is found swarming about ovens, kitchen fire-places, and localities of a similar nature. It makes its residence by cutting away the mortar with its powerful jaws, and so effectually will it do so, that it sometimes eats completely through the wall, opening communications between two or more houses. The manner in which it bears heat is wonderful, as it will live within a few inches of a fierce fire.

ACHĔTA.

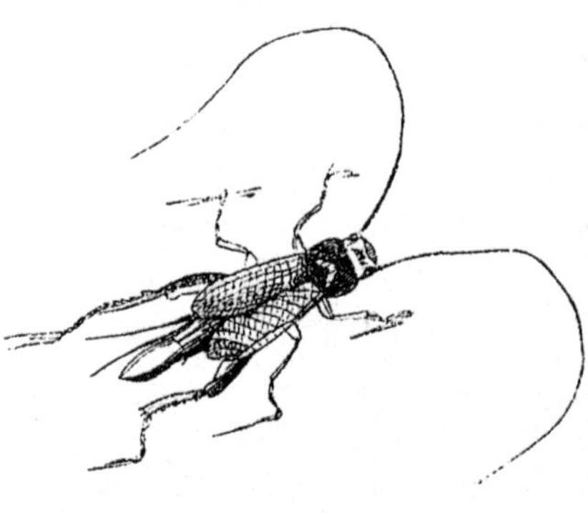

Domestĭcus (Lat. *domestic*), *the House Cricket.*

But the aridity and heat of the atmosphere in which it lives, render it very liable to thirst, and it consequently seeks every opportunity of quenching its thirst, by gnawing holes in wet linen, devouring any moist crumbs that may lie on the floor, or boldly climbing the milk-pan, in which latter case it gets a little too much liquid, and is generally "found drowned" next morning.

The wings of this insect, as well as those of the Field-Cricket, are very beautiful, and marked with an elegant pattern. The Cricket never appears to use them except at night, when it may be taken on the wing.

The MOLE-CRICKET.—The curious insect called the Mole-Cricket is not uncommon in England. It inhabits sandy banks, digging deep holes, and forming chambers, in which the eggs are laid. The fore legs closely resemble those of the mole, and are used for the same purpose. From its not unmusical cry it is called in some parts of England the Churr-worm, and near Oxford the rustics call it "Croaker."

GRYLLOTALPA.—(Lat. *Gryllus*, a Cricket *talpa*, a Mole.)

Vulgāris (Lat. *common*), *the Mole-Cricket.*

The LEAF INSECT is an inhabitant of South America. Not only does it resemble a leaf in shape, but even in colour, and its legs may be easily mistaken for dry twigs. Even the ramified veinings of the leaf are preserved on its wings. It is singular

PHYLLIA.*—(Gr. Φύλλον, a Leaf.)

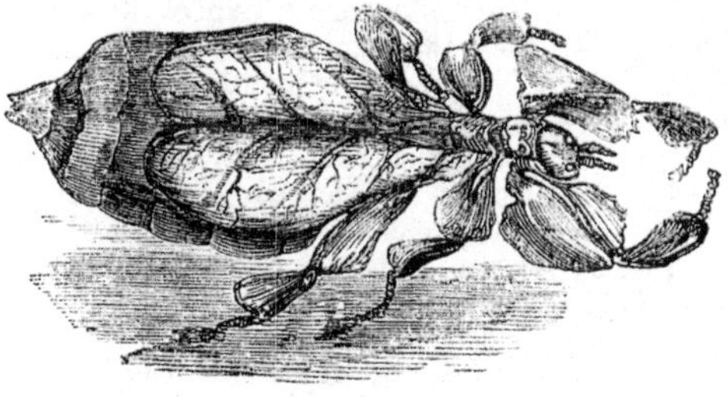

Foliāta (Lat. *like a leaf*), *the Leaf Insect.*

that while some insects closely resemble vegetables, some vegetables, as the Orchidaceæ, should as closely resemble insects. Nearly connected

* I have preferred to place these two insects in close proximity, as they both afford a curious instance of resemblance to another part of creation.

with this insect, is the Praying Mantis, so called from the curious manner in which it holds its fore legs. It is very voracious and exceedingly quarrelsome, fighting with the fore legs, which it uses like a sword. In China the inhabitants keep them in cages, and set them to fight as in other countries certain barbarians keep cocks for the same purpose.

Family, Blattĭdæ.—(Lat. *Blatta*, a Cockroach.)

BLATTA.

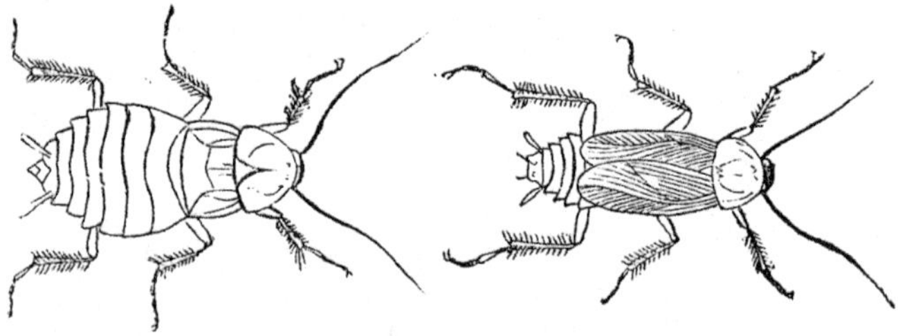

Orientālis (Lat. *Eastern*), *the Cockroach.*

The Cockroach.—Once upon a time, the French Academy were compiling a dictionary. Being determined to be quite accurate, they submitted each scientific word to some one skilled in that particular branch. One of these words was "Lobster," which the academicians defined as a little red fish that runs sideways. The word being zoological, came under Cuvier's notice, who, on reading the definition, observed that it would have been a very good one, but for three trifling circumstances, the first being that the lobster is not red until boiled, secondly that it is not a fish, and thirdly, that it does not run sideways.

In like manner the Cockroach has suffered under the hands of English housewives, who express their abhorrence of it under the name of "black beetle," a name egregiously false, as, in the first place, it is not black, and in the second place, it is not a beetle.

As is seen from the position of the insect, it belongs to the order Orthoptera, and its colour is a mahogany red. But, red or black, beetle or not, it is a very great plague, and fully deserves all the maledictions heaped upon it, which are not likely to be decreased by the fact that it is not even a good old English nuisance, but one of modern importation.

Its unpleasant character has caused innumerable plans to be laid for its destruction. Among these, strewing the ground with the peel of cucumber, or with red wafers, is said to be effectual in destroying the Cockroaches, but perhaps no plan is so successful as the glass pan with sloping sides, which lets the insects fall in, but prevents their escape altogether.

The eggs of the Cockroach are deposited, indeed, in little cases or purses, something like those of the shark, but without the strings.

Down one side a thick toothed ridge runs, and by this ridge the young escape when hatched.

The male Cockroach is furnished with very handsome wings, while the female is entirely destitute of these organs, and only possesses four little scales to mark their position. In the engraving, the right hand figure represents the male, the left being the female. These figures were drawn by the aid of the camera lucida, from specimens captured expressly for the purpose.

The COMMON MAY-FLY is so well-known an insect that it needs no long description. It is the fly so familiar to anglers under the name of the "Drake." It is to be found in swarms in the end of May and the beginning of June, rising and falling in the air in its peculiarly undulating manner.

Order IV. *NEUROPTĔRA.*
(Gr. Νεῦρον, a nerve; *πτέρον*, a wing.)
Family, Ephemerĭdæ.—(Gr. *ἐφήμερος, living for one day only.*)

EPHEMĔRA.

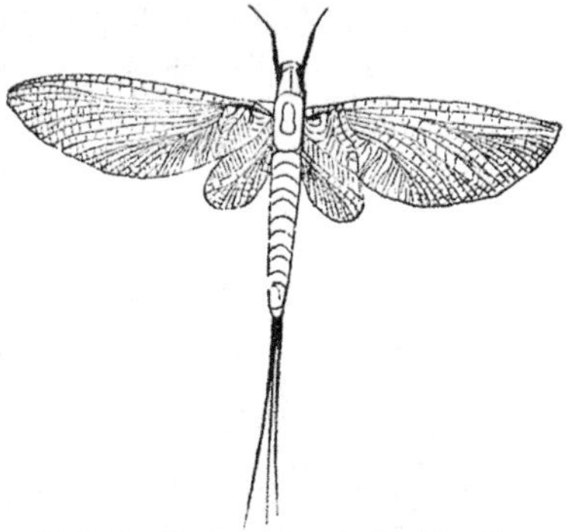

Vulgāta (Lat. *common*), *the May-fly.*

The May-fly spends the first portion of its existence in the water, under the shape of a longish grub, with leaf-like appendages to its tail. About May, the grubs may be seen to leave the water, and to crawl up the banks or climb the stems of aquatic plants. The skin then splits, and the May-fly creeps out. But it cannot immediately fly, as its wings are soft, and like two split peas. A short interval of exercise in the open air soon loosens them, and they are gradually shaken out until they have attained their full size, when the insect flies off. There is, however, another change yet. In a short time, the May-fly again settles, and sheds the entire skin a second time, even including the covering of the wings. These cast skins are often found sticking on the bark of willow-trees by the side of waters, and are mistaken for dead May-flies.

The DRAGON-FLY.—Well do the Dragon-flies deserve their name. Fierce, voracious, active, and powerful, they are a scourge to the insects. Few but the Coleoptera can escape them. They are on the wing nearly the whole day, seizing and devouring flies, spiders, and various insects; nor can even the broad-winged butterfly escape them: so voracious are they, that when held in the hand they will devour flies, &c., if held within their reach, and they have even been known, when their bodies have been severed in two, to eat flies, although they had no stomach to put them in. I once caught a dragon-fly in my net, and while holding

Family, Libellulĭdæ.—(Lat. *Libellula,* a Dragon-fly.)

LIBELLŬLA.

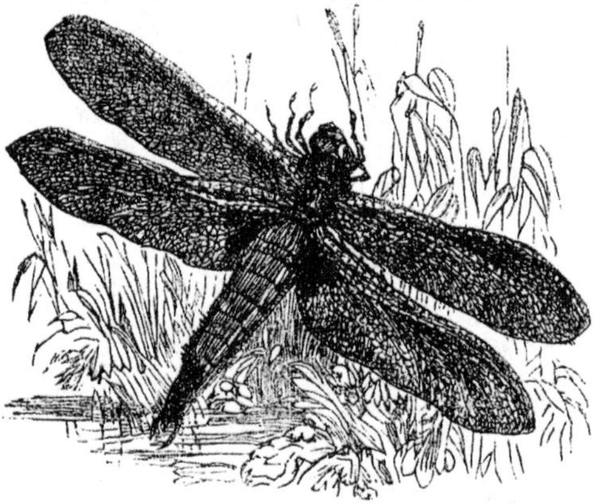

Depressa (Lat. *flattened*), *the Dragon-fly.*

it by the wings I presented to it no less than thirty-seven large flies in rapid succession, all of which it devoured, together with four long-legged spiders. It would probably have eaten as many more had I not been tired of catching flies for it.

A very great variety of these beautiful insects inhabit England. Some, the Agrionidæ, whose head resembles that of the hammer-headed shark, are of every vivid colour imaginable, floating in the air like beams of azure, emerald, and rosy light, while others have their wings marked with large indigo-coloured spots. The larva of the Dragon-fly inhabits the water, and is quite as voracious as in its perfect state. Affixed to its head is a curious set of organs, called the mask, which it can extend, and use for the purpose of seizing its prey, and holding it to its mouth.

The Ant-lion.—This insect in its perfect form, although it is very elegant, exhibits no peculiarity worthy of notice, but in its larva state its habits are so extraordinary as to have excited general attention. As it is slow and awkward in its movements, it has recourse to stratagem for capturing the agile insects on which it feeds. Choosing a light sandy soil, it digs for itself a conical pit, at the bottom of which it conceals itself, leaving only its jaws exposed. When an unwary insect approaches too near the edge of the pit, the sand gives way, and down rolls the insect into the very teeth of the concealed Ant-lion, who instantly

Family, Myrmelĕonĭdæ.—(Gr. Μύρμηξ, an Ant; λέων, a Lion.)

MYRMELĔON.*

Formicārum (Lat. *of ants*), *the Ant-lion.*

pierces its prey with its calliper-shaped fangs, and sucks out its juices through the jaws, which are hollow. Should, however, the Ant-lion miss its prey, and the insect endeavour to escape, its captor instantly makes such a turmoil by tossing up the sand with its closed jaws, and covering each side of the pit with the moving grains, that the insect is tolerably certain to be brought down to the bottom, and is seized by the Ant lion, who immediately drags it below the sand. When the insect is very strong and struggles hard to escape, the Ant-lion shakes it about as a dog does a rat, and beats it against the ground until it is disabled.

Family, Termitĭdæ.—(Lat. *Termes, properly a twig, also the insect.*)

TERMES.

Bellicōsus (Lat. *warlike.*)

The TERMITES, or WHITE ANTS as they are very erroneously called, belong to the Neuroptera, and are therefore not ants at all. These insects

* The winged Ant-lion is reduced one-half in size, and the figure on the right represents the larva.

live in large societies, and build edifices, sometimes of enormous size, and almost as hard as stone. Twelve feet in height is quite common, sc that were we to compare our works with theirs, St. Peter's in Rome, and St. Paul's in London, fall infinitely short of the edifices constructed by these little creatures. The common Termite inhabits Africa. Not only does it build these houses, but runs galleries underground, as, curiously enough, although blind, it always works either at night or in darkness. In each house or community, there are five different kinds of Termites:—1. the single male, or king, whose life is very short; 2. the single female or queen: these are the perfect insects, and have had wings, but have lost them soon after their admission into their cell; they also have eyes; 3. the soldiers or fighting men: these possess large jaws, do no work, but repel adversaries and watch as sentinels; 4. the pupæ, who resemble the workers, except that they possess the rudiments of wings; and 5. the larvæ, or workers. These do all the *work*, i. e. they collect food, attend to the queen, and watch over the eggs and young, and build and repair their castle. These are more numerous than all the other kinds.

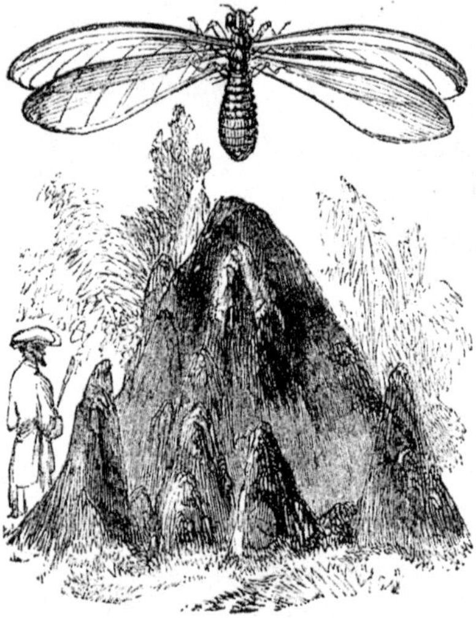

THE WHITE ANT.*

On the approach of the rainy season, the pupæ obtain wings and issue forth in swarms. Few, however, survive. Myriads are devoured by birds, reptiles, and even by man; and many are carried out to sea, and perish there. Those that do escape are speedily found by the labourers, who enclose a pair in a clay cell from which they never emerge. The male soon dies, but the female, after rapidly increasing to nearly three inches in length and one in breadth, continues to lay eggs unceasingly for a very long time. This cell becomes the nucleus of the hive, and round it all the other cells and galleries are built.

These insects are terribly destructive, as they eat through wooden beams, furniture, &c., leaving only a thin shell, which is broken down

* The upper figure is the Winged or King Termite, and below is represented one of the houses. In the cut on p. 421, the large figure is that of the female or Queen Termite; the left-hand one is the labourer, and the right-hand figure represents the soldier.

with the least extra weight, and many are the occasions when an unsuspecting individual, on seating himself on an apparently sound sofa or chair, finds himself, like Belzoni in the Pyramid, reposing among a heap of dust and splinters.

Mr. Cumming describes the habitations of the White Ant in these terms:—

"Throughout the greater part of the plains frequented by blesboks, numbers of the sunbaked hills or mounds of clay formed by the white ants occur. The average height of the ant-hills in these districts is from two to three feet. They are generally distant from one another from one to three hundred yards, being more or less thickly placed in different parts. These ant-hills are of the greatest service to the hunter, enabling him with facility to conceal himself on the otherwise open plain."

Order V. *TRICHOPTĔRA*.—(Gr. Θρίξ, hair; πτέρον, a wing.)

Family, Phryganidæ.—(Gr. Φρύγανον, a dry stick; alluding to their habitations.)

PHRYGANĔA.

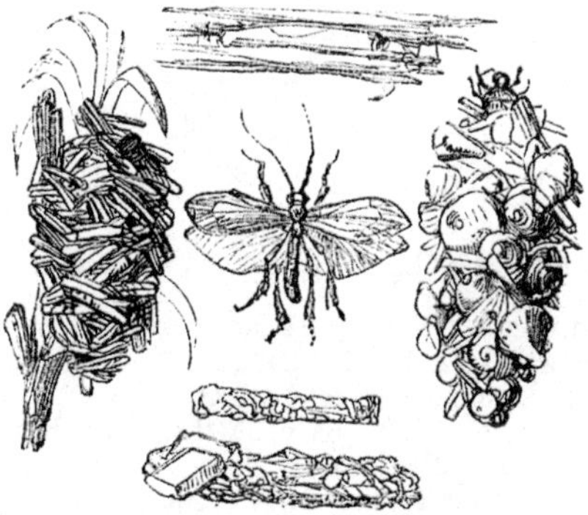

Grandis (Lat. *large*), *the Caddis-fly*.*

The Caddis-fly.—This fly is well known to every angler both in its larva and in its perfect state. The larva is a soft white worm, of which fishes are exceedingly fond, and it therefore requires some means of defence. It accordingly actually makes for itself a movable house of sand, small stones, straws, bits of shells, or even small living shells, in which it lives in perfect security, and crawls about in search of food,

* In this cut the cases of the Caddis-worm are of the natural size, but the insect in the centre is reduced one half.

dragging its house after it. When it is about to become a pupa, it spins a strong silk grating over the entrance of its case, so that the water necessary for its respiration can pass through, but at the same time all enemies are kept out. When the time for its change has arrived, the pupa bites through the grating, rises to the surface, and crawls out of the reach of the water, which would soon be fatal to it. The skin then splits down the back, and the perfect insect emerges.

The order is called Trichoptera, because the wings, instead of being covered with scales as are those of butterflies, are clothed with hairs.—There are many species of Caddís-flies.

Order VI. *HYMENOPTĔRA.*—(Gr. Ὑμήν, a Membrane; πτέρον, a wing.)
Family . . Ichneumonĭdæ.—(Gr. Ἰχνεύμων, a Hunter.)

PIMPLA.—(Gr. Πίμπλημι, to fill.)

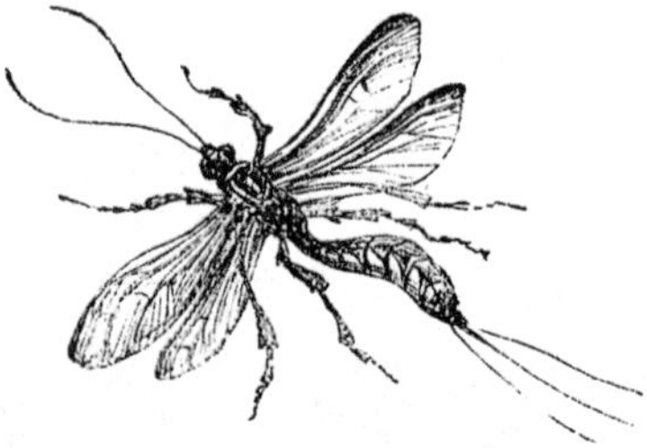

Manifestātor (Lat. *a pointer out*), *the Ichneumon-fly.*

The ICHNEUMON-FLY.—We have now reached a most important and interesting order. In it are contained the bees, wasps, ants, &c. This is the only order where the insects possess stings. The wings are four in number, with certain veinings upon them, the shape and number of which in many cases distinguishes the species.

The ICHNEUMONS form a very large section. They are most useful to mankind, as one ichneumon will destroy more caterpillars than a man could kill in his lifetime. They do not, as most other insects, deposit their eggs upon vegetable or dead animal substances, but they actually bore holes in other insects, while they are still in the larva state, and leave the eggs to hatch in their living receptacle. The most common ichneumon (*microgaster glomeratus*) is a very small insect, not so large as an ordinary gnat. This little creature may be seen searching for caterpillars. It generally selects the common cabbage caterpillar, and sitting upon it, pierces with its sting, or ovipositor as it is called, the skin of the caterpillar, and deposits an egg. After repeating this operation many times it flies off, and the caterpillar proceeds as before in the great business of its life, that is, eating, and continues in apparently perfect health until the time for its change into the chrysalis state occurs. The good condition of it, however, is merely deception, for the offspring of the little ichneumon have all this while been silently increasing in size, and feeding on the fat, &c. of the caterpillar, but cautiously avoiding any vital part, so that the plump appearance of the caterpillar is merely produced by the young ichneumons lying snugly under the skin. Just as the caterpillar commences its change, out come all the ichneumons, looking like little white maggots, and immediately each spins for itself a yellow oval case, frequently enveloping the form of the now emaciated caterpillar. In a few days a little lid on the top of each case is pushed open, and the perfect flies issue forth, and immediately commence their own work of destruction.

I have examined hundreds of caterpillars in the course of dissection, and have seldom found them free from ichneumons. I took out of one small goat caterpillar 137 of these insidious destroyers. I found them useful auxiliaries in dissection, as they had usually consumed all the fat, leaving the important organs ready cleared.

The remaining Hymenoptera are furnished with true stings, that is, with stings to which is attached a poison apparatus, like that belonging to the teeth of venomous snakes.

The Wood Ant is the largest of our British species. It is found principally in woods, and builds a large nest, which looks like a hillock of sand and earth, intermixed with bits of stick, leaves, &c. The interior of this hill is chambered out into a variety of apartments, and is traversed by passages. The so-called ants' eggs are not eggs at all, but the *pupa* cases of the insect, and if opened, the perfect insect is seen curled up inside. In the autumn, the ants burst forth by thousands, and may be seen hovering in clouds above the nest. Their beautiful wings do not last long, for when a female ant escapes, and founds an infant colony, her wings are soon lost, just as a highly accomplished young lady gives up her velvet painting and cross-stitchery when she marries and has a large family. Few *do* escape, as the birds find these living clouds a most agreeable and plentiful repast.

Family, Formicĭdæ.—(Lat. *Formīca*, an Ant.)

FORMĪCA.

Rufa (Lat. *red*), *the Wood Ant.*

Ants do not, as has been so frequently said, lay up stores of corn for the winter, for they are in a state of torpidity during the cold months, and require no food. Moreover, an ant would find as much difficulty in eating or digesting a grain of corn as we should in devouring a truss of straw.

In each nest are three kinds of ants,—males, females, and neuters, or workers.

Family, Vespĭdæ.—(Lat. *Vespa*, a Wasp.)

VESPA.

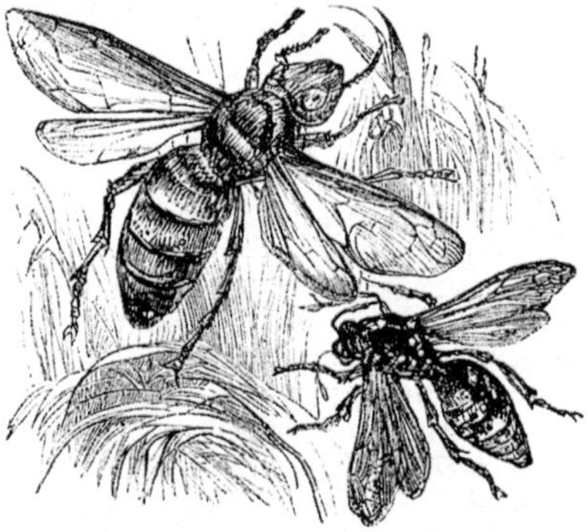

Crabro (Lat. *a Hornet*), *the Hornet.*

Vulgăris (Lat. *common*), *the Wasp.*

The Wasps.—Let us honour the Wasps as the first paper-makers, for of that material is the nest composed. The paper is rough and coarse, certainly, but it is still paper. The Wasp, in order to make this paper, rasps off fibres of decayed wood, which it afterwards mashes with its teeth into a pulp, and then spreads the pulp in layers, when it hardens and forms coarse paper.

The dreaded Hornet is usually found in woods, where it builds its nest in the hollows of trees. A deserted hut is a favourite spot, and when occupied by a full nest of hornets, is not particularly safe to enter, as the sting of this insect is peculiarly severe. In 1847, while on an entomological excursion in Bagley Wood, I saw five hornets sitting in a row, gnawing a dead branch. I was rather fearful of disturbing them, but at the same time, they were much wanted for a museum. They were all secured by tapping each in succession with a twig, and receiving it in my net as it flew off. Each bit a hole in the net, which had to be repaired before it could be used again with safety.

It feeds upon other insects, and even attacks and devours the formidable wasp.

The COMMON WASP builds its nest in the ground, usually in banks. The comb is laid horizontally, and not vertically like those of the bee. As the cells are made of paper, they will not hold honey, nor does the wasp endeavour to collect honey, although it is very fond of it, and never loses an opportunity of robbing a bee-hive, although its natural food is flies or other animal substances. Nor does it despise sugar, as every grocer's window testifies. Very few wasps survive the winter, and those who do, immediately set about forming a new nest. Only a few cells are made at first, but the number rapidly increases, until the nest is furnished with about sixteen thousand cells.

Some wasps build nests upon the branches of trees, and others suspend them from the branches.

The BEE.—This useful little creature is so well known that a lengthened description of it would be useless. A merely general sketch will be quite sufficient.

The cells of the bee are, as is well known, made of wax. This wax is secreted in the form of scales under six little flaps situated on the under

Family, Apĭdæ.—(Lat. *Apis,* a Bee.)

APIS.

Mellifĭca (Lat. *Mel,* honey; *facere,* to make), *the Honey Bee.**

side of the insect. It is then pulled out by the bee, and moulded with other scales until a tenacious piece of wax is formed. The yellow sub-

* In the cut, the upper figure is the Queen Bee; that on the left the Worker; and that on the right the Drone.

stance on the legs of the bees is the pollen of flowers. This is kneaded up by the bees, and is called bee-bread.

The cells are six-sided, a form which gives the greatest space and strength with the least amount of material, but the method employed by the bees to give the cells that shape is not known. The cells in which the drone or male bees are hatched, are much larger than those of the ordinary or worker bee. The edges of the cells are strengthened with a substance called propolis, which is a gummy material procured from the buds of various trees. This propolis is also used to stop up crevices and to mix with wax when the comb has to be strengthened.

The royal cells are much larger than any others, and are of an oval shape. When a worker larva is placed in a royal cell, and fed in a royal manner, it imbibes the principles of royalty, and becomes a queen accordingly. This practice is adopted if the queen bee should die, and there be no other queen to take her place.

The Queen Bee is lady paramount in her own hive, and suffers no other queen to divide rule with her. Should a strange queen gain admittance, there is a battle at once. which ceases not until one has been destroyed.

At the swarming time, the old queen is sadly put out by the encroachments of various young queens, who each wish for the throne, and at last is so agitated that she rushes out of the hive, attended by a large body of subjects, and thus the first swarm is formed. In seven or eight days, the queen next in age also departs, taking with her another supply of subjects. When all the swarms have left the original hive, the remaining queens fight until one gains the throne.

The old method of destroying bees for the sake of the honey was not only cruel but wasteful, as by burning some dry "puff-ball" the bees are stupefied, and shortly return to consciousness. The employment of a "cap" on the hive is an excellent plan, as the bees deposit honey alone in these caps, without any admixture of grubs or bee-bread. Extra hives at the side, with a communication from the original hive, are also useful.

The queen bee lays about eighteen thousand eggs. Of these about eight hundred are males or drones, and four or five queens, the remainder being workers.

The Swallow-tailed Butterfly.—We now arrive at the Haustellate Insects, so called, because they suck liquid food through an apparatus resembling the proboscis of an elephant. The first order of haustellate insects is the Lepidoptera, containing the butterflies and moths. The butterflies always fly by day, from which circumstance they are sometimes called Diurnal Lepidoptera. Most of the moths fly by night, and are called Nocturnal Lepidoptera. This is not a rule, however, as many moths fly by day, and some butterflies come out in the evening.

Butterflies are usually lighter in the body than moths, from which insects they are easily distinguished by the shape of the antennæ, which in the butterflies are slender, and terminate in a small knob, but in the moths terminate in a point, and are often beautifully fringed.

The name Lepidoptera is given to these insects because their wings are covered with myriads of minute scales, by which the beautiful colouring of the wings is produced. These scales vary in size and shape, according to the species, or the part of the wing from which they are taken. Under the microscope they are most exquisite objects, and well repay a long and careful examination.

Sub-class II. INSECTA HAUSTELLĀTA.—(Lat. *haurire,* to suck up.)
Order I. . . *LEPIDOPTĔRA.*—(Gr. Λεπίς, a scale; πτέρον, a wing.)
Family . . . Papilionĭdæ.—(Lat. *Papilio,* a Butterfly.)

PAPILĬO.

Machāon (Gr. *a proper name*), *the Swallow-tailed Butterfly.**

The Lepidoptera pass through three distinct changes before assuming their perfect form. They first exist in the larva state, in which state they are called caterpillars. They then pass to the pupa state, when they are known by the name of "aurelias" † or "chrysalides," ‡ both words being derived from words signifying gold, from the golden lustre of the pupa of certain butterflies. When they have remained in the pupa state during a time, varying from a few days to two years, they burst their shells and issue forth in their full and perfect beauty. This

* This figure is about one-third smaller than the insect. † Lat. *Aurum,* gold.
‡ Gr. Χρύσος, gold.

transformation has for many ages been used as an illustration of the resurrection after death.

The beautiful insect represented on p. 429 is not very uncommon in some parts of England, especially in the fenny parts of Cambridgeshire.*

It flies with exceeding rapidity, nearly in a straight line, and is very difficult to capture.

The colour of the wings is black, variegated most beautifully with yellow markings, and near the extremity of each hinder wing is a circular red spot, surmounted by a crescent of blue, and the whole surrounded by a black ring.

VANESSA.—(Lat. *a proper name.*) ARGYNNIS.—(Gr. *a proper name.*)

Adippe (*proper name*), *the Silver-spotted Fritillary.*
Atalanta (*proper name*), *the Red Admiral.*

The FRITILLARIES are well deserving of notice for the delicacy of their colouring, and the beauty of their markings. The Silver-spotted Fritillary is remarkable for the peculiar appearance presented by the under surface of the wings, which look as if they had been studded with pieces of burnished silver leaf. It is found mostly on thistles in woods, and is very common in Bagley Wood near Oxford, about the end of June, or during July.

The RED ADMIRAL is one of the most gorgeous of our butterflies. The colour of the wings is a deep black, relieved by a broad band of scarlet across each, and a series of semicircular blue marks edge each wing. It is usually found in woods and lanes, where there are nettles, as the larva feeds upon that plant. It appears about the middle of August.

* I once saw it in the water-meadows near Oxford.

The DEATH'S-HEAD MOTH.—This family is called Sphingidæ on account of the sphinx-like attitude that the caterpillars of some species assume. The larva of the Puss-moth (*Cerura vinula*) is particularly celebrated for this position. It holds the plants on which it feeds with its hinder feet, and raises the fore part of its body, just as the sphinx is represented. When in this position, it seems so remarkably self-satisfied, that the gardener of Rosel, a famous naturalist, was quite disconcerted, affirming that he never saw insects hold their heads so high.

Family, Sphingĭdæ.—(Gr. Σφίγξ, the Sphinx.)

ACHERONTĬA.—(Gr. Ἀχερόντιος, belonging to Acheron.)

Atrŏpos (Gr. *proper name; one of the Fates*), *the Death's-head Moth.*

The Death's-head Moth is the largest of the British Lepidoptera, as it not unfrequently measures nearly six inches across the wings. Its rather ominous name is derived from the singular marking in the thorax, which does not require much imagination to represent a skull and cross-bones.

Some naturalists have asserted that this moth makes its way into bee-hives, and robs the inhabitants of their honey, disarming their resentment by a curious squeaking noise which it has the power of producing.

The uneducated rustics have a great horror of this insect, and consider its appearance as a most disastrous omen. In a small village removed from the influence of railways, on one Sunday morning, as the inhabitants were going through the churchyard, a Death's-head Moth appeared on the path. Every one recoiled in dismay, and no one dared approach the dreaded object. Sundry heads were shaken at the evil omen, and various prophetic remarks made. At last, the blacksmith summoned up courage, and with a great jump, came down on the unfortunate moth, and happily destroyed it. The people were in blissful ignorance that, as there were several fields near planted with potatoes, on which vegetable the caterpillar generally feeds, there were probably a few hundred of Death's-head Moths in the vicinity.

I have this specimen now in my possession; it is of course mashed quite flat. It is a very singular fact, that those who, living so much in the open fields, would be supposed to have correct knowledge of natural phenomena, are really profoundly ignorant of facts that pass daily before their eyes. I have already mentioned the popular superstitions regarding efts.

In common with many other nocturnal insects, the eyes of the Death's-head Moth shine at night like two stars, which adds considerably to the terror inspired by its appearance.

The Humming-bird Moth.—This curious insect is called the Humming-bird Moth, because its appearance when on the wing exactly resembles that of a Humming-bird. It feeds on the wing as the bird does, hovering before each flower and sucking out the honey by means of its very long proboscis. It is very shy, and darts off if the slightest movement is made; but if the spectator remains perfectly quiet, the moth sees no danger, and will continue its meal within a yard of him. The moth appears to gain confidence if it is not disturbed, and in a few days will become almost tame, permitting the spectator to whom it is accustomed, to approach quite closely without appearing alarmed.

Family, Sesiïdæ.

Macroglossa.—(Gr. Μακρός, long; γλῶσσα, the tongue.)

Stellatārum, *the Humming-bird Moth.*

The Tiger-Moth.—This common but beautiful moth is found in the beginning of autumn. It runs on the ground with such swiftness as to be often mistaken for a mouse. I have more than once seen a kitten chasing a tiger-moth among the flowers in a garden, evidently deceived by its resemblance to a mouse. The larva is popularly called "the woolly bear." It is rather large, and is surrounded with tufts of long elastic hairs of a reddish brown colour, which serve as a defence against many enemies. When disturbed, it rolls itself round, just as a hedgehog does, and if on a branch, suffers itself to fall to the ground, when the long hairy covering defends it from being injured by the fall. When the caterpillar is about

Family, Arctiidæ.—(Gr. Ἄρκτος, a Bear; in allusion to the popular name of the larva.)

ARCTIA.

Caja (Lat. *proper name*), *the Tiger-moth.*

to change into a pupa, it spins a kind of hammock, and lies there until it comes forth as a moth.

The colour and markings of this moth vary considerably. The usual tints are, the thorax brown, the body red striped with black. The two anterior wings are cream colour, marked with bold patches of a deep brown: the posterior wings are bright red, spotted with bluish black.

The larvæ of the Geometridæ move in a very singular manner. When preparing to make a step, they hold firmly by their hinder legs to the substance on which they are moving, and then stretch out their body to the fullest extent, as if measuring their distance. After these preliminaries, they take a firm hold with the fore feet, and draw the hinder feet up to them, forming their body into an arch or loop. When at rest, the caterpillars often deceive an observer by their close resemblance to twigs, as they stretch themselves out motionless from the branch.

The family is very large, and contains many interesting species, but want of space compels me to omit all but the insect represented on page 434, the Swallow-tailed Moth. The caterpillar of this moth feeds principally on the elder, willow, and lime, and the moth appears in June and July. It is one of the largest of the British Geometridæ, as the spread of the wing considerably exceeds two inches. Its colour is a pale yellow,

Family, Geometrĭdæ.—(Gr. Γεομέτρης, a Land-measurer.)
Ouraptĕryx.—(Gr. Οὐρά, a tail; πτέρον, a wing.)
Family, Alucitĭdæ.
Alucĭta.—(*Linnean name.*)

Sambucarĭa (Lat. *Sambūcus*, the Elder-tree), *the Swallow-tailed Moth.*
Hexadactȳla (Gr. Ἕξ, six; δάκτυλος, a finger), *the Many-plumed Moth.*

and the lines across the wings are deep yellow. It derives its name from the shape of the hinder wings.

The Many-plumed Moth is found towards the close of autumn, usually running about windows. It is very small, measuring barely half an inch across the wings. The structure of the wings is very curious, each of the two anterior wings being divided into eight beautiful feather-like rays, and each of the posterior into four rays. Nearly allied to this are the common Feather Moths, the most common of which is the White-plumed Moth, whose wings measure nearly an inch across, and are divided into five feathered rays.

Diptera.—The insects of this order possess but two wings, the place of the others being supplied by two little organs something like drum-sticks, called "balancers." Without these the insect seems to be unable to direct its flight.

All are familiar with the Common Gnat. This pretty tormentor passes its larval existence in the water, in which state thousands may be seen in any uncovered water-butt, wriggling about with the most untiring energy, or reposing head downwards, only leaving the end of the tail at

Order II. *DIPTĔRA.*—(Gr. Δίς, twice; πτέρον, a wing.)

Family, Culicĭdæ.—(Lat. *Culex*, a Gnat.)

CULEX.

Family, Œstrĭdæ.—(Gr. Οἶστρος, the Gad-fly.)

ŒSTRUS.

Pipĭens (Lat. *humming*), *the Gnat.*

Bovis (Lat. *of the Ox*), *the Gad-fly.*

the surface. The reason for this is very curious. This larva breathes through its tail, and is moreover enabled by means of a fringe of hairs to carry air down with it.

It is a singular circumstance, that although the larva lives in the water, yet were either the eggs or the perfect insect to be submerged, they would be destroyed. The instinct of the Gnat in order to fulfil all three conditions is very beautiful. When the Gnat wishes to deposit its eggs, it rests on a leaf or twig on the surface of the water; it then takes each egg separately, and fastens them side by side in such a manner that they actually form a little boat, which will neither fill with water nor upset, however the water may be agitated. In a few days the eggs are hatched, when a little lid opens in the under end of each egg, and down tumbles the larva into the water.

After remaining in the water for some days it assumes the pupa form. In this state it floats at the surface with the back of the thorax uppermost. Soon this splits, and the insect emerges, standing on its own cast skin, which forms a raft for it until its wings are fully dry, when it takes to flight, leaving behind it the empty shell floating on the water. This change may be witnessed any warm day in summer.

The GADFLY has from the most ancient times been known as the terror of the herd. At the sound of its approach the cattle are driven almost mad with terror. The young gadflies are nourished under the skin, where they remain until they are fit to pass into the pupa state, when they bury themselves in the ground, and after a few days spent under the earth, issue forth in their perfect state.

Family, Bombylĭdæ (Gr. Βομβύλιος, *a Humble Bee.*)

BOMBYLĬUS.

Medĭus (Lat. *in the midst*), *the Humble Bee-fly.*

The HUMBLE BEE-FLY.—This very curious insect is found in the early days of spring, and may be seen hovering over the primroses and other spring flowers. It feeds in the same manner as the Humming-bird Moth, and much resembles that insect in many of its habits.

Order IV. *APHANIPTĔRA.*—(Gr. Ἀφανής, invisible; πτέρον, a wing.
Family . . Pulicĭdæ.—(Lat. *Pulex*, a Flea.)

PULEX.

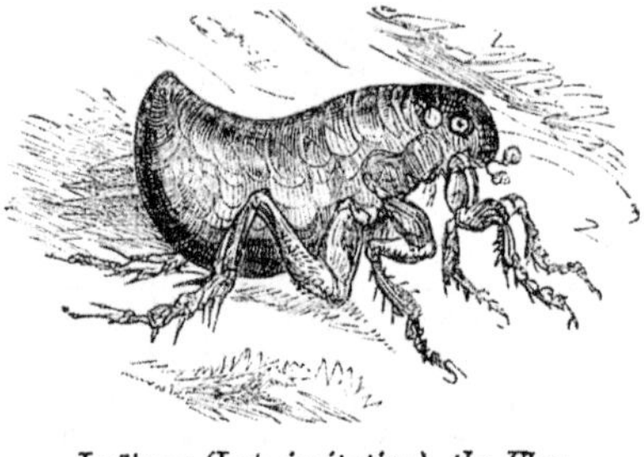

Irrītans (Lat. *irritating*), *the Flea.*

The FLEA.—The strength and agility of this curious but annoying little insect is perfectly wonderful. Many of my readers have doubtless seen the exhibition of the Industrious Fleas, who drew little carriages, and carried comparatively heavy weights with the greatest ease. The apparatus with which it extracts the blood of its victims is very curious, and forms a beautiful object under a microscope of low power. Its leap is tremendous in proportion to its size. This property it enjoys in common with many other insects, among which the Common Grasshopper, the

Frog-hopper, and the Halticas, or Turnip-flies are conspicuous. In all these insects the hinder pair of legs are very long and powerful.

I am here most reluctantly compelled to close this little work. Most willingly would I have entered into a sketch of the remaining Classes. These, however, are so numerous, and their habits are so different from those of the creatures whom we have already examined, that even a very slight description would consume too much space and time.

Here, then, I take my leave of the reader, with a sincere hope that the perusal of this volume will not only have proved interesting, but will also have given him some insight into the beautiful order of the animated world.

INDEX.

ABDOMINALIA, 360.
Abramis, 362.
Acanthopterygii, 351.
Acaridæ, 406.
Accentor, 212.
——— Hedge, 212.
Accentorinæ, 212.
Accipiter, 181.
Accipitres, 164.
——— Diurni, 164.
——— Nocturni, 183.
Accipitrinæ, 180.
Acherontia, 431.
Acheta, 416.
Achetidæ, 416.
Acipenser, 380.
Acipenseridæ, 380.
Adder, Puff, 333.
Adippe, 430.
Adjutant, 295.
Admiral, Red, 433.
Agamidæ, 329.
Agouti, 95.
Alauda, 248, 249.
Alaudinæ, 248.
Albatros, Wandering, 319.
Alca, 314.
Alcedinidæ, 196.
Alcedininæ, 196.
Alcedo, 196.
Alces, 135.
Alcidæ, 313.
Alcinæ, 313.
Alligator, 344.
Alligatoridæ, 344.
Alucita, 434.
Alucitidæ, 434.
Ampelidæ, 224.
Ampelinæ, 224.
Ampelis, 224.
Amphibia, 345.
Anas, 308.
Anatidæ, 304.
Anatinæ, 308.
Anchovy, 373.
Angler, 359.
Anguilla, 376.
Anguis, 327.
Anobium, 411.
Anseres, 304.
Anserinæ, 305.
Ant, Wood, 425.
Ant-Eater, Great, 160.
——— Middle, 161.
——— Little, 162.
Ant-Lion, 421.
——— White, 421, 422.
Anthropoides, 291.
Anthus, 216.
Aphaniptera. 436.
Apidæ, 427.
Apis, 427.
Apoda, 376.
Apteryginæ, 283.
Apteryx, 283.
Aquila, 171.
Aquilinæ, 171.
Arachnida, 403.
Araneidæ, 403.
Arctia, 433.
Arctiidæ, 433.
Arctomys, 102.
Ardea, 289.
Ardeidæ, 288.
Ardeinæ, 289.
Argala, 295.
Argonaut, 389.
Argonauta, 389.
Argus, 267.
——— Pheasant, 267.
Argynnis, 430.
Aricia, 393.
Armadillo, 159.
Arvicola, 91.
Arvicolina, 91.
Asinus, 137.
Ass, 137, 138, 139.
Astacidæ, 401.
Astacus, 402.
Astur, 180.
Ateles, 17.
Athene, 184.
Auk, Great, 314.
Aves, 164.
Avocet, 298.
Axis, 132.
Babyroussa, 147.
Badger, 58.
Balæna, 78.
Balænidæ, 77.
Barbel, 361.
Barbus, 361.
Bat, Long-eared, 24.
—— Vampire, 21.
Batrachia, 345.
Beagle, 44.
Bear, 61.
—— Grizzly, 62.
—— Polar, 64.
Beaver, 82.
Bee, Honey, 427.
—— eater, 197.
Beetle, Burying, 409.
—— Cockchaffer, 410.
—— Dor, 410.
—— Ground, 408.
—— Musk, 412.
—— Rove, 413.
—— Stag, 410.
—— Tiger, 407.
Bernicla, 305.
Bernicle, 398.
Bernicle Goose, 305.
Bison, 108.
Bittern, 292.
Blackbird, 220.
Blatta, 418.
Blattidæ, 418.
Blindworm, 327.
Bloodhound, 42.
Boa, 336.
Boar, 147.
Boidæ, 336.
Bombylidæ, 436.
Bombylius, 436.
Booby, 322.
Bos, 103, 105.
Boselaphus, 115.
Botaurus, 292.
Bovidæ, 103.
Bovina, 103.
Bower-bird, Satin, 240.
Bradypidæ, 156.
Bradypus, 156.

Bream, 362.
Bubalus, 106, 107
Bubo, 185.
Buboninæ, 185.
Buccinidæ, 393.
Buccinum, 393.
Buceros, 251.
Bucerotidæ, 251.
Bufo, 346.
Buffalo, 106.
——— Cape, 107.
Bulldog, 46.
Bullfinch, 250.
Bull-frog, 346.
Bull-head, 352.
Bunting, 247.
——— Yellow, 247.
Burying-beetle, 409.
Bustard, Great, 286.
Buteo, 174.
Buteoninæ, 174.
Butterfly, Atalanta, 430.
——— Fritillary, 430.
——— Swallow-tailed, 429.
Buzzard, 174.
——— Honey, 175

Cacatua, 256.
Cachalot, 83.
Caddis-fly, 423.
Calamodyta, 204.
Callithrix, 19.
Camel, 124.
——— Bactrian, 126.
Camelina, 124.
Camelopardalis, 121.
Camelopardina, 121.
Camelus, 124, 126.
Canary, 246.
Cancer, 399.
Canceridæ, 399.
Canis, 41, 51, 52.
Canina, 41.
Capercaillie, 274.
Capra, 119.
Caprimulgidæ, 188.
Caprimulginæ, 188.
Caprimulgus, 188.
Capybara, 95.
Carabidæ, 408.
Carabus, 408.
Caracal, 36.
Carduelis, 246.
Carp, 360.
Carrier, 265.
Cassowary, 281.
Castor, 92.
Castorina, 92.
Casuarius, 281.
Cat, 35.
—— Civet, 39.
—— Tiger, 34.
Catoblepas, 112.
Cebidæ, 17.
Cephalopoda, 388.
Cerambycidæ, 411.
Cerambyx, 411.
Cerastes, 334.
Certhia, 202.
Certhidæ, 202.
Certhinæ, 202.
Cercoleptes, 66.
Cercoleptina, 65.
Cervina, 129.
Cervus, 129, 130, 131.
Cete, 78.
Chaffinch, 243.
Chameleon, 330.
Chameleonidæ, 330.
Chamois, 118.
Charadridæ, 287.
Charadriinæ, 288.
Chatterer, Waxen, 224.
Chelidon, 193.
Chelonia, 339, 341.
Cheloniadæ, 341.
Chenopis, 307.
Chetah, 38.
Chicken, Mother Cary's, 317.
Chiff-chaff, 208.
Chimpansee, 11.
Chinchilla, 98.
Chinchillina, 98.
Chough, 238.
Chub, 365.
Cicindela, 407.
Cicindelidæ, 407.
Ciconia, 294.
Ciconinæ, 294.
Circinæ, 182.
Circus, 182.
Cirrhopoda, 398.
Civet Cat, 39.
Clotho, 333.
Clupea, 371, 372.
Clupeidæ, 371.
Coaita, Spider Monkey, 17.
Coati Mondi, 65.
Cobitis, 365.
Cobra de Capello, 337.
Coccothraustes, 242.
Cockatoo, Great Sulphur, 256.
Cockchaffer, 410.
Cockroach, 418.
Cod, 373.
Coleoptera, 407.
Colubridæ, 337.
Colubrina, 336.
Columba, 261, 262.
Columbæ, 261.
Columbidæ, 261.
Columbinæ, 261.
Colymbidæ, 311.
Colymbinæ, 311.
Colymbus, 311.
Conchifera, 394.
Condor, 167.
Cone, 392.
Conger, 377.
Coniidæ, 392.
Conirostres, 228.
Conus, 392.
Coot, 303.
Coracia, 238.
Coracias, 194.
Coraciidæ, 194.
Coraciinæ, 194.
Cormorant, 323.
Corncrake, 302.
Corvidæ, 228.
Corvinæ, 229.
Corvus, 232, 234, 235, 236, 237.
Cotile, 193.
Cottidæ, 352.
Cottus, 352.
Coturnix, 273.
Cowry, 393.
Crab, 399.
—— Hermit, 400.
Cracticornis, 297.
Crane, 288.
——— Demoiselle, 291.
Crangon, 402.
Crangonidæ, 402.
Cray-fish, 401.
Creeper, 202.
Creophilus, 413.
Cricetus, 91.
Cricket, Field, 416.
——— House, 416.
——— Mole, 417.
Crocodile, 342.
Crocodilidæ, 342.
Crocodilus, 342.
Cross-bill, 250.
Crossopus, 69.
Crotalidæ, 331.
Crow, 236.
——— Hooded, 237.
——— Nutcracker, 229.
Crustacea, 399.
Cuckoo, 260.
Cuculidæ, 260.
Cuculinæ, 260.
Cuculus, 260.
Culex, 435.
Culicidæ, 435.
Curlew, 297.
——— Stone, 287.
Cuttle-fish, 388.
Cyclobranchiata, 394.
Cyclosaura, 326.
Cyclostomi, 386.
Cyclothurus, 162.
Cygninæ, 305.
Cygnus, 305, 306.
Cynocephalus, 16.
Cypræidæ, 393.
Cyprinidæ, 360.
Cyprinus, 360, 361.
Cypselinæ, 190.
Cypselus, 190, 191.

Dabchick, 313.
Dace, 363, 364.
Dactylophori, 351.
Dama, 133.
Dasypidæ, 158.
Dasypina, 159.
Dasyprocta, 95.
Dasyproctina, 95.
Dasypus, 159.
Death's Head Moth, 431.
Death-Watch, 411.
Decapoda, 399.
——— Anomoura, 400.
——— Brachyura, 399.
——— Macroura, 401.
Deer, Fallow, 133.
—— Musk, 128.
—— Red, 130.
—— Rein, 134.
Delphinidæ, 84.
Delphinus, 84.
Dendrosaura, 330.
Dentirostres, 204.
Dermaptera, 414.
Dicotyles, 148.
Didelphina, 74.
Didelphys, 74.
Didinæ, 284.
Didus, 284.
Diodontidæ, 379.
Diomedea, 319.
Dipina, 99.
Dipper, 217.
Diptera, 435.
Dipus, 99.
Diver, Great Northern, 311.
Dodo, 284.
Dog, Bull, 46.
—— King Charles's, 42.
—— Newfoundland, 41.
—— Pug, 47.
—— Shepherd's, 48.
Dog-fish, Spotted, 381.
Dolphin, 84.
Dormouse, 99.
Dory, John, 356.
Dove, Ring, 261.
——— Stock, 262.
——— Turtle, 262.
Draco, 329.
Dragon, Flying, 329.
Dragon-fly, 420.
Dromaius, 282.
Duck, Wild, 308.
——— Eider, 310.
Dzigguetai, 138.

Eagle, Golden, 171.
——— Whiteheaded, 173.
Earwig, 414.
Echeneidæ, 358.
Echeneis, 358.
Ectopistes, 263.
Eel, Sharp-nosed, 376.
—— Electric, 377.
Eider Duck, 310.
Eland, 115.
Elanoides, 176.
Elephant, Indian, 140.
——— African, 144.
Elephantidæ, 140.
Elephantinæ, 140.
Elephas, 140, 144.
Eleutheropomi, 380.
Elk, 135.
Emberiza, 247.
Emberizinæ, 247, 248.
Emu, 282.
Emydosauri, 342.
Engraulis, 373.
Entellus, 16.
Ephemera, 419.
Ephemeridæ, 419.
Ephialtes, 185.
Equidæ, 136.
Equus, 136.
Erinaceus, 70.
Erinacina, 69.
Ermine, 56.
Erythacinæ, 210.
Erythacus, 211.
Esocidæ, 366.
Esox, 366.
Exocœtus, 367.

Falco, 177, 178.
Falcon, Gyr, 177.
——— Peregrine, 178.
——— Stone, 287.
——— Swallow-tailed, 176.
Falconidæ, 171.
Falconinæ, 177.
Felidæ, 26.
Felina, 26.
Felis, 35.
Feræ, 26.
Fieldfare, 218.
Fishing-frog, 359.
Fissirostres Diurnæ, 190.
——— Nocturnæ, 188.
Flamingo, 304.
Flea, 436.
Fly-catcher, Spotted, 223.
Flying-fish, 367.
Forficula, 414.
Forficulidæ, 414.
Formica, 425.
Formicarinæ, 217.
Formicidæ, 425.
Fowl, domestic, 269.
Fox, 53.
Foxhound, 44.
Fratercula, 313.
Fregata, 324.
Frigate Pelican, 325.
Fringilla, 243, 244, 245.
Fringillidæ, 242.
Fritillary, Silver-spotted, 430.
Frog, 345.
—— Bull, 346.
—— Fishing, 359.
—— Tree, 346.
Fulica, 303.
Fuligulinæ, 310.

Gadfly, 435.
Gadidæ, 373.
Gallinæ, 266, 269.
Gallinula, 302.
Gallinulinæ, 302.
Gallus, 269.
Gannet, 322.
Garrulinæ, 228.
Garrulus, 228.
Gasteropoda, 391.
Gasterosteus, 356.
Gazella, 118.
Gazelle, 118.
Gecko, 328.
Geckotidæ, 328.
Geissosaura, 326.
Genet, 40.
Genetta, 40.
Geometridæ, 434.
Geotrupes, 410.
Geotrupidæ, 410.
Gibbon, Agile, 14.
Giraffe, 121.
Glires, 88.
Glow-worm, 411.
Glutton, 58.
Gnat, 435.
Gnoo, 112.
Gnu, 112.
Goat, 119.
Goat-sucker, 188.
Gobio, 362.
Goldfinch, 244.
Gold-fish, 361.
Goose, Bernicle, 305.
——— Solan, 322.
Goshawk, 180.
Gradientia, 347.
Grallæ, 287.
Grebe, Crested, 312.
——— Little, 313.
Greenfinch, 245.
Greyhound, 49.
Grosbeak, 242.
Grouse, Red, 275.
——— Black, 275.
Gruinæ, 288.
Grus, 288.
Gryllotalpa, 417.
Gudgeon, 362.
Gueparda, 38.
Guillemot, 316.
Guinea-fowl, 271.
Guinea-pig, 95.
Gull, Black-backed, 320.
Gulo, 58.
Gurnard, 351.
Gymnotidæ, 377.

Gymnotus, 377.
Gypaëtidæ, 164.
Gypaëtus, 164.
Gyps, 169.
Gyrfalcon, 177.

Hag-fish, Glutinous, 387.
Haliaetus, 173.
Hamster, 91.
Hare, 96.
——— Alpine, 97.
Harvest-bug, 406.
Hawfinch, 242.
Hawk-owl, 183.
——— Sparrow, 181.
Hedgehog, 70.
Helicidæ, 391.
Helix, 391.
Hen Harrier, 182.
—— Water, 302.
Heron, 289.
Herpestes, 40.
Herring, 372.
Hippocampus, 357.
Hippopotamina, 154.
Hippopotamus, 154.
Hirundinidæ, 190.
Hirundininæ, 192.
Hirundo, 192.
Hobby, 178.
Holodactyli, 352.
Hominidæ, 1
Homo, 1.
Hoopoe, 198.
Hornbill, Rhinoceros, 251.
Hornet, 426.
Horse, 136.
——— Sea, 357.
Hound, Blood, 42.
——— Fox, 44.
——— Grey, 40.
Howler, Ursine, 18.
Humble Bee-fly, 436.
Humming-bird, Bar-tailed, 200.
——— Cora, 200.
——— Double-crested, 200.
——— Gould's, 200.
——— Moth, 432.
——— Ruby-throated, 199.
Hydrobata, 217.
Hydrochœrina, 95.
Hydrochœrus, 95.
Hyæna, 39.
Hyenina, 39.
Hylobates, 14.
Hymenoptera, 424.
Hypotriorchis, 178, 179.
Hystricidæ, 94.
Hystricina, 94.
Hystrix, 94.

Ibex, 119.
Ibis, Sacred, 296.
Ichneumon, 40.
Ichneumon-fly, 424.
Ichneumonidæ, 424.
Icterinæ, 241.
Icterus, 241.
Iguana, 329.
Iguanidæ, 329.
Insecta, 407.
Insecta Haustellata, 429.
——— Mandibulata, 407.
Invertebrata, 388.

Jacana, 301.
Jacchus, 19.
Jackal, 52.
Jackdaw, 235.
Jaguar, 33.
Jay, 228.
Jerboa, 99.
Jerboidæ, 98.
John Dory, 356.

Kahau, 15.
Kangaroo, 72.
Kestrel, 179.
Kingfisher, 196.
Kinkajou, 66.
Kite, 175.
Koodoo, 114.

Lacertinidæ, 326.
Lagopus, 275, 276.
Lämmergeyer, 164.
Lampern, 387.
Lampetra, 387.
Lamprey, 386.
Lampyris, 411
Lampyridæ, 411.
Landrail, 302.
Lanidæ, 225.
Laninæ, 225.
Lanius, 225, 227.
Lapwing, 288.
Laridæ, 320.
Larinæ, 320.
Lark, Sky, 248.
—— Tit, 216.
—— Wood, 249.
Larus, 320.
Leaf Insect, 417.
Lemur, 20.
Lemuridæ, 20.
Leo, 26.
Leopard, 31.
——— Hunting, 38.
Leopardus, 31, 32, 33, 34.
Lepidoptera, 429.
Leporidæ, 96.
Leptoglossæ, 326.
Leptoptilos, 295.
Leptus, 406.
Lepus, 96, 97.
Leuciscus, 363, 365.
Libellula, 420.
Libellulidæ, 420.
Limacidæ, 391.
Limax, 391.
Limosinæ, 297.
Limpet, 394.
Linnet, 244.
——— Green, 245.
Lion, 26.
Lizard, 326.
Llama, 127.
Loach, 365.
Lobster, 401.
Locust, 414.
Locusta, 414.
Locustidæ, 414.
Lophiidæ, 359.
Lophius, 359.
Loris, 21.
Loxia, 250.
Loxina, 250.
Lucanidæ, 410.
Lucanus, 410.
Luscinidæ, 204.
Luscininæ, 204.
Luscinia, 205.
Lutra, 60.
Lynx, 37.

Macaco, 20.
Macaw, Blue and Yellow, 254.
Mackarel, 353.
Macrocercus, 254.
Macroglossa, 432.
Macropidæ, 72.
Macropina, 72.
Macropus, 72.
Magpie, 230.
Malacopterygii, 360.
Malacostraca, 399.
Mallard, 308.
Mammalia, 1.
Man, 1.
Mandrill, 16.
Manina, 158.
Manis, 158.
Mareca, 308.
Martin, 193.
——— Chimney, 192.
——— Sand, 193.
Marmoset, 19.
Marmot, 102.
Marmotta, 102.
Marten, Pine, 55.
Martes, 55.
Mastiff, 46.
May-fly, 419.
Meantia, 349.
Megapode, Mound-making, 278.
Megapodidæ, 277.
Megapodius, 278.
Meleagrina, 397.
Meleagrinæ, 270.
Meleagrinidæ, 397.
Meleagris, 270.
Meles, 58.

Mellivora, 57.
Melolontha, 410.
Melolonthidæ, 410.
Menurinæ, 204.
Merlin, 179.
Meropidæ, 197.
Meropinæ, 197.
Merops, 197.
Micromys, 90.
Miller's Thumb, 352.
Milvinæ, 175.
Milvus, 175.
Mocking Bird, 221.
Mole, 67.
Mollusca, 388.
Monkey, Entellus, 16.
——— Proboscis, 15.
——— Spider, 17.
Monodon, 86.
Morrhua, 373.
Morse, 76.
Morunga, 76.
Moschina, 128.
Moschus, 128.
Motacilla, 215, 216.
Motacillinæ, 215.
Moth, Death's-head, 431.
——— Humming-bird, 432.
——— Many-plumed, 434.
——— Swallow-tailed, 434.
——— Tiger, 433.
Mouse, 89.
——— harvest, 90.
Mullingong, 162.
Murænidæ, 376.
Murex, 394.
Muricidæ, 394.
Muridæ, 88.
Murinæ, 88.
Mus, 88, 89.
Muscicapa, 223.
Muscicapidæ, 223.
Muscicapinæ, 223.
Mussel, Edible, 397.
Mustela, 56, 57.
Mustelina, 55.
Mycetes, 18.
Mygale, 403.
Myoxina, 99.
Myoxus, 99.
Myrmecophaga, 160.
Myrmecophagina, 160.
Myrmeleon, 421.
Myrmeleonidæ, 421.
Mytilidæ, 397.
Mytilus, 397.
Myxine, 387.
Myxinidæ, 387.

Naja, 337.
Narwhal, 86.
Nasua, 65.
Natrix, 338.
Nautilus, Paper, 389.
Necrophagus, 409.
Neuroptera, 419.
Newt, 347.
Nightingale, 205.
Nucifraga, 229.
Numenius, 299.
Numida, 271.
Nutcracker, 229.
Nuthatch, 203.
Nyctea, 184.
Nyctisaura, 328.
Nylghau, 113.

Ocelot, 34.
Octopidæ, 388.
Octopus, 388.
Œdicnemidæ, 287.
Œdicneminæ, 287.
Œdicnemus, 287.
Œnanthe, 210.
Œstridæ, 435.
Œstrus, 435.
Ophidia, 331.
Opossum, 74.
Orang Outan, 13.
Oriole, Baltimore, 241.
——— Golden, 223.
Oriolinæ, 223.
Oriolus, 223.
Ornismya, 200.
Ornithorhynchina, 162.
Ornithorhynchus, 162.
Orpheus, 221.
Orthagoriscus, 379.
Orthoptera, 414.
Ortolan, 248.
Ortygometra, 302.
Oryx, 116.
Osprey, 172.
Ossei, 351.
Ostrea, 396.
Ostrich, 279.
Otinæ, 286.
Otter, 60.
Otus, 286.
Ounce, 32.
Ourapteryx, 434.
Ouzel, Water, 217.
Ovibos, 111.
Ovis, 120.
Owl, Barn, 186.
—— Burrowing, 184.
—— Great-eared, 185.
—— Hawk, 183.
—— Scops-eared, 185.
—— Snowy, 184.
Ox, 103.
— Musk, 111.
Oyster, 396.
——— Pearl, 397.

Pachyglossæ, 328.
Paguridæ, 400.
Pagurus, 400.
Palæmonidæ, 402.
Palæmon, 402.
Palæornis, 255.
Palamedeidæ, 301.
Pandion, 172.
Pangolin, 158.
Panther, 31.
Papilio, 429.
Papilionidæ, 429.
Paradise, Emerald Bird of, 239.
Paradisea, 239.
Paradiseidæ, 239.
Parinæ, 213.
Parra, 301.
Parrakeet, Ringed, 255.
Parrinæ, 301.
Partridge, 272.
Parus, 213, 214.
Passer, 246.
Passeres, 188.
Patella, 394.
Patellidæ, 394.
Pavo, 266.
Pavoninæ, 266.
Peacock, 266.
Peccary, 148.
Pecten, 395.
Pectinidæ, 395.
Peewit, 288.
Pelias, 335.
Pelecanidæ, 321.
Pelecaninæ, 322.
Pelecanus, 324.
Pelican, Frigate, 325.
——— White, 324.
Penguin, Cape, 315.
Pentalasmis, 398.
Perca, 352.
Perch, 352.
Percidæ, 352.
Perdicinæ, 272.
Perdix, 272.
Pernis, 175.
Petrel, Fulmar, 317.
——— Stormy, 317.
Petromyzon, 386.
Petromyzonidæ, 386
Pettichaps, 208.
Phaeton, 321.
Phaetoninæ, 321.
Phalacrocorax, 323.
Phatagin, 158.
Phasianidæ, 266.
Phasianinæ, 267.
Phasianus, 268.
Pheasant, 268.
——— Argus, 267.
Philomachus, 300.
Phoca, 75.
Phocæna, 85.
Phocidæ, 75.
Phocina, 75.
Phœnicopterinæ, 304.
Phœnicopterus, 304.
Phryganea, 423.
Phryganidæ, 423.

Phyllostomina, 21
Phyllia, 417.
Physeter, 83.
Pica, 230.
Picidæ, 257.
Picinæ, 257.
Picus, 257, 258.
Pigeon, Domestic, 265.
——— Passenger, 263.
Pike, 366.
Pilchard, 371.
Pimpla, 424.
Pipit, Meadow, 216.
Pisces, 351.
———Chonderopterygii, 380.
——— Malacopterygii, 360.
——— Ossei, 351.
Platalea, 293.
Plecotus, 24.
Plectognathi, 379.
Pleuronectidæ, 375.
Podicepinæ, 312.
Podiceps, 312, 313.
Poephagus, 110.
Pointer, 45.
Polecat, 56.
Porcupine, 94.
Porpoise, 85.
Portax, 113.
Potamobius, 401.
Pouter, 264.
Poultry, 269.
Prawn, 402.
Presbytes, 15, 16.
Primates, 1.
Pristidæ, 383.
Pristis, 383.
Proboscis Monkey, 15.
Procellaria, 317.
Procellaridæ, 317.
Procellarinæ, 317.
Procyon, 65.
Procyonina, 65.
Proteidæ, 349.
Proteus, 349.
Psetta, 375.
Psittacidæ, 254.
Ptarmigan, 276.
Pteromys, 101.
Ptilonorhynchinæ, 240.
Ptilonorhynchus, 240.
Ptinidæ, 411.
Puff Adder, 333.
Puffin, 313.
Pulex, 436.
Pulicidæ, 436.
Pulmobranchiata, 391.
Pulmonaria, 403.
Puma, 33.
Putorius, 56.
Pyrrhocoracinæ, 238.
Pyrrhula, 250.
Pyrrhulinæ, 250.

Quadrumana, 10.
Quagga, 139.
Quail, 273.
Querquedula, 310.

Rabbit, 97.
Racoon, 65.
Raia, 385.
Raidæ, 384.
Rallidæ, 302.
Rallinæ, 302.
Ram, 120.
Ramphastidæ, 253.
Ramphastidinæ, 253.
Ramphastos, 253.
Rana, 345.
Rangifer, 134.
Rat, 88.
—— Water, 91.
Ratel, Honey, 57.
Rattle-snake, 331.
Raven, 232.
Recurvirostra, 298.
Recurvirostrinæ, 298.
Red Admiral, 507.
Redbreast, 211.
Redstart, 211.
Regulus, 209.
Reindeer, 134.
Remora, 358.
Reptilia, 326.
Rhinaster, 151.
Rhinocerina, 150.
Rhinoceros, 150, 151.
————— two-horned, 151.
Rhinoceros Hornbill, 252.
Ring-dove, 261.
Roach, 363.
Roebuck, 129.
Roller, 194.
Rook, 234.
Rove-beetle, 413.
Ruff, 300.
Rupicapra, 118.
Ruticilla, 211.

Sable, 55.
Salamandridæ, 347.
Salientia, 345.
Salmo, 368, 369.
Salmon, 368.
Salmonidæ, 368.
Sandpiper, 297.
Sarcorhamphidæ, 167.
Sarcorhamphos, 167, 168.
Saura, 326.
Saw-fish, 383.
Saxicola, 210.
Scalaria, 392.
Scallop, 395.
Scansores, 253.
Scincidæ, 327.
Sciurinæ, 100.
Sciurus, 100.
Scolopacidæ, 297.
Scolopacinæ, 299.
Scolopax, 299.
Scomber, 253.
Scomberidæ, 353.
Scops-eared Owl, 185.
Scorpio, 405.
Scorpion, 405.
Scorpionidæ, 405.
Scyllidæ, 381.
Scyllium, 381.
Sea-horse, 357.
Seal, 75.
—— Elephant, 76.
Secretary Bird, 182.
Serpentarius, 182.
Sesiidæ, 432.
Shark, White, 382.
——— Hammer-headed, 383
Sheep, 120.
Shrew, Mouse, 69.
——— Water, 69.
Shrike, Great Grey, 225.
——— Red-backed, 227.
Shrimp, 402
Silphidæ, 409.
Simia, 13.
Simiadæ, 11.
Siskin, 245.
Sitta, 203.
Sittinæ, 203.
Skate, Thornback, 385.
Skylark, 248.
Sloth, 156.
Slowworm, 327.
Slug, 391.
Snail, 391.
Snake, Rattle, 331.
——— Ringed, 338.
Snipe, 299.
Solan Goose, 322.
Sole, 375.
Solea, 375.
Somateria, 310.
Sorex, 69.
Spaniel, Water, 42.
Sparrow, 246.
———— hawk, 181.
———— hedge, 212.
Spheniscinæ, 315.
Spheniscus, 315.
Sphingidæ, 431.
Sphyrnias, 383.
Spider, Bird, 403.
Spoonbill, White, 293.
Springbok, 117.
Squalidæ, 382.
Squalus, 382.
Squirrel, 100.
———— Flying, 101.
Stag, 130.
Stag Beetle, 410.
Staphylinidæ, 413.
Starling, 242.
Steinbok, 119.
Sterna, 321.
Sterninæ, 321.

Stickleback, 356.
Stoat, 56.
Stockdove, 262.
Stone Curlew, 287.
Stork, 294.
Strepsiceros, 114.
Strigidæ, 183.
Striginæ, 186.
Strix, 186.
Strobilosaura, 329.
Struthio, 279.
Struthiones, 279.
Struthionidæ, 279.
Struthioninæ, 279.
Sturgeon, 380.
Sturnidæ, 240.
Sturninæ, 242.
Sturnus, 242.
Sub-brachiata, 373.
Sucking-fish, 359.
Suina, 147.
Sula, 322.
Sun-fish, 379.
Surnia, 183.
Surninæ, 183.
Sus, 147.
Swallow, Esculent, 194
Swan, 305.
—— Black, 307.
—— Whistling, 306.
Swift, 191.
—— Alpine, 190.
Sword-fish, 355.
Sylvia, 206, 207, 208, 209.
Syngnathidæ, 357.

Tajaçu, 148.
Talegallus, 277.
Talpa, 67.
Talpidæ, 67.
Talpina, 67.
Tamandua, 161.
Tantalinæ, 296.
Tapir, 146.
Tapirina, 146.
Tapirus, 146.
Tarantula, 405.
Teal, 310.
Tee-Tee, Collared, 19.
Tench, 363.
Tenuirostres, 198.
Termes, 421.
Termitidæ, 421.
Tern, Common, 321.
Terrier, English, 47.
—— Scotch, 47.
Testudinidæ, 339.
Testudo, 339.
Tetrao, 274, 275.
Tetraonidæ, 272.
Tetraoninæ, 274.
Thalarctos, 64
Thalassidroma, 317.
Thrush, Misseltoe, 217.
—— Song, 219.
Thynnus, 354.
Tiger, 29.
—— Cat, 34.
Tiger Moth, 434.
—— Beetle, 407.
Tigris, 29.
Tinca, 363.
Tinnunculus, 179.
Titlark, 216.
Titmouse, Blue, 214.
—— Great, 213.
—— Long-tailed, 214.
Toad, 346.
Torpedo, 384.
Tortoise, 339.
Totaninæ, 297.
Toucan, Toco, 253.
Trachearia, 406.
Trematopnei, 381.
Trichicina, 76.
Trichicus, 76.
Trichoptera, 423.
Trigla, 351.
Triglidæ, 351.
Tringinæ, 300.
Tringoides, 297.
Triton, 347.
Trochilidæ, 199.
Trochilus, 199.
Troglodytes, 11, 204.
Trogon, 195.
Trogonidæ, 195.
Tropic Bird, 321.
Trout, 369.
Tumbler, 264.
Tunny, 354.
Turbinidæ, 392
Turbot, 375.
Turdidæ, 217.
Turdinæ, 217.
Turdus, 217, 218, 219, 220
Turkey, 270.
—— Brush, 277.
Turtle, 341.
Turtle-dove, 262.
Turtur, 262.

Ungulata, 103.
Upupa, 198.
Upupidæ, 198.
Upupinæ, 198.
Uria, 316.
Urinæ, 316.
Uropsophus, 331.
Ursidæ, 61.
Ursina, 61.

Vampire Bat, 21.
Vampirus, 21.
Vanellus, 288.
Vanessa, 430.
Vertebrata, 1.
Vespa, 426.
Vespertilionidæ, 21.
Vespertilionina, 24.
Vespidæ, 426.
Viper, 335.
Viperina, 331.
Viperidæ, 333.
Viverra, 39.
Viverrinæ, 39.
Vulpes, 53.
Vulture, Griffon, 169
—— King, 168.
Vulturidæ, 169.
Vulturinæ, 169.

Wagtail, Pied, 215.
—— Yellow, 216.
Walrus, or Morse, 76.
Wapiti, 131.
Warbler, Blackcap, 207.
—— Dartford, 206.
—— Garden, 208.
—— Grasshopper, 204.
Wasp, 426.
Water-hen, 302.
Waxwing, Bohemian, 224.
Weasel, 57.
Wentletrap, Royal Staircase, 392.
Whale, 78.
—— Spermaceti, 83.
Wheatear, 210.
Whelk, 393.
Whitethroat, 206.
Widgeon, 308.
Wolf, 51.
Wolverene, 58.
Wood Ant, 425.
Woodcock, 299.
Woodcock, Thorny, 394.
Woodlark, 249.
Woodpecker, Green, 258.
—— Spotted, 257.
Wren, 204.
—— Golden-crested, 209.
—— Willow, 209.
Wryneck, 259.

Xiphias, 355.

Yak, 110.
Yellowhammer, 247.
Yuncinæ, 259.
Yunx, 259.

Zebra, 139.
Zebu, 105.
Zeidæ, 356.
Zeus, 356.
Zootoca, 326.

R. Clay, Son, and Taylor, Printers,
Bread Street Hill, London.

Printed in Dunstable, United Kingdom

91220867R00268